CYTOTECHNOLOGY
细胞工程技术

郭华荣　主编

中国海洋大学出版社
·青岛·

图书在版编目(CIP)数据

CYTOTECHNOLOGY 细胞工程技术：英汉对照/郭华荣主编. —青岛：中国海洋大学出版社，2011.10
ISBN 978-7-81125-907-0

Ⅰ.①C… Ⅱ.①郭… Ⅲ.①植物－细胞工程－英、汉②动物－细胞工程－英、汉 Ⅳ.①Q943②Q952

中国版本图书馆 CIP 数据核字(2011)第 203874 号

出版发行	中国海洋大学出版社			
社　　址	青岛市香港东路 23 号	邮政编码	266071	
出 版 人	杨立敏			
网　　址	http://www.ouc-press.com			
电子信箱	WJG60@126.com			
订购电话	0532—82032573(传真)			
责任编辑	魏建功	电　话	0532—85902121	
印　　制	青岛海大印务有限公司			
版　　次	2011 年 10 月第 1 版			
印　　次	2011 年 10 月第 1 次印刷			
成品尺寸	185 mm×260 mm			
印　　张	21.5			
字　　数	503 千字			
定　　价	48.00 元			

Fig. 2.1. Laminar flow hoods for aseptic procedures. (By Van Staaveren bv, Rijsenhout, The Netherlands)

Figure 2.2. A walk-in controlled environment room for incubation of *in vitro* cultures. (By SBW, Rijsenhout Department, The Netherlands)

Figure 2.3. Shakers for the aeration of liquid cultures. (By P+S PLANTLAB b.v., The Netherlands)

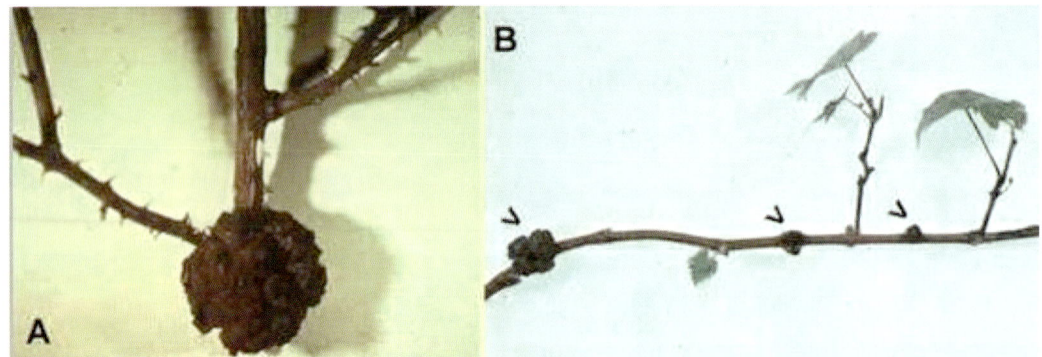

Figure 16.1. Crown Gall (*Agrobacterium tumefaciens*) in grapes.

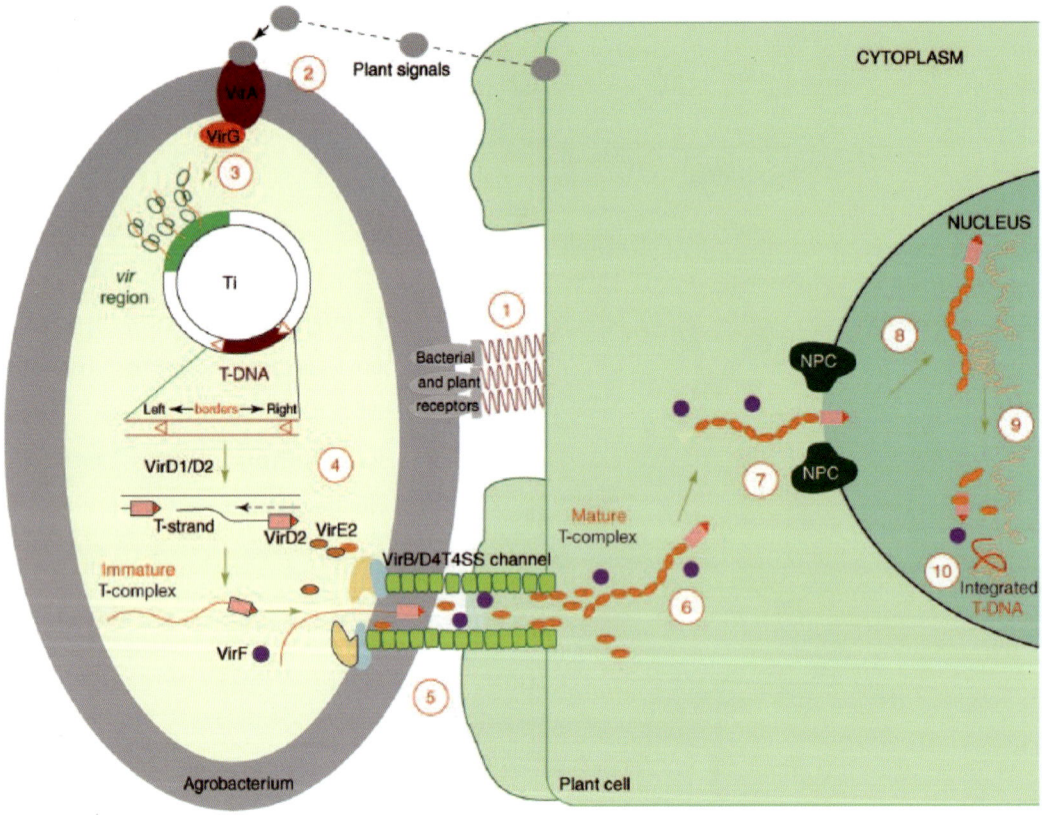

Fig. 16.3. A model for Agrobacterium-mediated transformation. The transformation process comprises 10 major steps and begins with recognition and attachment of the Agrobacterium to the host cells ① and the sensing of specific signals by the Agrobacterium VirA/VirG two-component-signal-transduction system ②. Following activation of the vir gene region ③, a mobile copy of the T-DNA is generated by the VirD1/VirD2 protein complex ④ and delivered as a VirD2-DNA complex (immature T-complex), together with several other vir proteins, into the host cell cytoplasm ⑤. Following the association of the virE2 with the T-strand, the mature T-complex forms, travels through the host cell cytoplasm ⑥ and actively imported into the host cell nucleus ⑦. Once inside the nucleus, the T-DNA is recruited to the point of integration ⑧, stripped of its escorting proteins ⑨ and integrated into the host genome ⑩.

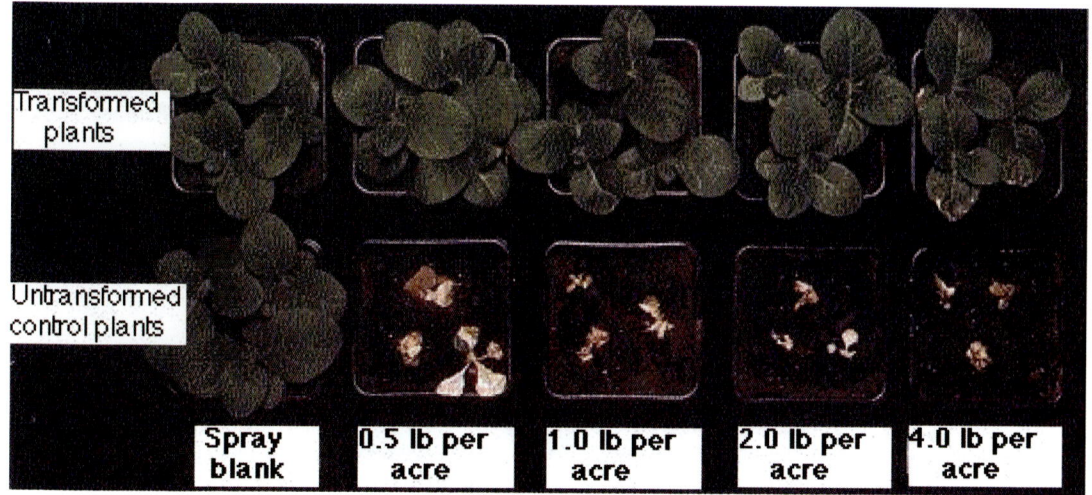

Figure 16.5. Effect of the herbicide bromoxynil （溴苯腈）on tobacco plants transformed with a bacterial gene whose product breaks down bromoxynil (top row) and control plants (bottom row). "Spray blank" plants were treated with the same spray mixture as the others except the bromoxynil was left out. (By Calgene, Davis, CA.)

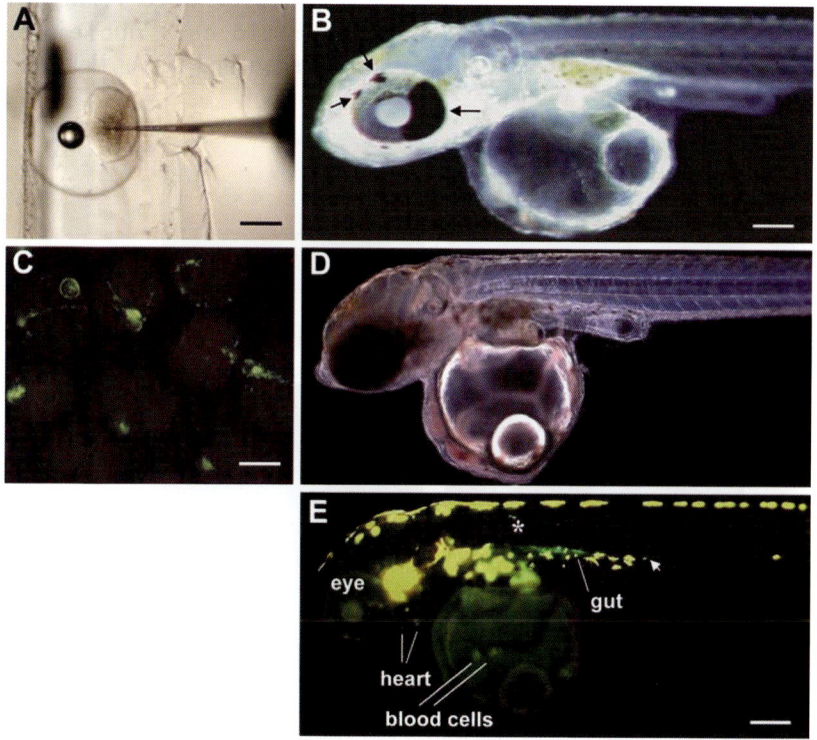

Fig. 31.8. Fish chimera. A. Injection of SaBE-1c cells into a recipient embryo (bar = 250 2m). B. Pigmented chimeric fry showing the contribution of MES1 derivatives to melanocytes (black areas, arrows) in the developing eye (bar = 200 μm). C. Group of GFP positive chimeras 24 hrs post fertilization (bar =600 μm). D, E. Chimeric fry showing the contribution of GFP expressing MES1 derivatives to the trunk (star), somite (arrow) and other cell types. Dark-field D. and merge (E) between dark field and fluorescent optics (bar = 200 μm). (By M. C. Alvarez et al., 2007)

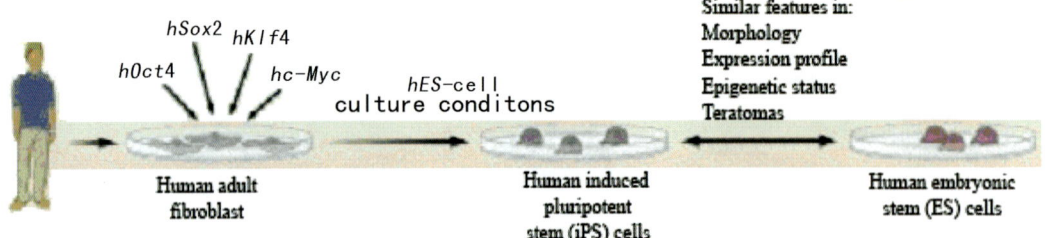

Fig. 32.5. Transcription factor-induced pluripotency. Adult fbroblasts from human donors were exposed to retroviral vectors expressing a cocktail of four transgenes encoding the human factors hOct4, hSox2, hKlf4, and hc-Myc (Takahashi et al., 2007). Thirty days after transduction and further cultivation under human ES cell growth conditions, human induced pluripotent stem (iPS) cell colonies (among others) that could be propagated and expanded further were isolated. Comparative analysis of human iPS cells and human ES cells using assays for morphology, surface-marker expression, gene expression profling, epigenetic status, and in vitro and in vivo differentiation potential revealed a remarkable degree of similarity between these two pluripotent stem cell types. (Modified from H. Zaehres and H. R. Schöler1, 2007))

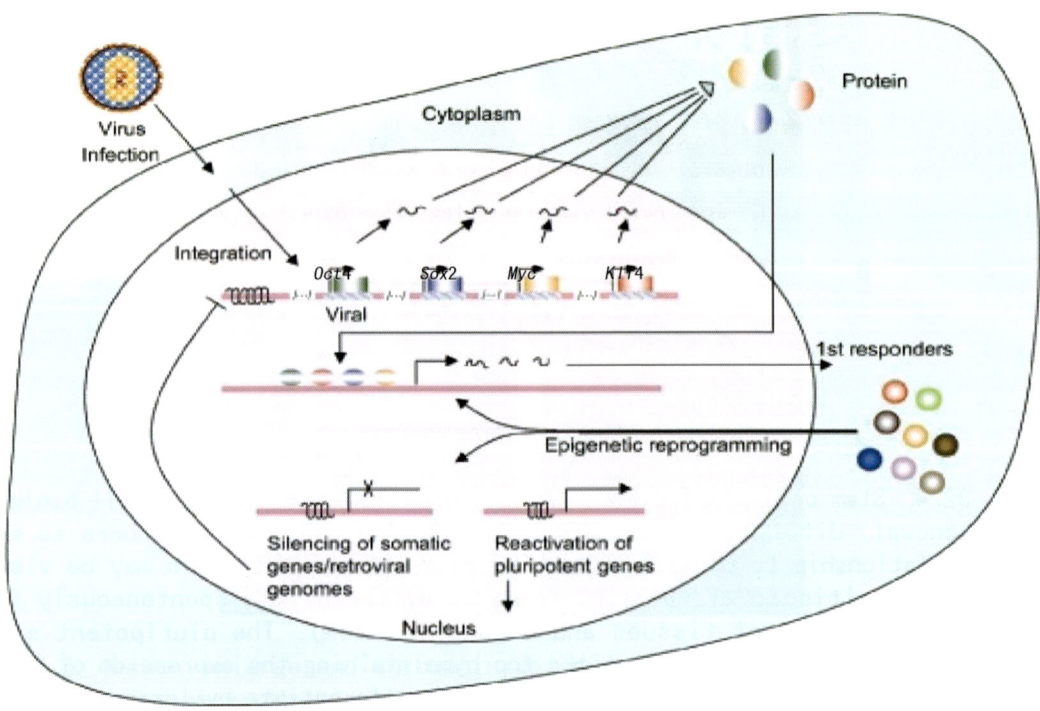

Fig. 32.6. Mechanical steps involved in the reprogramming of somatic cells into pluripotent ones by Oct4/Sox2/Klf4/Myc. (By D. Pei, 2009)

前　言

细胞工程是细胞水平上的生物技术,是一门既有悠久历史,又有崭新内容,不断发展更新的学科。近年来,国内已有多种版本的《细胞工程》教材出版,但尚缺乏具有相应教学内容的英文原版教材,从而阻碍了该课程双语教学内容的开展。我在10多年的教学和科研实践的基础上,经过精心筛选和提炼,从英文原版专著或公开发表的论文中,收集了细胞工程相关的教学内容并编辑成书。书中主要专业词汇均标注了汉语释义,以帮助学生理解书中内容。

全书分为植物细胞工程和动物细胞工程两大部分。植物细胞工程部分的第1~15章内容主要来源于"Dodds J H and Roberts L W, 1995. Experiments in plant tissue culture. 3rd edition, Cambridge University Press, New York, USA"。为减少篇幅,本书中省略了其中的参考文献。动物细胞工程部分的第17~30,33~35章内容主要来源于"Freshney R I, 2000. Culture of animal cells — a manual of basic technique. 4th edition. Wiley-Liss Inc, New York, USA"。其他各章内容是在参考了多篇公开发表的中外学术论文和论著的基础上编撰完成的。

本书的内容力求精炼,重点阐述细胞工程技术的原理和方法,可读性好,实用性强,适合作为高等院校本科生和研究生的细胞工程双语教材,以及科研工作者的实验工具书。作者尽管已倾注了大量心血,限于水平,难免有不当之处,恳请读者批评指正。

编者

2011年8月

CONTENTS

PART I CYTOTECHNOLOGY IN PLANTS（植物细胞工程）

CHAPTER 1 INTRODUCTION FOR PLANT CELL, TISSUE AND ORGAN CULTURE（植物细胞、组织和器官培养简介） ………… 3
CHAPTER 2 LABORATORY FACILITIES（实验室装备） ………… 11
CHAPTER 3 ASEPTIC TECHNIQUES（无菌操作技术） ………… 14
CHAPTER 4 MEDIA COMPOSITION AND PREPARATION（培养基组成和制备） ………… 23
CHAPTER 5 INITIATION AND MAINTENANCE OF CALLUS（愈伤组织的诱导和培养） ………… 36
CHAPTER 6 ORGANOGENESIS（器官发生） ………… 42
CHAPTER 7 CELL SUSPENSIONS（细胞悬浮培养） ………… 47
CHAPTER 8 SOMATIC EMBRYOGENESIS（体细胞胚胎发生） ………… 52
CHAPTER 9 CULTURE OF ISOLATED ROOTS（离体根的培养） ………… 58
CHAPTER 10 MICROPROPAGATION BY BUD PROLIFERATION（芽的快繁） ………… 61
CHAPTER 11 ANTHER AND POLLEN CULTURES（花药和花粉的培养） ………… 65
CHAPTER 12 ISOLATION AND CULTURE OF PROTOPLASTS（原生质体的分离和培养） ………… 73
CHAPTER 13 PROTOPLAST FUSION AND SOMATIC HYBRIDIZATION（原生质体的融合和体细胞杂交） ………… 79
CHAPTER 14 CRYOPRESERVATION OF GERMPLASM（种质资源的低温保存） ………… 84
CHAPTER 15 PRODUCTION OF SECONDARY METABOLITES（次生代谢产物的生产） ………… 87
CHAPTER 16 TRANSGENIC PLANTS（转基因植物） ………… 90

PART II CYTOTECHOLOGY IN ANIMALS（动物细胞工程）

CHAPTER 17 INTRODUCTION TO ANIMAL CELL CULTURE（动物细胞培养简介） ………… 107

CHAPTER 18	BIOLOGY OF CULTURED CELLS（培养细胞的生物学特性）…	114
CHAPTER 19	EQUIPMENTS FOR ANIMAL CELL CULTURE（动物细胞培养器材）………………………………………………………………	120
CHAPTER 20	MEDIA FOR ANIMAL CELLS（动物细胞培养基）…………	128
CHAPTER 21	PREPARATION AND STERILIZATION OF MEDIA（培养基的配制和灭菌）……………………………………………………	147
CHAPTER 22	SERUM-FREE MEDIA（无血清培养基）……………………	152
CHAPTER 23	PRIMARY CELL CULTURE（原代细胞培养）………………	158
CHAPTER 24	CELL LINES（细胞系）………………………………………	171
CHAPTER 25	CELL CLONING AND SELECTION（细胞克隆和筛选）…	180
CHAPTER 26	TRANSFORMATION OF CULTURED CELLS（培养细胞的转化）………………………………………………………………	191
CHAPTER 27	CHARACTERIZATION OF CELL LINES（细胞系的鉴定）……	201
CHAPTER 28	CONTAMINATION OF CULTURED CELLS（培养细胞的污染）………………………………………………………………	210
CHAPTER 29	CRYOPRESERVATION OF CUL TURED CELLS（培养细胞的低温冻存）…………………………………………………	221
CHAPTER 30	CYTOTOXOCITY（细胞毒性）………………………………	227
CHAPTER 31	EMBRYONIC STEM CELLS（胚胎干细胞）………………	235
CHAPTER 32	NUCLEAR REPROGRAMMING AND INDUCED PLURIPOTENT STEM CELLS（细胞核重编程与诱导性多能干细胞）……	252
CHAPTER 33	ORGAN CULTURE, HISTOTYPIC CULTURE AND ORGANOTYPIC CULTURE（器官培养、组织型培养和器官型培养）…	264
CHAPTER 34	ANIMAL CELL FUSION（动物细胞融合）…………………	270
CHAPTER 35	ANIMAL CELL TRANSFECTION（动物细胞转染）………	279
CHAPTER 36	TRANSGENIC ANIMAL（转基因动物）……………………	286
CHAPTER 37	CLONED ANIMAL（克隆动物）……………………………	311

GLOSSARY …………………………………………………………………… 322

REFERENCES ………………………………………………………………… 331

PART I
CYTOTECHNOLOGY IN PLANTS
(植物细胞工程)

CHAPTER 1　INTRODUCTION FOR PLANT CELL, TISSUE AND ORGAN CULTURE(植物细胞、组织和器官培养简介)

　　The idea of experimenting with the tissues and organs of plants in isolation under controlled laboratory conditions arose during the latter part of the nineteenth century, finding its focus in the work of the great German plant physiologist Haberlandt (1902). Haberlandt's vision was of achieving continued cell division in explanted tissues on nutrient media-that is, of establishing true, potentially perpetual (长期) tissue cultures. In this, he was himself unsuccessful, and about 35 years were to elapse(逝去)before the goal was attained-as it could be only after the discovery of the auxins(生长素). Gautheret, Nobecourt, and White were the pioneers in this second phase. The research they set in train was at first mainly concerned with establishing the conditions in which cell division and growth would take place in explants, and in exploring the nutritional and hormonal requirements of the tissues. But this quickly gave place to a period during which cultured tissues were used as a research tool, in studying more general problems of plant cell physiology and biochemistry and the complex processes of differentiation and organogenesis. The achievements were considerable; but above all, the finding that whole plants could be regenerated from undifferentiated tissues-even single cells-in culture gave the method enormous power. In an extraordinary way this has meant that at one time the entity-a plant-can be handled like a microorganism and subjected to the rigorous procedures of molecular biology, and at another called almost magically back into existence as a free-living, macroscopic organism. If genetic engineering, involving the direct manipulation of the stuff of heredity, is ever to contribute to that part of man's welfare that depends on his exploitation of plants, the procedures adopted will inevitably depend ultimately upon the recovery of "real" plants from cultured components. No wonder, then, that the technology has escaped from the confines of the university laboratory to become part of the armory(军械库)of industry and agriculture.

HISTORY

Early Attempts, 1902-1939

　　The concept that the individual cells of an organism are totipotent (全能性) is implicit in the statement of the cell theory. Schwann (1839) expressed the view that each living cell of a multicellular organism should be capable of independent development if provided with the proper external conditions. A totipotent cell is one that is capable of developing by regeneration into a whole organism. The basic problem of cell culture was clearly stated by White (1954): If all the cells of a given organism are essentially identical and totipotent, then the cellular differences observed within an organism must arise from responses of those cells to their microenvironment and to other cells within the organism. It should be possible to restore suppressed functions by isolating the cells from those organismal influences responsible for their suppression. If there has been a loss of certain functions so that the cells in the intact organism are no longer totipotent, then isolation would have no effect on restoring the lost activities. The use of culture techniques enables the scientist to segregate cells, tissues, and organs from the parent organism for subsequent study as isolated biological units. The attempts to reduce an organism to its constituent cells, and subsequently to study these cultured cells as elementary organisms, is therefore of fundamental importance.

Several plant scientists performed experiments on fragments of tissue isolated from higher plants during the latter part of the 19th century. Wound callus, formed on isolated stem fragments and root slices, was described. 'Callus' refers to a disorganized proliferated mass of actively dividing cells. Rechinger (1893) examined the "minimum limits" of divisibility of isolated fragments of buds, roots, and other plant material. He found that pieces thicker than 1.5 mm were capable of further growth on sand moistened with water, but isolated fragments thinner that 1.5 mm not, although no nutrients were used in these experiments. Rechinger reported that the presence of vessel elements appeared to stimulate growth of the fragments. Unfortunately, he did not pursue this clue, since his observations suggested the ability of cambial tissue(形成层组织) to proliferate was associated with vascular tissues(导管组织).

Haberlandt (1902) originated the concept of cell culture and was the first to attempt to cultivate isolated plant cells in vitro on an artificial medium. A tribute to Haberlandt's genius, with a translation of his paper "Experiments on the culture of isolated plant cells," has been published. Unlike Rechinger, Haberlandt believed that unlimited fragmentation would not influence cellular proliferation. The culture medium consisted mainly of Knop's solution, asparagine(天冬酰胺), peptone(蛋白胨), and sucrose. Although the cultured cells survived for several months, they were incapable of proliferation. Haberlandt's failure to obtain cell division in his cultures was, in part, due to the relatively simple nutrients and to his use of highly differentiated cells. Since Haberlandt did not use sterile techniques, it is difficult to evaluate his results because of the possible effects of bacterial contamination. As example of his genius, Haberlandt suggested the utilization of embryo sac fluids and the possibility of culturing artificial embryos from vegetative cells. In addition, he anticipated the paper-raft technique. Following his lack of success with cell cultures, Haberlandt became interested in wound healing. Experiments in this area led to the formulation of his theory of division hormones. Cell division was postulated as being regulated by two hormones: One was "leptohormone"(韧皮部激素), which was associated with vascular tissue, particularly the phloem(韧皮部). The other was a wound hormone released by the injured cells. Subsequent research investigators verified the association of hormones with vascular tissues.

Early in the twentieth century interest shifted to the culture of meristematic tissues (分生组织) in the form of isolated root tips. These represented the first aseptic organ cultures. Robbins (1922) was the first to develop a technique for the culture of isolated roots and Kotte (1922), a student of Haberlandt's, published similar studies independently. These cultures were of limited success. Robbins and Maneval (1923), with the aid of subcultures, maintained maize(玉米) roots for 20 weeks. White (1934), experimenting with tomato roots, succeeded for the first time in demonstrating the potentially indefinite culture of isolated roots. According to White (1951), two difficulties hampered the development between 1902 and 1934 of a successful method for culturing excised plant material: (1) the problem of choosing the right plant material, and (2) the formulation of a satisfactory nutrient medium.

With the introduction of root tips as a satisfactory experimental material, the crucial problem became largely one of organic nutrition. White's early success with tomato roots can be attributed to his discovery of the importance of B vitamins, plus the fact that indefinite growth was achieved without the addition of any cell division factor to the liquid medium.

It is important at this point to make a distinction between organ culture and tissue culture. In the case of excised roots as an example of organ culture, the cultured plant material maintains its morphological identity as a root with the same basic anatomy and physiology as in the roots of the parent plant. There are some exceptions, and slight changes in anatomy and physiology may occur during the culture period. According to Street (1977), the term "tissue culture" can be applied to any multicellular culture growing on a solid medium (or attached to a substratum and nurtured with a liquid medium) that consists of many cells in protoplasmic continuity. Typically, the culture of an explant, consisting of one or more tissues, results in a callus that has no structural or functional counterpart with any tissue of the normal plant body.

The first plant tissue cultures, in the sense of long-term cultures of callus, involved explants of cambial tissues isolated from carrot and tobacco tumor tissue from the hybrid *Nicotiana glauca* × *N. langsdorffii*. The latter tumor tissue requires no exogenous cell-division factor. Results from these three laboratories, published independently, appeared almost simultaneously. Fortunately, plant physiologists working in other areas had discovered some of the hormonal characteristics of indole-3-acetic acid, IAA (吲哚乙酸), and the addition of this auxin to the culture medium was essential to the success of the carrot cultures maintained by Nobecourt and Gautheret. According to Gautheret (1939), the carrot cultures required Knop's solution supplemented with Bertholot's salt mixture, glucose, gelatine(明胶), thiamine(维生素 B1), cysteine-HCl(半胱氨酸盐酸盐), and IAA. The goal at that time was to demonstrate the potentially unlimited growth of a given culture, by repeated subcultures, with the formation of undifferentiated callus. The workers were fascinated by the apparent immortality of their cultures and devoted much effort to determining the nutritional requirements for sustained growth.

Basic Studies on Nutrition and Morphogenesis, 1940-1978

Because of the lull(暂停) in botanical research during the war years (1939-1945), relatively little was accomplished until a resurgence of interest in the early 1950s. Some fundamental studies, however, were undertaken. White and Braun (1942) initiated experiments on crown gall(冠瘿瘤) and tumor formation in plants. Probably the most significant event leading to advancement in the next decade was the discovery of the nutritional quality of liquid endosperm(胚乳) extracted from coconut(椰子). Following the success of Van Overbeek, Conklin, and Blakeslee (1941) with the culture of isolated *Datura*(蔓陀罗) embryos on a medium enriched with coconut milk, other workers rapidly adopted this natural plant extract. The combination of coconut milk and 2,4-D had a remarkable effect on the proliferation of cultured carrot and potato tissues. Although it was first thought that a single substance, termed the "coconut-milk factor," was involved as a growth stimulant, several constituents were later found responsible for its activity. Steward's group at Cornell University made numerous contributions in technique, nutrition, quantitative analyses of culture growth, and morphogenesis(形态发生). The regeneration of carrot plantlets from cultured secondary-phloem cells of the taproot(主根) clearly demonstrated the totipotency of plant cells. The phenomenon of somatic embryogenesis(体细胞胚胎发生) in carrot cultures was discovered at approximately the same time by Steward (1958) and Reinert (1959).

The discovery of cytokinins(细胞分裂素) stems from Skoog's tissue culture investigations at the University of Wisconsin. During attempts to induce unlimited callus production from mature tobacco pith(髓) cells, numerous compounds were tested for possible activity in stimulating cell division. Although coconut water or yeast extract plus IAA promoted cell division, efforts were made to locate a specific cell-division factor. Since adenine(腺嘌呤), in the presence of auxin, was found to be active in stimulating callus growth and bud formation in tobacco cultures, nucleic acids were then examined. Skoog's group eventually located a potent cell-division factor in degraded DNA preparations. It was isolated, identified as 6-furfurylaminopurine (呋喃甲基氨基嘌呤), and named "kinetin"(激动素). The related analogue, 6-benzylaminopurine (6-苄氨基腺嘌呤, 6-BA), was then synthesized, and it too stimulated cell division in cultured tissues.

The generic term "cytokinin" was given to this group of 6-substituted aminopurine compounds that stimulate cell division in cultured plant tissues and behave in a physiological manner similar to kinetin. Later it was discovered that zeatin(玉米素), isopentyl adenine(异戊烯腺嘌呤), and other cytokinins are naturally occurring plant hormones. Often these compounds are attached to ribose sugar ("ribosides") or to ribose and phosphate ("ribotides"). The stimulatory properties of coconut water are partly due to the presence of zeatin riboside. Skoog and Miller (1957) advanced the hypothesis that shoot and root initiation in cultured callus can be regulated by varying the ratio of auxin and cytokinin in the medium. In addition to the cytokinins, other endogenous cell-division factors may exist in plant tissues.

It was found that callus fragments, transferred to a liquid medium and aerated on a shaker, gave a suspension of single cells and cell aggregates that could be propagated by subculture. Steward's group made extensive use of carrot suspension cultures, and it became evident that this technique offered much potential for studying many facets of cell biology and biochemistry. Street and co-workers have pioneered the development of various procedures for the culture of cell suspensions (e. g., chemostat(恒化培养) and turbidostat(恒浊培养)).

Muir (1953) succeeded in developing a technique for the culture of single isolated cells. Single cells were placed on squares of filter paper, and the lower surface of the paper was placed in contact with an actively growing "nurse" culture. This paper-raft nurse technique(滤纸看护培养技术) provided the isolated cells not only with nutrients from the medium via the older culture, but also with growth factors synthesized by the nurse tissue. Although Muir's experiments involved bacteria-free crown gall (*Agrobacterium tumefaciens*) tumor cells, single-cell clones were produced later from normal cells. In another approach, a single cell was suspended as a hanging drop in a microchamber. The agar-plating method of Bergmann (1960) involved separating a single-cell fraction by filtration, mixing the cells with warm agar, and then plating the cells as a thin layer in a Petri dish. Although Muir and his colleagues reported in 1954 that single isolated cells exhibited cell division, this claim was contested by De Ropp (1955). Subsequent investigations provided the necessary evidence that single cells are capable of proliferation. The question of totipotency was completely resolved by Vasil and Hildebrandt (1965) by the demonstration that a single isolated cell can divide and ultimately give rise to a whole plant.

Plant tissue cultures have been used extensively for the study of cytodifferentiation, particularly the formation of tracheary elements(导管成分)in cultured tissues. A variety of different techniques have been employed. Wedges(楔子) of agar containing auxin and sucrose "grafted" to a block of callus induced tracheary element formation. Primary explants from many different plant tissues are capable of producing tracheary elements during culture on agar or in a liquid medium.

Many of the early investigators employed either herbaceous(草本的)or woody dicot(双子叶植物) tissues as sources of primary explants, although other groups of plants were also used as experimental material. Morel (1950) successfully cultured monocot(单子叶植物)tissues with the aid of coconut water. Ball (1955) cultured tissues of the gymnosperm(裸子植物)*Sequoia sempervirens*(北美红杉) and Tulecke prepared haploid cultures from the pollen of *Taxus*(紫杉)(1959) and *Ginkgo biloba*(银杏)(1953, 1957). Harvey and Grasham (1969) published media requirements for establishing cultures of 12 conifer(松类)species. Tissue culture procedures have been used in developmental anatomy studies involving excised shoot apices(顶端, apex 的复数)of lower plants (e.g., ferns(蕨类), *Selaginella* (卷柏), and *Equisetum* (问荆)).

During the 1960s it was shown that cultured pollen and the microsporogenous tissue of anthers (花药)have the potential to produce vast numbers of haploid embryos. Later, with a technique developed by C. Nitsch, it became possible to culture microspores of *Nicotiana* (烟草)and *Datura*(曼陀罗), to double the chromosome number of the microspores, and to collect seeds from the homozygous diploid plants within a five-month period. Although there are technical problems associated with this technique, haploid cultures have been used successfully in China for the selection of improved varieties of crop plants.

Another important development during the 1960s was the enzymatic isolation and culture of protoplasts. This method involves removing the cell wall with purified preparations of cellulase(纤维素酶)and pectinase(果胶酶), while regulating protoplast expansion with an external osmoticum(渗透剂). The cultured protoplasts regenerate new cell walls, form cell colonies, and ultimately form plantlets(小苗). Some of the experimental approaches currently employed include:(1) protoplast fusion within species, between species, between monocots and dicots, and even between plants and animals;(2) introduction of mitochondria and plastids (质体) into protoplasts;(3) uptake of blue-green algae, bacteria, and viruses by protoplasts; and (4) the transfer of genetic information into isolated protoplasts.

Cells containing foreign organelles are termed "cytoplasmic hybrids"(胞质杂交体) or "heteroplasts"(异质体) whereas cells containing transferred nuclei are referred to as "heterokaryons"(异核体)or "heterokaryocytes"(异核细胞). The fusion of the two nuclei in heterokaryons produces hybrid cells.

Plant tissue culture techniques have been widely used for the commercial propagation of plants. Most of the applications are based on the characteristic of cytokinins to stimulate bud proliferation in the cultured shoot apex. Ball (1946) demonstrated the possibility of regenerating plants from isolated explants of angiosperm(被子植物)shoot apices. Later, Wetmore and Morel (1951) regenerated whole plants from shoot apices measuring 100-250 μm in length and bearing one or two leaf primordia（原基）. Modifications of Morel's (1960) shoot-apex technique have been used for orchid propagation (Morel, 1964). Premixed culture media, specifically formulated for the propagation of certain plants, are commercially available.

One of the earliest applications of plant tissue culture involved the study of plant tumor physiology. White and Braun (1942) reported the growth of bacteria-free crown gall tissue. Gautheret (1946) observed that a callus culture of *Scorzonera hispanica*（菊牛蒡）, which originally required auxin in the medium for growth, often developed outgrowths of callus that would grow indefinitely on an auxin-deficient medium. The term "habituation" refers to inherited changes in nutritional requirements arising in cultured cells, especially changes involving plant hormones. For example, an auxin-habituated culture has lost its original requirement for exogenous auxin. These cultures are important in investigations of plant cancer. The grafting of auxin-and cytokinin-habituated tissues into healthy plants produces tumors. Tissue culture techniques have also been used to produce pathogen-free plants via apical meristem cultures.

Emergence of A New Technology, circa 1978 to the present

By the late 1970s it had become evident that plant tissue culture technology was beginning to make significant contributions to agriculture and industry. In agriculture, the major areas are haploid breeding, clonal propagation, mutant cultures, pathogen-free plants, production of secondary products, and genetic engineering. In addition, the cryopreservation of plant tissue cultures and the establishment of in vitro gene banks have attained considerable interest.

The greatest success has been achieved with in vitro clonal propagation. In vitro techniques have revitalized the orchid industry. Murashige (1977) estimated that over six hundred species of ornamental plants（观赏植物）have been cloned. Cloning has been extended to forest trees, fruit trees, oil-bearing plants, vegetables, and numerous agronomic crop plants（农作物）. Clonal propagation of potato plants has been achieved on a large scale by regeneration from isolated leaf-cell protoplasts（原生质体）. By 1982 more than a hundred tissue culture facilities were engaged in the commercial propagation of plants. In other areas, plant tissue culture has had only limited economic success. Haploid breeding has produced relatively few established cultivars, mainly because of the low frequency of the appearance of new agriculturally important genotypes by this method. Progress is being made, however, particularly in China. The application of mutagenic agents to cultures, followed by suitable screening techniques, has led to the regeneration of mutant plants showing disease or stress resistance. Several pathogen-free plants have been developed, and tissue culture technology now plays an important role in plant pathology. For example, protoplasts are currently employed for the study of virus infection and biochemistry. The prospects of success with the genetic manipulation of plants have created considerable public interest, and advances to date are now considerable. Several approaches have been considered. One technique involves the transfer of genetic information by the fusion of protoplasts isolated from two different organisms. This method provides the opportunity of producing hybrids between related sexually incompatible species. Melchers, Sacristan, and Holder (1978) produced a somatic hybrid plant from the fusion of potato and tomato protoplasts, both members of the Solanaceae family (茄科). It appears unlikely, however, that this technique will yield any plants of economic importance in the near future. Another technique concerns the insertion of foreign genes attached to a plasmid vector into the naked protoplast. The transfer of the plasmid DNA into the protoplast is usually accomplished by means of liposomes-A major problem, however, is whether the inserted gene will be integrated into the host genome, transcribed, and expressed in the mature plant. Several research groups have produced transformed tobacco plants following single-cell transformation (i. e., gene insertion). The use of Agrobacterium-mediated plant transformation (根癌农杆菌介导的植物转化技术), and the use of the so-called gene gun (基因枪) literally to shoot DNA into plant cells, has led to the development of many novel plants.

Industrial applications involve large-scale suspension cultures capable of synthesizing significant amounts of useful compounds. These secondary products of industrial interest include antimicrobial compounds, antitumor alkaloids, food flavors, sweeteners, vitamins, insecticides, and enzymes. A major problem has been the genetic instability of the cultures, as well as engineering problems associated with this technique. Nevertheless, progress has been made at the industrial level, especially by the Japanese.

知识要点

植物的组织培养开始于19世纪后半叶，Haberlandt是第一个用人工培养基对分离的植物细胞进行培养的人，并提出了细胞培养、激素作用和植物细胞全能性等概念，被誉为植物组织培养之父。Gautheret、Nobecourt和White改进了植物培养基，尤其是发现了植物生长激素和B族维生素在植物细胞生长和分裂中的重要作用，被誉为植物组织培养基的奠基人。1958年美国的Steward和德国的Reinert分别由培养的胡萝卜细胞诱导形成了胚状体，1965年由Vasil和Hildebrandt用分离的单个细胞培养获得了再生植株，从而使植物细胞全能性的理论真正得到了科学的证实。20世纪60年代，Guha和Maheshwari，Rourgin和Nitsch进行植物花粉培养，成功获得了烟草和胡萝卜的单倍体植株，为单倍体育种奠定了基础。Cocking等(1967)利用纤维素酶和果胶酶成功获得了烟草细胞原生质体，为植物细胞融合创造了条件。植物组织培养已经变成了一种常规的实验技术，广泛应用于植物的脱毒、快繁、基因工程、细胞工程、次生代谢物质的生产、工厂化育苗等多个方面。

CHAPTER 2 LABORATORY FACILITIES(实验室装备)

A laboratory devoted to in vitro procedures with plant tissues must have adequate space for the performance of several functions. According to White (1963), it must provide facilities for: (a) media preparation, sterilization, cleaning, and storage of supplies; (b) aseptic manipulation of plant material; (c) growth of the cultures under controlled environmental conditions; (d) examination and evaluation of the cultures; and (e) assembling and filing of records.

The grouping of functions will vary considerably from one laboratory to another. The ideal organization will allow a separate room for each of the following functions: media preparation, aseptic procedures, incubation of cultures, and general laboratory operations. If one has the opportunity to plan an in vitro laboratory in advance, the component facilities should be arranged as a production line. The area involved with washing and storage of glassware should lead to the facilities for oven sterilization and media preparation. Materials should then move from autoclave sterilization to the aseptic transfer facility. After the aseptic operations, the cultures are transferred to incubators or controlled-environment chambers. The cultures should be in close proximity to the laboratory containing microscopes and facilities for evaluation of the results. Discarded and contaminated cultures are transferred back to the washing area. It is of utmost importance to give careful consideration to the arrangement of the aseptic procedures. Because most laboratories do not have a separate sterile room, some type of laminar flow cabinet or hood is required (彩页 Fig. 2.1).

DISHWASHING

Discarded cultures, as well as contaminated ones, are autoclaved briefly in order to liquefy the agar and to kill any contaminants that may be present. The culture glassware is easier to wash after the spent medium has been liquefied and removed. After scrubbing with a brush in a hot detergent bath, the glassware is rinsed repeatedly with tap water, and then given two or three rinses in distilled water. If an automatic dish-washer machine is used, a final rinse with distilled or demineralized water should be used to remove any possible traces of detergent. After washing, the glassware is oven dried prior to storage. New glassware may release chemicals that are toxic to the cultured tissues.

MEDIA PREPARATION

Although media preparation requires a balance sensitive to milligram quantities for weighing hormones and vitamins, a less sensitive scale may be used for weighing agar and carbohydrates. The media reagents should be shelved near the balance for convenience. A refrigerator and a freezer in the media room are necessary for storing stock solutions (母液) and chemicals that degrade at room temperature. A combination hot plate and magnetic stirrer is a time saver for dissolving inorganic reagents. Either a pH meter or pH indicator paper is required for adjusting the final pH of the medium. Relatively large quantities of single and double-distilled water must be available in the media room.

Sterilization equipment is an integral part of media preparation. A commercial electric stove is the most economical type of oven sterilization. Wet-heat sterilization involves either an autoclave or a pressure cooker. Some hormones and vitamins are sterilized by ultrafiltration at room temperature. After sterilization of the culture vessels by dry heat and autoclaving the medium, the culture tubes or Petri dishes are poured in the transfer chamber. It is also useful to have a microwave oven to dissolve agar and melt medium.

INCUBATION OF THE CULTURES

The freshly prepared cultures are grown under carefully regulated environmental conditions; that is, temperature, light, and humidity. This is accomplished with an incubator, plant growth chamber, or controlled environment room (彩页 Fig. 2.2). If cell suspensions are cultured, some type of shaker or aeration equipment will be necessary (彩页 Fig. 2.3). Several engineering aspects should be considered in designing a culture room: safety and convenience of the electrical system, air flow for uniform temperature regulation, arrangement of the shelving, and elimination of airborne contaminants. Optimal environmental conditions will vary depending on the species and the purpose of the experiment, and consideration should be given to diurnal (每日的) temperature variations, light intensity, light quality, and photoperiod (i. e., relative length of light-dark cycles). Fluorescent lamps have certain advantages over incandescent sources(白炽光源): a better spectral quality, a more convenient shape, and a lower heat output. When constructing a growth room, it is good to keep the light bulbs outside, as they generate a lot of heat. Some cultures, however, appear to show the best growth in the presence of a mixture of both types of illumination. Experiments conducted by Murashige (1974) with *Asparagus*(芦笋), *Gerbera*(大丁草), *Saxifrage*(虎耳草), and *Bromeliads*(凤梨) indicated an optimum light intensity of 1,000 lux during culture initiation and shoot proliferation. A higher optimum of 3,000-10,000 lux was required for the establishment of plantlets. It is advisable to equip the culture room with a clock-operated timing switch for the regulation of photoperiods. Although some investigators may want to expose the cultures to thermoperiodic cycles, most experiments are conducted with constant temperatures set at approximately 25-27℃. Some morphogenetic responses are evidently sensitive to temperature fluctuations.

ADDITIONAL NEEDS

General laboratory requirements vary considerably, depending on the type of data required. A suitable hand lens and dissection microscope are important for the macroexamination of the cultures, and a compound microscope equipped with photomicrograph accessories is available in most laboratories. A chemical hood should be used with maceration（浸泡）procedures involving chromic acid（铬酸）, as well as for the storage of volatile and potentially dangerous chemicals. Protoplast purification requires a benchtop centrifuge. Also, an inverted microscope is helpful. All plant tissue culture laboratories should be equipped with a fire extinguisher and a first aid kit.

知识要点

理想的植物组织培养实验室应该配备有基本操作室、无菌操作室、培养室、鉴定分析室和温室等独立的功能分区。基本操作室用于洗涤、灭菌、培养基的配制和贮藏；无菌操作室用于培养材料的接种、传代和分化培养操作，一般要配备有缓冲区，并安装有紫外灯和超净工作台，以便能长期处于无菌状态；培养室用于植物外植体的培养，一般配备有暗培养箱、恒温恒湿光照培养箱、培养架和摇床等；鉴定分析室用于实验材料的研究和观察分析；温室用于炼苗和移栽。

CHAPTER 3   ASEPTIC TECHNIQUES(无菌操作技术)

The importance of maintaining a sterile environment during the culture of plant tissues is absolutely necessary and cannot be overemphasized. A few simple precautions to avoid contamination will save valuable time in not repeating experiments. The best example of aseptic technique can be found in the operating room of a modern hospital.

The single most important factor in the selection of a suitable working area is the possible flow of unfiltered air over the disinfected working area. Air currents must be avoided because of airborne spores of contaminating microorganisms. An interior room, similar in layout to a photographic darkroom, is an excellent choice for aseptic procedures. Because opening the door creates a draft, post a NO ADMITTANCE sign on the door during aseptic procedures. If precautions are taken, an open laboratory bench can be used in a draft-free room. Under these conditions it may be prudent to use either a face mask or a plastic biohazard shield.

Most plant scientists using tissue culture procedures conduct sterile operations within some type of transfer chamber, bacteriological glove box, or laminar flow cabinet. With a laminar flow cabinet (see Fig. 2.1) air is forced through a dust filter and then passed through a high-efficiency particulate air (HEPA) filter. Depending on the type of cabinet, the air is directed either downward or outward over the working area. The gentle flow of sterile air is designed to prevent any spore-laden(含有孢子的) unfiltered air from entering the cabinet. Since ethanol is highly inflammable(易燃的), flaming instruments in a laminar flow cabinet should be done with caution: The air flow from the cabinet would direct a flash fire toward the worker. A few minutes before the start of any sterile procedure, the working area should be thoroughly scrubbed with a tissue soaked with ethanol or isopropanol(异丙醇,70% v/v).

Aseptic cabinets and transfer rooms are often equipped with one or more germicidal lamps emitting ultraviolet (UV) light. This type of radiation is useful in eliminating airborne contaminants and for surface disinfection. The emission at 254 nm is slowly germicidal, but UV does not penetrate surfaces; dust and shadowed areas protect contaminants from its effects. In the presence of plastics, the radiation may produce inhibitory substances in culture media. The use of UV radiation should be kept to a minimum since it is not a substitute for cleanliness. If such a lamp is used, switch it on about 30 min prior to culture time. Do not leave the UV lamp on for several hours or overnight because the accumulation of toxic ozone(臭氧) in a confined space creates a hazard.

An extremely important point about aseptic procedure, and one of the leading causes of contamination, is unclean hands. Simply rinsing the hands with water is insufficient; it is necessary to scrub them *vigorously* with soap and hot water for several minutes. Attention must be given to the fingernails and to any part of the forearm that extends into the working area. After a hot-water rinse, blot(吸干)the skin partially dry with paper towels. It is not advisable to use strong disinfectants that could produce a skin rash (皮疹). The hands may be dipped in a dilute solution of ethanol or isopropanol, although this can cause skin dryness and must not be used indiscriminately(无差别地) around an open flame.

According to Biondi and Thorpe (1981), the seed coat can be removed manually from surface-sterilized seeds without contamination by dipping the fingers repeatedly in ethanol (40%-70% v/v) prior to the operation. Hexachlorophene (六氯酚,2,2'methylenebis [3,4,6trichlorophenol]) has been used as an antibacterial agent in soaps. This chemical penetrates the intact skin and can cause brain damage.

Several techniques are used for the sterilization of glassware, surgical instruments, liquids, and plant material. The term "sterilization" is an absolute one that implies the total inactivation of all forms of microbial life in terms of the ability of the organisms to reproduce. "Disinfection", on the other hand, means the reduction of bacterial numbers to some arbitrary "acceptable" level. Ethanol is a bacteriostatic agent and disinfects a working area, but the treated area is not sterile. The methods of sterilization can be classified as follows: dry heat, wet heat, microwave, microfiltration, and chemical.

DRY HEAT

This method is used only for glassware, metal instruments, and other materials that are not charred(烧焦) by oven temperatures. Strips of tape should not be on any glassware, and the cutting edges of surgical blades may be dulled by the prolonged oven heat. Cotton, paper, and plastic should never be placed in the oven for sterilization. Although laboratory drying ovens may be used, the oven of a gas or electric kitchen stove will serve the same purpose. It is highly important not to use a laboratory oven that has previously been used for paraffin embedding(石蜡包埋).

Objects to be sterilized are wrapped in heavy-duty aluminum foil before being placed in the oven. Take care not to pack the foil-wrapped packages in the oven too tightly, but to leave some space between them. In calculating the time required for dry-heat sterilization, three time periods must be considered:

1. Approximately 1 h (heating-up period) is allowed for the entire load to reach the sterilization temperature of 180℃ (356°F).

2. A minimum of 2 h at this temperature is required to kill all organisms including

the spore formers.

3. Finally, a cooling-down period is advisable in order to prevent the glassware from cracking due to a rapid drop in temperature.

WET HEAT

This procedure employs an autoclave operated with steam under pressure (Fig. 3.1). For the sterilization of paper products, glassware, instruments, and liquids the standard procedure is for the autoclave to operate at a steam pressure of 15 lb/in. (103.4 kPa) and a chamber temperature of 121℃ (250°F). For anything other than liquids, the time required for sterility is 15 min after the chamber has reached the sterilization temperature of 121℃. The time required for the sterilization of liquids varies considerably depending on the volume: The greater the volume, the longer it will take for the contents to reach sterilization temperature. Unfortunately, there are evidently some heat-resistant species of *Bacillus*(杆菌) that can withstand recommended sterilization times as well as instrument flaming.

Figure 3.1　Portable electric autoclave for wet sterilization of media and equipment

All of the residual air in the chamber must be removed so that the steam is in direct contact with all the materials in the autoclave. If a pressure cooker is used, do not close the escape valve until a steady stream of pure steam is evident. At the end of the sterilization period, the pressure must be permitted to return to the atmospheric level slowly because rapid decompression will cause the liquids to boil out of the vessels. Prolonged autoclaving must be avoided because it results in the degradation of certain components of the medium. For example, one group reported that 5% of the sucrose in a liquid medium was hydrolyzed during autoclaving. If the autoclave does not have a drying cycle, paper products should be placed in a drying oven (<60℃) briefly in order to evaporate the condensed moisture. Steam in the autoclave chamber must penetrate the materials; a temperature of 121℃ will not by itself achieve sterilization. With the exception of the flasks, instruments and other materials should be wrapped in unwaxed kraft paper(无蜡牛皮纸). Although aluminum foil is commonly used as a wrapping, it is impermeable to the steam vapors and therefore is not recommended. Demineralized water should be used in boilers of autoclaves that generate their own supply of steam, as well as in pressure cookers. Steam generated by external power plants often contains contaminates that may be absorbed by the materials in the autoclave. Caution should be used in subjecting plastic labware to steam heat.

Another form of wet-heat sterilization is a boiling-water bath. The bath is filled with distilled water and heated to 100℃. Instruments, placed directly into the water, should remain in the boiling water for a minimum of 20 min. Although microorganisms in the vegetative state are destroyed by this treatment, spores will be unaffected. Also, a boiling-water bath is ineffective as a device for sterilization above 6,000 ft elevation.

MICROWAVE

Liquid and agar media can be sterilized using a household-type microwave oven. The required microwave treatment is somewhat empirical since it depends on the energy produced by the magnetron(磁控管), vessel type, volume of medium, and the presence of energy-sink water reservoirs. The problem of the agar media boiling over in test tubes was diminished by placing in the oven two 1-liter Pyrex borosilicate(一种耐热玻璃) glass bottles containing distilled water; liquid media can be microwaved without the use of the energy-sink water bottles.

MICROFILTRATION

Microfiltration is the process of removing contaminants in the range of 0.025-10 μm from fluids by passage through a microporous medium, such as a membrane filter. This is necessary for media components that are degraded during autoclaving and must be sterilized at room temperature. A relatively small volume can be sterilized by passage through a filtration unit attached to a hypodermic syringe(皮下注射器). For example, the Swinney (stainless steel) and the Swinnex (polypropylene) are reusable units equipped with membrane filters, whereas the Millex units are disposable (Millipore Corporation). Millex units are available in 4-, 13-, 25-, and 50-μm diameters. The appropriate volume of the sterile liquid is added directly to the autoclaved medium with a graduated syringe. If an agar medium is used, this is done while the agar is still hot (approximately 45℃) and in the liquid state. For larger volumes special units are equipped for either vacuum or pressure operation.

Most of the membrane filters are screens consisting of a uniform continuous mesh of polymeric material with pore size precisely determined by the manufacturing process. For the sterilization of hydrophobic solvents, such as dimethyl sulfoxide(二甲基亚砜, DMSO)solutions, a Fluoropore (Millipore) filter constructed of polytetrafluoroethylene (多聚四氟乙烯) is recommended. Disposable filtration units composed of polystyrene (聚苯乙烯) with 0.1-or 0.2-μm cellulose acetate membranes are practical for infrequent filtrations. These units have been presterilized by radiation and may be purchased with a variety of filter pore sizes and filling and receiving containers. Although some workers use a 0.45-m pore size, the pore diameter should measure at least 0.22 μm for the complete removal of all bacteria, yeasts, and molds.

CHEMICALS

The working area is generally disinfected with either ethanol or isopropanol (70% v/v). Although acidified alcohol (70% v/v, pH 2.0) may be a more effective disinfectant, it has a corrosive effect on metal instruments.

The surface sterilization of plant material may be accomplished with an aqueous solution of either sodium hypochlorite(次氯酸钠, NaOCl) or calcium hypochlorite(次氯酸钙, Ca[OCl]$_2$). Most workers use common household bleach such as Clorox. These commercial products contain about 5% NaOCl as the active agent. These are usually used at concentrations from 10%-20% (v/v), that is, containing about 0.5%-1.0% NaOCl. This concentration is adequate for the surface sterilization of pith parenchyma (髓实质) explants. There is no general agreement in regard to NaOCl concentrations, and higher levels are often used. The sterilization may be enhanced by agitating the solution and by the addition of a wetting agent such as Tween 20 or Tween 80.

The following procedure-preparation of a potato tuber(块茎) for explant removal-may be used as an example. Scrub the unpeeled tuber surface with water containing a detergent. Rinse the tuber under running tap water for a few minutes. Immerse the tuber for 30 min in a 20% bleach solution (200 cm^3 bleach: 800 cm^3 distilled water). Add 10 drops of Tween 20 (polyoxyethylene sorbitan monolaurate) to the bleach solution. After protecting the hands with surgical gloves, peel the tuber leaving only the inner parenchymatous tissue. Care must be taken to remove all traces of eyes (buds), and surface blemishes(污点,伤口). Cut rectangular blocks of tissue and surface sterilize them in a 10% bleach solution for 15 min. Gently agitate the solution on an orbital shaker, with a magnetic stirrer(磁力搅拌器), or by hand. The tissue blocks must then be thoroughly rinsed with three changes of sterile DDH$_2$O in order to remove all traces of the hypochlorite (次氯酸盐). Subsequently, cylindrical primary explants are prepared with a sterile cork borer(穿孔器). A similar procedure was used in studies on potato callus. Because of the corrosive effect on metal instruments within the chamber, the bleach solution and the rinse water should be discarded immediately after use.

Several agents were tested by Sweet and Bolton (1979) for the surface sterilization of seeds. Calcium hypochlorite was one of the most effective and least injurious agents. Sodium ions (i. e., in sodium hypochlorite) can induce abnormal development in some seedlings. In addition to $Ca(OCl)_2$, the mixture contained a phosphate buffer giving a final pH of 6.0, and a 1.0% solution of either Triton or Tween 80 as a wetting agent. The seeds were immersed in the hypochlorite mixture for 10 min and then rinsed in three changes of sterile distilled water. Other workers have used 9%-10% calcium hypochlorite solutions for periods ranging from 5 to 30 min.

Some workers prefer to treat seed surfaces with a combination of hypochlorite and ethanol. One worker suggests treating barley（大麦）seeds in the following manner: Seeds were immersed in a 20% bleach solution containing a detergent and agitated 15-20 min with a magnetic stirrer. They were transferred aseptically to a Büchner funnel and rinsed with sterile DDH_2O. Then the seeds were covered with 70% ethanol for 1 min, and finally rinsed with sterile DDH_2O.

Some plant tissues pose a problem if they contain microorganisms within the tissue sample. Surface treatments will obviously be completely ineffective. If a fleshy organ examined for explant preparation shows any localized discoloration（变色）, it should be rejected. The goal in surface sterilization is to remove all of the microorganisms with a minimum of damage to the plant cells to be cultured. In some cases, the achievement of this goal is empirical and the worker must be flexible in the approach to the problem. In the case of seeds, the use of higher concentrations of chemicals or longer periods of treatment does not appear to improve the decontamination without reducing the percentage of germination.

Leifert and Waites (1990) have reviewed the problem of contaminants of plant tissue cultures. The most difficult plant material for surface sterilization included plant tissues exposed to or near the soil, field-grown plants in tropical climates, and plants receiving overhead irrigation. Faulty（错误）sterile technique is a major cause of culture contamination. Approximately 50% of the bacteria found after numerous subcultures are unlikely to have been present in the primary explant. The identification of *Staphylococcus* spp.（金黄色葡萄球菌）, *Micrococcus* spp.（微球菌）, and *Candida albicans*（白色念珠菌）in contaminated cultures clearly signifies human error since these microorganisms are obligate human parasites.

ANTIBIOTICS

Although antibiotics are used routinely in animal cell cultures, they have not been widely used in plant tissue cultures. The early botanists were aware that these natural products may alter the growth and development of plant tissues cultured in vitro. Antibiotics should be regarded as a form of prophylaxis(预防方法). Prophylactic use of antibiotics in medicine depends on three conditions, according to Falkiner (1990): A single known pathogen is targeted; the pathogen remains sensitive to the drug; and the period of exposure to the drug is short and limited to the period of maximum risk. These criteria are helpful, to some extent, to plant scientists. Identification of the contaminants is highly important, and it would be prudent to limit the treatment to the stages of culture when prophylaxis would be most beneficial. The effectiveness of antibiotics is questionable, and their usage is not a substitute for the strict adherence to proper sterile techniques. No known antibiotic is effective against all microorganisms that might cause contamination. In one study eight agents were tested, and none was capable of completely eradicating(根除) bacterial contamination. These agents, or their degradation products, may be metabolized by plant tissues with unpredictable results. Xylogenesis in explants of lettuce(莴苣) pith and Jerusalem artichoke(洋姜,鬼子姜) tuber was strongly inhibited in the presence of gentamycin sulfate(硫酸庆大霉素), although this effect was observed within the concentration range (50-100 $\mu g/cm$) recommended by the manufacturer for use in tissue cultures. If antibiotic usage is deemed necessary, the agents of choice are those that act specifically within bacterial cell walls and bacterial membranes. Selection of an effective agent cannot be made until the contaminants are known.

HEALTH HAZARDS

Hypochlorite solutions should be used with care. Inhalation can produce severe bronchial irritation and skin contact can be harmful. One should never use the mouth to pipette a hypochlorite solution, or any other chemical solution used in the tissue culture laboratory: Pipette fillers are available and inexpensive.

A fire danger exists if the student, after flaming an instrument, reinserts the hot instrument into the alcohol dip. Ethanol is highly inflammable. One must be extremely careful about spilling ethanol or other alcohol in the vicinity of an open flame. The test tube containing the 80% (v/v) ethanol should be kept in a metal can to avoid the release of burning ethanol in case of glass breakage.

Ultraviolet radiation poses some serious health risks. One should never look at a live tube with the naked eye. UV burns to the eye are very painful, although not normally of lasting effect. A glass barrier between the eyes and the UV source provides complete protection. UV radiation can produce irritation to unprotected skin, so avoid placing the hands in the cabinet when the lamp is on. Another problem is the formation of ozone (O_3) resulting from the photochemical reaction with atmospheric oxygen. This explosive gas is a powerful oxidizing agent, and high concentrations can cause severe irritation to the respiratory tract and eyes. The UV lamp should never be left on for long periods of time with the hood closed.

Mercuric chloride (HgCl$_2$) has been used for surface sterilization, but this chemical is an *extremely* dangerous poison. An aqueous solution of HgCl$_2$ is slightly volatile at room temperature, and this has resulted in mercury poisoning to laboratory workers.

Although Lysol (来苏尔, 3.5%, v/v) kills vegetative cells, it is ineffective in eliminating spore contamination. Lysol is a preparation of cresols (甲酚) and phenol and consequently leaves an oily film on surfaces. In addition, it can produce a severe burn to the skin. If residual Lysol is autoclaved, it will result in the chemical contamination of the contents of the autoclave plus the autoclave itself. Because Lysol is a common constituent of industrial floor cleaners, one should make certain that such chemicals are not used in the tissue culture laboratory. Volatile phenols can have an adverse effect on cultures.

Gas sterilization should not be used in a classroom laboratory without *strict* supervision by an instructor trained in the handling of toxic gases. Small plastic objects can be sterilized by subjecting them to a saturated atmosphere of ethylene oxide(氧化乙烯) for several hours in a sealed container at room temperature. Ethylene oxide is violently explosive in nearly all mixtures with air, and the gas is toxic at concentrations not detected by smell. It is highly irritating to the eyes and mucous membranes, and high concentrations can cause pulmonary edema(肺部水肿).

Special care must be taken in handling certain antibiotics. Vancomycin(万古霉素) and chloramphenicol(氯霉素) are toxic to humans, and some of the β-lactams(β-内酰胺) are allergenic in humans.

AVOIDING CONTAMINATION

One important factor in preventing contamination is the elimination of drafts that carry airborne microorganisms into the working area. Keep all doors and windows closed during aseptic procedures. It is best not to have any other person in the room during aseptic transfers. Avoid breathing or coughing into the working area. A surgical face mask is often worn by those working in laminar flow hoods. It is of the utmost importance that your hands, wrists, and forearms be scrupulously(小心谨慎地) clean, and *never* pass your hand or arm directly over a sterile exposed surface (e.g., a water rinse or an open agar plate). All sterile open surfaces should be placed as far back in the hood as conveniently possible. When pouring sterile liquids, grasp the flask at the base, and keep the hands as far as possible from the open tube or Petri dish receiving the liquid. In opening a sterile Petri dish, hold the lid with the thumb and middle finger on opposite sides, and gently pull the lid back; that is, do not permit the fingertips to pass over the sterile bottom half of the plate. A common practice in microbiology is to flame the mouth of culture tubes and flasks during transfers. Unfortunately, this practice introduces ethylene into the vessels. This gas, a production of combustion, is a plant hormone and therefore can influence the growth and morphogenesis of plant tissue cultures. At the conclusion of each step of the procedure, remove all unnecessary glassware, instruments, aluminum foil, and other materials that have been used.

Finally, it is important to remove all contaminated cultures from incubators and plant growth chambers. They will ultimately produce spores and place the entire facility in jeopardy(危险). With a plastic squeeze bottle, add a few cubic centimeters of ethanol (70%, v/v) to each of the contaminated cultures

QUESTIONS FOR DISCUSSION

1. Can you think of ways to avoid contamination that were not mentioned in this chapter?

2. What are the most important hazards to your safety as a worker using plant tissue culture techniques?

3. Why is it important to identify contaminating microorganisms?

4. Can you think of some reasons why antibiotics are not used more commonly in plant tissue cultures?

知识要点

无菌的培养环境是实现植物细胞、组织和器官成功离体培养的前提,因此实验材料、培养基和培养器皿等都要进行彻底的消毒灭菌。无菌操作技术包括清洗、消毒灭菌和无菌操作等环节。紫外灯可进行表面消毒,烤箱可对耐热和不易燃的器具器皿进行干热灭菌,高压灭菌锅可对耐高温和耐腐蚀的器具器皿和溶液进行湿热灭菌,抽滤装置可对不耐高温的生物溶液进行灭菌。超净工作台可为我们提供一个无菌的实验操作环境。用75%酒精消毒操作台面、手以及用酒精灯烧灼瓶口等是必须要养成的好习惯。

CHAPTER 4 MEDIA COMPOSITION AND PREPARATION(培养基组成和制备)

Because there is a division of labor by different organs of the plant in the biosynthesis of organic metabolites, we know less about the nutritional requirements of the individual organs and tissues of the plant than we do of the whole plant. A tomato plant growing in the garden requires only an external supply of mineral elements for the successful completion of its life cycle. An isolated root of this tomato plant has different requirements for normal growth and development. In addition to the essential mineral elements, the isolated root also requires certain organic compounds because in the whole plant the root system was provided with these compounds. That is, essential organics were synthesized elsewhere in the plant and transported down to the root system. The requirement for a particular organic supplement could be due either to the inability of the culture to produce it or to a new requirement resulting from a shift in metabolism. In 1934 White discovered that isolated tomato roots had the potential for unlimited growth if they were provided with a liquid medium containing a mixture of inorganic salts, sucrose, thiamine(维生素B1), pyridoxine(维生素B6), nicotinic acid(烟酸), and glycine.

The components of a plant tissue culture medium include macronutrients(大量营养成分), micronutrients(微量营养成分), a separate iron supplement, vitamins, a carbon source, and usually plant growth regulators. Amino acids and various nitrogenous compounds may be present in the vitamin mixture. Other topics discussed in this chapter include complex organic additives, charcoal(活性炭), osmotica(渗透剂, osmoticum 的复数形式), water, medium matrix, and high-temperature degradation of media components.

MACRONUTRIENTS

Cultured plant tissues require a continuous supply of certain inorganic chemicals. Aside from carbon, hydrogen, and oxygen, the essential elements required in relatively large amounts are termed macronutrient elements. The macronutrients are nitrogen(氮), phosphorus(磷), potassium(钾), calcium(钙), magnesium(镁), and sulfur(硫). Nitrogen, added in the largest amount, is present as either a nitrate (NO_3^-) or an ammonium (NH_4^+) ion, or a combination of these ions. Magnesium sulfate ($MgSO_4 \cdot 7H_2$

O) satisfies both the Mg and S requirements. Sulfur may also be present in the form of Na_2SO_4. Phosphorus can be represented by $NaH_2PO_4 \cdot H_2O$, KH_2PO_4, or $(NH_4)H_2PO_4$. Potassium, the cation (阳离子) found in the largest amount, is given as KCl, KNO_3, or KH_2PO_4. The calcium requirement involves $CaCl_2 \cdot 2H_2O$, $Ca(NO_3)_2 \cdot 4H_2O$, or anhydrous(无水) forms of these salts.

MICRONUTRIENTS

Traces of certain mineral elements are required by all plant cells. Because these quantities are exceedingly small for some of the elements, a more concentrated stock solution (except for iron) is prepared in advance. (It is also convenient to prepare in advance stock solutions of macronutrients, vitamins, and certain plant growth regulators. These solutions can be stored for limited periods of time in glass containers at 4℃.) Micronutrient elements essential for all higher plant cells include iron, manganese(锰), zinc(锌), boron(硼), copper(铜), molybdenum(钼), and chlorine(氯).

Certain plants have unique requirements for some micronutrients; media nurturing explants taken from these plants could have similar requirements. Sodium could be a micronutrient for the culture of certain halophytes(盐土植物), plants with C4 photosynthetic pathways, and plants with Crassulacean acid metabolism(景天酸代谢,简称 CAM). Since nickel(镍) is essential for the structure and functioning of urease(脲酶), cultures of jackbean (刀豆,*Canavalia ensiformis* L.) and soybean (黄豆,*Glycine max* L.) may require a trace of nickel in the medium. Although iodine(碘)in the form of KI is a constituent of several media, the necessity of this element remains questionable. A trace of cobalt(钴) is found in several media, and yet this element is not known to have any function in higher plants. One unresolved problem is that even the purest chemical reagents will contain traces of inorganic contaminants, and these always constitute a hidden source of micronutrients. Agar is a source of numerous minerals, possibly traces of vitamins, and even toxic substances. Little critical work has been done on the micronutrient requirements of media designed for special purposes.

Owen and Miller (1992) have published the correct concentrations and chemical formulations of inorganic constituents for the following media: White, Murashige and Skoog (MS), B5, Nitsch and Nitsch, Anderson, and Lloyd and McCown's woody plant medium.

IRON SUPPLEMENT

An iron stock is prepared separately because of the problem of iron solubility. This element requires acidic conditions for solubility. Usually the iron stock is prepared in a chelated form(螯合形式)as the sodium salt of ferric(三价铁) ethylenediaminetetraacetic acid (NaFeEDTA). EDTA itself, however, may have side effects on certain enzyme systems and on morphogenesis in cultures. Some inhibitory effects may be due to light-induced degradation of EDTA.

VITAMINS

Vitamins have catalytic functions in enzyme systems and are required only in trace amounts. Thiamine (vitamin B_1) may be the only essential vitamin for nearly all plant tissue cultures, whereas nicotinic acid (niacin) and pyridoxine (vitamin B6) may stimulate growth. Thiamine is added as thiamine-HCl in amounts varying from about 0.1 to 10.0 mg/L. The need for thiamine is particularly evident at low levels of cytokinins. In the presence of fairly high concentrations of cytokinins (0.1-10.0 mg/L), tobacco cells grew without the addition of exogenous thiamine. Presumably tobacco cultures can develop the capability of synthesizing thiamine.

Both nicotinic acid and pyriodixine are required for the culture of *Haplopappus gracilis* (纤细单冠葡). Nicotinic acid is an essential growth factor for the optimal growth of sugarcane(甘蔗) suspension cells.

Some other vitamins that have been used in plant tissue culture media include *p*-aminobenzoic acid (PABA; vitamin B_x), ascorbic acid (vitamin C), tocopherol (vitamin E), biotin (vitamin H), choline chloride (氯化胆碱), cyanocobalamin (vitamin B_{12}), folic acid (vitamin B_c), calcium pantothenate (泛酸钙, vitamin B_5), and riboflavin (vitamin B_2).

CARBON SOURCES

All plant tissue culture media require the presence of a carbon and energy source. Sucrose or D-glucose is usually added in concentrations of 20,000-30,000 mg/L. Nearly all cultures appear to give the optimum growth response in the presence of the disaccharide(二糖)sucrose, whereas there can be considerable variability in growth when other disaccharides or monosaccharides(单糖)are substituted for sucrose. There are, however, some photoautotrophic(光自养)cultures that are grown in the absence of an exogenous organic carbon source. In these cultures CO_2 fixation via photosynthesis provides sufficient carbohydrate. Although many laboratories autoclave sucrose with the remainder of the nutrient medium, sucrose is heat labile(不稳定的), and the result is a combination of sucrose, D-glucose, and D-fructose(D-果糖). Such an autoclaved medium may give completely different results compared to a medium containing filter-sterilized sucrose.

The cyclitol *myo*-inositol(肌醇) is added to some vitamin supplements as a growth factor at a concentration of 100 mg/L, although higher concentrations may be present in special media. Although this compound is a carbohydrate, it has special functions mainly in the form of phosphoinositides(磷酸肌醇) and phosphatidylinositol(磷脂酰肌醇). Inositol bisphospholipids(肌醇二磷酸酯) may have a role in the calcium messenger system and the *IAA-myo-*inositol conjugate is thought to play roles in the storage, transport, and release of auxin.

The choice and concentration of the sugar to be used depend mainly on the tissue to be cultured and the purpose of the experiment. A recent study demonstrated that the optimum concentration of sucrose for the induction of xylogenesis(木质部形成) in lettuce pith explants was 0.2% (w/v), although xylem formation was stimulated by the addition of as little as 0.001% (1.0 mg/100 cm^3) sucrose.

The question of the purity of the carbohydrate reagents has been raised because the occlusion(吸收) of a variety of organic substances, particularly traces of amino acids, occurs during the crystallization(结晶化) of sucrose.

PLANT GROWTH REGULATORS

The growth regulator requirements for most callus cultures are some combination of auxin and cytokinin (Fig. 4.1). The term "hormone" should be reserved for naturally

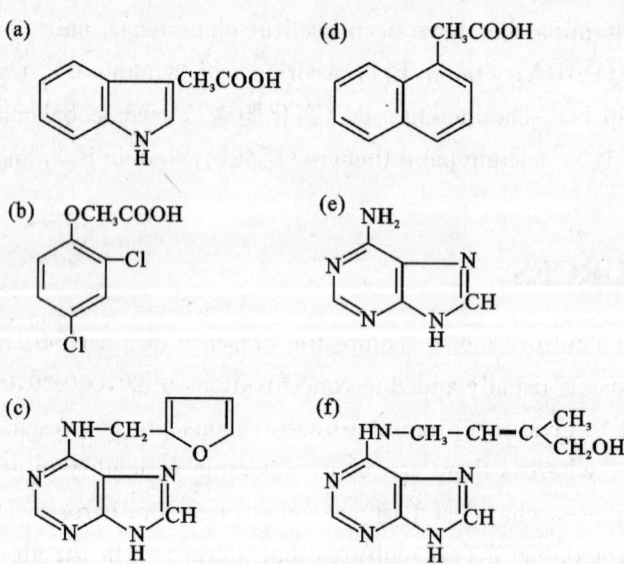

Figure 4.1 Structural formulae of some auxins and cytokinins. Auxins include (a) indole-3yl-acetic acid (IBA), (b) α-naphtha leneacetic acid (α-NAA), and (c) 2,4-dichlorophenoxyacetic acid (2,4-D) are auxins. (d) adenine, (e) kinetin, and (f) *trans*-zeatin are cytokinins.

occurring plant growth regulators. Synthetic growth regulators, such as 2,4-D and kinetin, are not considered to be plant hormones. Auxins, a class of compounds that stimulate shoot cell elongation, resemble IAA in their spectrum of activity. As a supplement they are useful to stimulate the formation of adventitious roots(不定根), inhibit bud formation, and to play a role in embryogenesis. Cytokinins, which promote cell division in plant tissues under certain bioassay conditions and only in the presence of auxin, regulate growth and development in the same manner as kinetin (6-furfurylaminopurine). In addition to callus initiation, cytokinins stimulate bud proliferation and inhibit rooting. Cytokinins are mainly N^6-substituted aminopurine derivatives(氨基嘌呤衍生物), although there are some exceptions.

Auxin-cytokinin supplements are instrumental in the regulation of cell division, cell elongation, cell differentiation, and organ formation. Gibberellins(赤霉素, GA) are rarely added to culture media, although GA_3 has been used in apical meristem cultures (顶端分生组织培养)and studies on vascular differentiation. Increasing attention has been given to ethylene in the initiation of buds and tracheary element differentiation. Relatively few studies have used abscisic acid (脱落酸, ABA) as a supplement. Some auxins employed in culture media include IAA, α-NAA, 2,4-D, and picloram(毒莠定). Both the α and β isomers of NAA are commercially available, but the (isomer is always used in media. The β isomer is a weak auxin with relatively little physiological activity. Indole-3-butyric acid (吲哚丁酸, IBA), p-chlorophenoxyacetic acid (对氯苯氧基乙酸, 4-CPA), and 2,4,5-trichlorophenoxyacetic acid (2,4,5-三氯苯氧乙酸, 2,4,5-T) are also effective auxins. IBA is a particularly effective rooting agent. IAA is a naturally occurring auxin, but unfortunately it is rapidly degraded by light and enzymatic oxidation. Because IAA oxidase may be present in cultured tissues, IAA is added to media in a relatively high concentration (1-30 mg/L). The synthetic α-NAA is not subject to the same enzymatic oxidation as IAA, and it may be effective in a lower concentration (0.1-2.0 mg/L). One of the most effective auxins for callus proliferation is 2,4-D (10^{-7}-10^{-5} M), often employed in the absence of any exogenous cytokinin. This herbicide is a powerful suppressant of organogenesis and should not be used in media involving root and shoot initiation. Picloram(毒莠定) offers certain advantages over 2,4-D: It is water soluble, effective at lower concentrations than 2,4-D, may be less toxic to plant tissue cultures at optimum levels, and offers the potential for direct regeneration of plants from calli (callus 的复数).

The most widely used cytokinins in media are kinetin, 6-benzylaminopurine (6-苄氨基嘌呤, N^6-benzyladenine, 6-BA), and zeatin (玉米素). Kinetin and 6-BA are synthetic compounds, whereas zeatin occurs naturally in plants. Other naturally occurring cytokinins, which are considerably less expensive than zeatin, are 6-[γ,γ-dimethylallylaminojpurine (2iP) or N^6-[Δ^2-isopentyljadenine (IPA). Also, 1,3-diphenylurea (1,3 二苯脲) exhibits cytokinin-like responses in some bioassays. Another chemical with potent cytokinin activity is thidiazuron (苯基噻二唑基脲, 噻苯隆, TDZ), a substituted phenylurea. Kinetin is typically added at a concentration of 0.1 mg/L, in combination with a source of auxin, for the induction of callus. A preparation of autoclaved coconut water can be added to culture media as a cytokinin source for a final concentration in the medium of 10-15% (v/v).

There are exceptions to this dual requirement for auxin and cytokinin. Some cultures require no exogenous auxin. Although some explants initially may have high endogenous auxin, cultured tissues apparently can develop auxin biosynthetic pathways. The terms "anergy（无反应力）" and "habituation（适应）" have been given to this autonomous condition, originally studied in cultures of tumor tissue. Some cultures require the addition of auxin, but not cytokinin. The conversion of cultured tobacco cells to a cytokinin-habituated phenotype occurs in response to cytokinin or high temperatures. Although the habituated state is highly stable, reversion occurs when cloned lines are induced to form plants. An interesting example of what appears to be a combined auxin-cytokinin habituation was reported for callus production from cultured leaf disks of sugar beet（甜菜）cultivars. Callus was initiated on a hormone-free medium after an average of 96.7 days, and subsequently organogenesis occurred on the same medium in some populations.

AMINO ACIDS AND OTHER NITROGENOUS ADDITIVES

With the exception of glycine, which is a component of several media, amino acids are not generally added to plant nutrient media. If a mixture of organic nitrogen is necessary, the medium can be enriched with either casein hydrolysate or casamino acids（水解酪蛋白，0.05%-0.1% w/v）. Casein, a bovine milk protein, consists of an ill-defined mixture of at least 18 different amino acids. Assuming this supplement has a beneficial effect, additional experiments can be made substituting various amino acids and amides（氨基化合物）for the hydrolysate. Ultimately one may be successful in identifying the specific organic nitrogen requirement. Some of the nitrogen compounds that are used most frequently include L-aspartic acid(天冬氨酸,asp), L-asparagine（天冬酰胺,asn）, L-glutamic acid(谷氨酸,glu), L-glutamine(谷氨酰胺,gln), L-arginine(精氨酸,arg), and L-tyrosine（酪氨酸,tyr）. Traces of L-methionine(蛋氨酸), added to the medium for the enhancement of ethylene biosynthesis, have a stimulatory effect on xylogenesis（木质部发生）. Often growth inhibition occurs following the addition of a combination of amino acids, a phenomenon that has been attributed to competitive interactions among the various amino acids.

COMPLEX ORGANIC SUPPLEMENTS

The trend in plant tissue culture has been to attempt to define all of the constituents of a given medium and to eliminate the use of crude natural extracts. Such products as peptone(蛋白胨), yeast extract, and malt extract(麦芽抽提液) are seldom used today. Although this attitude is commendable from a scientific viewpoint, the use of natural extracts should not be ignored when chemically defined media fail to produce the desired results. Fruit juices are also useful supplements. Explants from several *Citrus* spp. (橘) were stimulated in growth by the addition of orange juice to the medium. Tomato juice (30% v/v) has been used effectively. Banana powder and coconut water are available commercially from Sigma.

CHARCOAL

Activated charcoal (AC) will absorb many organic and inorganic molecules from a medium. Although the precise effects of AC are unknown, there are several possible modes of operation. It may remove contaminants from agar and secondary products secreted by the cultured tissues, or regulate the supply of growth regulators. Some of the effects of AC may be due to darkening of the support matrix and thus approximate soil conditions more closely. As a nutrient supplement AC has been reported to stimulate embryogenesis. On the other hand, AC can inhibit growth and morphogenesis. The type of AC used is important because the adsorptive characteristics and pH are dependent on the manufacturing process. Wood charcoal is much higher in carbon content in comparison to bone charcoal, and the latter preparation contains ingredients that may adversely affect plant tissue cultures.

OSMOTICA(渗透)

The uptake of water by plant cells is governed by the relative water potential values between the vacuolar sap(液泡液) and the external medium. The major components of the nutrient medium that influence water availability are the concentration of the agar, the amount of sugar present, and any nonmetabolite added as an osmoticum(渗透调节物质). One colloidal(胶状) characteristic of the gel state of agar is the imbibitional (吸入) retention of water within the micelles (微胶粒)of the gel. Carbohydrates not only function as a carbon source, but they play an important role in the regulation of the external osmotic potential. Often a weakly metabolized sugar for example, mannitol(甘露糖)or sorbitol(山梨糖)-is used as an external osmoticum. Polyethylene glycol (PEG) has been used as an osmoticum in protoplast fusion experiments and in the cryopreservation of cultures.

WATER

The water used in tissue culture media, including the water employed in the preparation of the explant, should be double distilled or of equivalent purity. In addition to glass distillation, other water purification methods include screen and depth filters, electrodialysis(电透析), carbon adsorption, resin-based deionization, ultrafiltration, and reverse osmosis. Only distillation and reverse osmosis have the capability of removing all the sources of contamination (inorganics, organics, bacteria, pyrogens, and particles). One should be cautioned against the prolonged storage of redistilled water in polyethylene containers since these receptacles(容器)release substances that may be toxic to the cultures. The lengthy storage of sterile water is unwise, and this has been a problem in some of our hospitals.

MEDIUM MATRIX

Unless the culture is suspended in an aqueous medium, it is grown on a semisolid gel or solid (porous) matrix. Many early experiments were based on agars from Difco Laboratories, and these evidently contained unknown contaminants; consequently, the matrix itself was a nutritional supplement. Today we have a wide range of relatively pure gelling agents. The choice of the appropriate gelling agent, and the concentration to be employed, may be as important as the selection of the ideal nutrient mixture. Here are a few specific examples: An agarose preparation termed Sea Plaque (FMC Corporation) is a suitable matrix. Phytagel (Gelrite, Merck) is an agar substitute synthesized from gellan gum. Agargel (Sigma), a blend of agar and Phytagel, was devised to control vitrification(玻璃化)of cultures. Transfergel (Sigma) is a carrier gel, supplemented with a complete nutritional medium, for nurturing propagules(繁殖体)such as somatic embryos, microcuttings, and shoot tips.

In addition to gels, other materials have been tried. Filter-paper platforms were introduced by Heller (1965), and filter-paper disks impregnated (添加) with nutritives have been used. Glass fiber filters are useful supports for cultures, although they should be pretreated to remove contaminants and to saturate the cation exchange sites on the glass fibers with specific ions. Filter paper has been used to separate contiguous(相邻) cultures of two different origins, that is, as a "nurse" culture. To measure the growth of a culture with a minimum of disturbance, a thin layer of cells can be separated from the agar medium by a filter-paper disk. The disk and the cultured cells can be periodically removed, weighed aseptically(无菌地), and replaced on the medium without sacrificing the cells.

HIGH-TEMPERATURE DEGRADATION OF MEDIA COMPONENTS

Several chemicals employed in plant tissue culture media degrade on exposure to steam sterilization. Gibberellins(赤霉素)are rapidly degraded by high temperatures, and the biological activity of a freshly prepared solution of GA_3 was reduced by more than 90% as a result of autoclaving. The auxins IAA, (-NAA, and 2,4-D are relatively thermostable depending on the inorganic basal medium and supplements. Aqueous solutions of kinetin, zeatin, and 2iP have been chromatographed on thin-layer silica-gel chromatograms before and after prolonged autoclaving with no breakdown products detected. On the other hand, biologically inactive 1, 3-or 9-substituted purine molecules were converted into callus-inducing N^6-substituted purines by autoclave treatment. Crude plant extracts that possibly contain inactive purine molecules should be filter sterilized. Heat sterilization apparently has no effect on the isomers of abscisic acid.

Vitamins have varying degrees of thermolability. Most workers autoclave the vitamins with the remainder of the medium. Nicotinic acid(烟酸), pyridoxine(Vitamine B_6), and thiamine(维生素 B_1)in an MS liquid medium (pH 5.5-5.6) showed no signs of degradation after autoclaving; nevertheless, thiamine is rapidly destroyed if the pH of the medium is much above 5.5. Calcium pantothenate(泛酸钙)cannot be autoclaved without destruction. If the research study involves vitamin activity, then the vitamins should be sterilized by microfiltration. Since those employed in most of our experiments are apparently thermostable, the vitamin supplement will be added to the medium prior to autoclaving.

One of the most frequently employed carbohydrates in media is sucrose. This disaccharide decomposes, to some extent, on autoclaving to release a mixture of D-glucose and D-fructose. A recent study indicated that 5 percent of the sucrose in an MS liquid medium was hydrolyzed during autoclaving. This degradation can be inhibitory to some cultured tissues. Presumably the toxicity is due to the degradation products of D-fructose. Steam sterilization may also catalyze reactions within the media between carbohydrates and amino acids.

SOME SUGGESTIONS ON THE SELECTION OF MEDIUM

The choice of a particular medium depends on the species of the plant, the tissue or organ to be cultured, and the purpose of the experiment. If the plant material has been cultured successfully in other laboratories, it is always best to start with published information. A suitable starting point for the initiation of callus from a dicot tissue explant would be the preparation of the MS basal medium. One characteristic of this medium is its relatively high concentration of nitrate, potassium, and ammonium ions in comparison with other formulations. The B5 medium is another effective basal mixture. In addition to the basal mineral salts, it is recommended to add the MS vitamin mixture, myoinositol (100 mg/L), and sucrose (2%-3% w/v). A possible modification of the MS vitamin mixture would be to increase the thiamine content. The B5 medium contains 10 mg/L thiamine in comparison to 0.1 mg/L in the MS medium. For callus formation the addition of 2,4-D (0.2-2.0 mg/L) serves as an effective auxin, and the addition of kinetin or 6-benzylaminopurine (0.5-2.0 mg/L) is advised. If these combinations fail to produce the desired result, then a supplement of amino acids or some natural plant extract might be considered.

Aside from considerations of callus formation, the initiation of various morphogenetic events in vitro often requires special adjustments in the concentration of the components of the medium. Embryogenesis responds favorably to high levels of potassium in some systems. In addition to the proper ratio of growth regulators, shoot formation may require a medium either high in phosphate or low in ammonium nitrate. The total salt concentration may be a factor of some importance.

QUESTIONS FOR DISCUSSION

1. What are sources of inorganic and organic contaminants that are unwittingly added to our cultures?

2. What evidence indicates that cultured plant tissues can experience changes in certain biochemical pathways over time?

3. What are possible substitutes for agar as a medium matrix? Discuss their advantages and disadvantages.

APPENDIX

Preparation of Murashige and Skoog (MS) Stocks

The formulation of Murashige and Skoog's (1962) medium is given in Table 4.1.

Iron stock ($20\times$; Table 4.1B). Dissolve $FeSO_4 \cdot 7H_2O$ in 40 cm^3 of warm DDH_2O in a 100-cm^3 beaker. In a separate beaker dissolve $Na_2EDTA\ 2H_2O$ in 40 cm^3 of warm DDH_2O. Mix the two solutions and transfer to a 100-cm^3 volumetric flask. Add DDH_2O to the final volume. The iron stock should be protected from light by storing the solution in an amber bottle (琥珀瓶), or wrap the entire flask with aluminum foil (锡箔纸). Store at room temperature since precipitation may occur at chilling temperatures. Pipette 5 cm^3 of iron stock for 1 liter of MS medium.

Micronutrient stock ($100\times$; Table 4.1C). Add approximately 400 cm^3 DDH_2O to to a 1-liter beaker. Weigh and dissolve each of the salts to a 1-liter volumetric flask, and add DDH_2O to the final volume. Store under refrigeration. Pipette 10 cm^3 of the micronutrient stock for 1 liter of MS nutrient medium.

Vitamin stock ($100\times$; Table 4.1D). Add about 50 cm^3 DDH_2O to a 100-cm^3 beaker. Weigh and dissolve each of the vitamins indicated. Transfer the vitamin mixture to a 100-cm^3 volumetric flask, and add DDH_2O to the final volume. Store under refrigeration. Pipette 1 cm^3 of vitamin stock for 1 liter of MS medium.

Note: Do not pipette directly from stock bottles, and do not return any unused stock solutions to the stock bottles. Label all stock solutions and include the concentration, your initials, and the date of preparation. Although inorganic salts are relatively stable in solution under refrigeration, vitamin stock should be discarded after 30 days. Also, vitamin stock should be visually examined periodically for any signs of microorganisms.

Table 4.1 Medium for Nicotiana tabacum stem callus

Ingredient	Concentrations*	
	Stock	MS medium
(A) Macronutrients		(mg/L)
$(NH_4)NO_3$		1,650
KNO_3		1,900
$CaCl_2 \cdot 2H_2O$		440
$MgSO_4 \cdot 7H_2O$		370
KH_2PO_4		170
(B) Iron	$(mg/100 cm^2)(20\times)$	(mg/L) (5 cm^3 stock gives)
$Na_2EDTA \cdot 2H_2O$	744	37.2
$FeSO_4 \cdot 7H_2O$	556	27.8
(C) Micronutrients	$(mg/L)(100\times)$	(mg/L) (10 cm^3 stock gives)
$MnSO_4 \cdot 4H_2O$	2,230	22.3
$ZnSO_4 \cdot 7H_2O$	860	8.6
H_3BO_3	620	6.2
KI	83	0.83
$Na_2MoO_4 \cdot 2H_2O$	25	0.25
$CuSO_2 \cdot 5H_2O$	2.5	0.025
$CoCl_2 \cdot 6H_2O$	2.5	0.025
(D) Vitamins	$(mg/100 cm^3)(100\times)$	(mg/L) (1 cm^3 stock gives)
glycine	200	2.0
nicotinic acid	50	0.5
pyridoxine · HCl	50	0.5
thiamine · HCl	10	0.1
(E) Auxinpicloram		(mg/L) 3.0
myo-inositol		(mg/L)
sucrose		100
Gelrite (0.2% w/v) pH 5.7		30,000

* In the source publication, ranges of concentrations were used for IAA (1-30 mg/L) and kinetin (0.04-10mg/L). Since picloram will be the sole plant growth regulator for the first experiment involving potato callus, it will be used in this formulation instead of IAA and kinetin. A casein hydrolysate preparation was given as optional (1.0 mg/L). Corrections have been made involving changes in water of hydration for Na_2EDTA and $ZnSO_4$ as given by Owen & Miller (1992).

Source: Murashige & Skoog (1962).

Preparation of the Complete MS Medium

1. Add approximately 400 cm^3 DDH$_2$O to a 1-liter beaker. Weigh and dissolve each of the macronutrient salts given in Table 4.1A using a magnetic stirrer.

2. Pipette the following from the stock solutions: 5 cm^3 iron, 10 cm^3 micronutrients, and 1 cm^3 vitamins.

3. Weigh 100 mg myo-inositol and dissolve it in the medium mixture.

4. Weigh 3.0 mg picloram(毒莠定), dissolve in a few drops of DDH$_2$O, and transfer it to the medium mixture.

5. Add DDH$_2$O until the total volume of liquid is about 800 cm^3. While agitating the solution with a magnetic stirrer, adjust the pH to 5.7 with droplets of 1 N NaOH or 1 N HCl with separate Pasteur pipettes.

6. Transfer the medium to a 1-liter volumetric flask and add DDH$_2$O to the final volume. Store under refrigeration. Label, initial, and give the date of preparation.

Final Procedure

7. Sterilize the Petri dishes or culture tubes in advance with dry heat.

8. Weigh 0.2 g Gelrite (固化剂) and 3.0 g reagent-grade sucrose, and transfer them to a 250-cm^3 flask. Add 100 cm^3 of the MS medium (step 6). Seal the flask with an aluminum foil cap and sterilize the medium with wet heat.

9. While the medium is in the autoclave, clean the interior of the hood with a tissue soaked in 70% (v/v) ethanol. Arrange the sterile Petri dishes or culture tubes to receive the autoclaved medium.

10. After the sterilized medium is removed from the autoclave, the flasks are swirled for a few minutes to ensure the dissolution of the sucrose and to mix the Gelrite with the remainder of the medium prior to pouring into the culture tubes. The flasks that contained the medium should be washed immediately after use, that is, before the residual gel has solidified.

11. After the gel in the tubes has cooled, replace them in the storage jar and wrap the jar in aluminum foil. Store the units in the refrigerator until 1 h before culture time.

Note on the Importance of pH

The pH of the nutrient medium is highly important since it influences the uptake of various components of the medium as well as regulating a wide range of biochemical reactions occurring in plant tissue cultures. Although media are usually adjusted to pH 5.2-5.8 with NaOH and HCl before autoclaving, most media are poorly buffered, and the pH drifts during the course of an experiment. Numerous workers have reported that the high temperature of autoclaving causes the pH to be altered from the preset value. Postautoclave pH may be influenced by the type of carbohydrate employed, the brand of the gelling agent, and the presence and type of activated charcoal.

知识要点

培养基是离体材料赖以生存的营养基质。植物培养基主要是由水、大量无机营养成分、微量无机营养成分、铁添加物、维生素、碳水化合物、氨基酸和植物生长调节类物质等组成。植物细胞的悬浮培养采用液体培养基,愈伤组织的诱导和分化则采用固体培养基。固体培养基的配制需要添加琼脂等固化剂。植物培养基的 pH 值一定要调节到 5.2～5.8,pH 值太高或太低,都不利于培养基的配制和植物材料的生长。为了方便培养基的配制和减少微量元素称量的误差,一般先将培养基中各种成分按照添加成分的性质和量,分别配制成母液,需要时再进行稀释。在培养基配制过程中,要注意生长调节类物质的溶解性、热稳定性和光稳定性,对于不溶于水的物质,需要先用 NaOH、HCl 或乙醇溶解后,再用水稀释。

CHAPTER 5 INITIATION AND MAINTENANCE OF CALLUS(愈伤组织的诱导和培养)

A callus consists of an amorphous mass of loosely arranged thin-walled parenchyma cells arising from the proliferating cells of the cultured explant. Frequently, as a result of wounding, a callus is formed at the cut end of a stem or root. The term "callus" should not be confused with "callose(胼胝质)," another botanical term. The latter refers to a polysaccharide associated primarily with sieve elements. Although the major emphasis has been on angiosperm(被子植物) tissues, callus has been observed in gymnosperms(裸子植物), ferns(蕨类植物), mosses(苔藓), and liverworts(地钱).

The stimuli involved in the initiation of wound callus are the endogenous hormones auxin and cytokinin. In addition to mechanical injury, callus may be produced in plant tissues following an invasion by certain microorganisms or by insect feeding. Plant material typically cultured includes vascular cambia(形成层,cambium 的复数), storage parenchyma, pericycle(中鞘柱) of roots, cotyledons(子叶), leaf mesophyll(叶肉), and provascular tissue(维管束原组织). In fact, all multicellular plants are potential sources of explants for callus initiation.

In 1939 the first successful prolonged cultures of experimentally induced callus were achieved almost simultaneously at the research laboratories of Gautheret in Paris, Nobecourt in Grenoble(格勒诺布尔,法国), and White in Princeton. These cultures were originally derived from explants of cambial tissue of carrot and tobacco. The most important characteristics of callus, from a functional viewpoint, is that it has the potential to develop normal roots, shoots, and embryoids that can form plants and, in addition, can be used to initiate a suspension culture.

Establishment of a callus from an explant can be divided roughly into three developmental stages: induction, cell division, and differentiation. During the initial induction phase metabolism is stimulated prior to mitotic activity. The length of this phase depends on the physiological status of the explant cells as well as the cultural conditions. Subsequently, there is a phase of active cell division as the explant cells revert to a meristematic state(分生组织状态). The third phase involves the appearance of cellular differentiation and the expression of certain metabolic pathways that lead to the formation of secondary products.

The growth characteristics of a callus involve a complex relationship among the plant material used to initiate the callus, the composition of the medium, and the environmental conditions during the incubation period. Some callus growths are heavily lignified（木质化）and hard in texture, whereas others break easily into small fragments. Fragile growths that crumble（易碎）readily are termed "friable cultures." Callus may appear yellowish, white, green, or pigmented with anthocyanin（花青素）. Pigmentation may be uniform throughout the callus or some regions may remain unpigmented.

There is considerable variability in the anatomy of callus cultures. A homogeneous callus consisting entirely of parenchyma cells is rarely found, although exceptions have been reported for *Agave*（龙舌兰）and *Rosa*（蔷薇）cultures. Cytodifferentiation occurs in the form of tracheary elements（导管成分）, sieve elements（筛管成分）, suberized cells（栓细胞）, secretory cells, and trichomes（毛状体）. Small nests of dividing cells form vascular nodules (meristemoids) that may become centers for the formation of shoot apices, root primordia, or incipient embryos（早期胚胎）. Vascular nodules typically consist of discrete zones of xylem and phloem separated by a cambium. The orientation of the xylem and phloem with respect to the cambial zone is influenced by the nature of the original tissue.

The hormonal requirements for the initiation of callus depend on the origin of the explant tissue. Juice vesicles from lemon fruits, and explants containing cambial cells, exhibit callus growth without the addition of any exogenous growth regulators. Most excised tissues, however, require the addition of one or more growth regulators in order to initiate callus formation. Explants can be classified according to their exogenous requirements, in the following manner: (a) auxin, (b) cytokinin, (c) auxin and cytokinin, and (d) complex natural extracts.

After the callus has been grown for a while in association with the original tissue, it becomes necessary to subculture the callus to a fresh medium. Growth on the same medium for an extended period will lead to a depletion of essential nutrients and to a gradual desiccation of the gelling agent. Metabolites secreted by the growing callus may accumulate to toxic levels in the medium. The transferred fragment of callus must be of a sufficient size to ensure renewed growth on the fresh medium. If the transferred inoculum is too small, it may exhibit a very slow rate of growth or none at all. Street (1969) recommended that the inoculum be 5-10 mm in diameter and weigh 20-100 mg. Successive subcultures are usually performed every four to six weeks with cultures maintained on an agar medium at 25℃ or above. Passage time, however, is somewhat variable and depends on the rate of growth of the callus. A friable callus can be subdivided with a thin spatula（刮勺）or scalpel（解剖刀）, transferred directly to the surface of a sterile Petri dish, and sliced into fragments with a scalpel. Only healthy tissue should be transferred, and brown or necrotic tissue must be discarded. Interest has been shown in developing alternative methods for long-term maintenance of tissue cultures, for example, freeze preservation.

What are some of the best plant materials for the initiation of a callus culture? Young healthy tissues that are rich in nutrients, and possibly endogenous hormones, are the best choices for the induction of cell division; for example, storage organs and cotyledons (子叶) of seeds. These include tissues from potato tuber (*Solanum tuberosum*), storage roots of turnip (芫菁, *Brassica rapa*), sweet potato (*Ipomea batatas*), and carrot (*Daucus carota*). Also, callus is easily started from the cotyledons of soybean (*Glycine max*). Stem pith parenchyma from lettuce (*Lactuca sativa*) and tobacco (*Nicotiana tabacum*) readily divides in the presence of auxin and cytokinin.

Woody plant material is generally a poor choice. Plant tissues that are high in oxidase activity pose a special problem since enzymatic browning retards cell division. The browning results from the activity of wound-induced copper oxidases (polyphenoloxidase, 多酚氧化酶). This may be suppressed, to some extent, by the use of an antioxidant mixture.

The purpose of the following experiment is to acquaint (使熟悉) the student with the technique of inducing callus formation in explants excised from a potato tuber (*Solanum tuberosum* L.). Sluggish growth (生长缓慢) has been a common problem in the early initiation of callus from a potato tuber. Hagen and his colleagues (1990) found that the synthetic auxin picloram (毒莠定) was highly effective in initiating and maintaining potato callus. Picloram, the active ingredient in the herbicide Tordon, was effective as the only growth regulator in the medium of Hagen's group. For most potato cultivars a concentration of 10 mM (2.41 mg/L) was optimal, although one cultivar responded better to 20 mM.

PROCEDURE

Prepare 1 liter of MS medium supplemented with picloram (3.0 mg/L), sucrose (3.0% w/v), and Gelrite (脱乙酰吉兰糖胶, 固化剂, 0.2% w/v). Prepare to culture 30 potato tuber explants, which will require the preparation of 30 culture tubes of medium (10 cm^3 each).

1. Prepare a tuber for explant removal. Wash the tuber with water containing a detergent, followed by running tap water. The tuber is given a pretreatment by immersion in a liter of 20% bleach solution containing Tween 20 (10 drops) for 30 min. It is advisable to wear surgical gloves, and care must be taken not to splash any of the hypochlorite solution on the skin or clothing. Rinse the tubers briefly with tap water. With a paring knife and vegetable scraper remove the outer 1-2 mm of suberized periderm (栓化外皮) and carefully remove all traces of buds ("eyes") and surface discoloration (变色).

2. Disinfect the working area within the hood with a tissue soaked with ethanol (70% v/v). Place the tissue blocks in one of the sterile 600-cm^3 beakers. Add the 10% (v/v) hypochlorite solution to the beaker, and set the timer for 10 min. If the hood is e-

quipped with a UV lamp, *do not* turn on the lamp; otherwise, UV-induced degradation of the hypochlorite solution might occur.

3. After about 8 min, preparations can be started for rinsing the tissue blocks. Thoroughly wash your hands with soap and hot water before starting the aseptic procedure. Unwrap the remaining three beakers (600 cm^3), and add about 300 cm^3 sterile DDH_2O to each beaker. Following the 10-min sterilization period, remove the tissue blocks from the hypochlorite solution with the forceps, and rinse them successively for 30 sec to 1 min in each of the three rinse beakers. Withdraw from the hood the beakers containing the hypochlorite solution and the first two rinses. Light the methanol lamp and place the scalpel in the ethanol dip. Open the foil packet containing the cork borer (木塞穿孔器)and the packet of empty Petri dishes. Place two of the dishes in the rear of the hood, fill each dish about halfway with sterile DDH_2O, and partially remove the lid of one of the dishes. Remove the lid of the third empty Petri dish, exposing the sterile inner surface of the lower half (boring platform, 钻孔平台).

4. Flame the forceps and transfer one of the tissue blocks from the final rinse beaker to the boring platform. Steady the tissue with the forceps, and make a single vertical boring with the cork borer through the center of the tissue block. The cork borer must be inserted all the way so that it cuts completely through the tissue block. Lift the block with the borer still inserted in it, and hold the block directly over the DDH_2O in the partially opened Petri dish. Gently exert pressure on the metal rod. This slight force should eject the tissue cylinder into the pool of water. Return the tissue block to the boring platform. Place the arms of the forceps on each side of the borer and withdraw the borer from the tissue. Repeat the process with the other blocks of tissue until you have prepared six or seven tissue cylinders. Finally, withdraw from the hood the final rinse beaker, cork borer, remains of the potato tissue, and the boring platform.

5. Open the packet containing the explant cutting guide. If this device is unavailable, the bottom half of a sterile Petri dish can be used for slicing explants from the cylindrical tissue borings. Arrange the following three Petri dishes in the rear of the hood. The nearest dish should be the explant cutting guide, another dish will contain the cylinders of tissue, and the third dish contains sterile DDH_2O for explant rinsing (prepared in step 3). Partially open the two plates in the rear of the hood and completely remove the lid of the explant cutting guide.

6. Flame the forceps and scalpel, and transfer a tissue cylinder with the forceps to the cutting guide. The flamed instruments should be permitted to cool briefly before bringing them in contact with living plant tissues. Trim and discard approximately 2 mm of tissue from each end of the cylinder. Slice the remaining cylinder into segments of 2-3 mm in length. Each explant, therefore, will measure about 8 mm in diameter and 2-3 mm in thickness. Each cylinder should yield a minimum of 5 explants. With the flat

blade of the scalpel, transfer the explants to the plate containing the DDH_2O rinse. Repeat the cutting operation until 30 explants have been prepared. Flame the forceps and scalpel several times during the course of the slicing operation. Remove from the hood the Petri plate that contained the cylinders of tissue and the cutting guide.

7. Fill an empty Petri dish halfway with DDH_2O. Flame the forceps and transfer the explants to the plate containing the rinse water. Remove from the hood the plate that formerly contained the explants. Open the paper bag containing the Petri dishes with the Whatman No. 1 filter paper. Arrange the culture tubes to receive the explants.

8. Flame the forceps and transfer the explants one at a time to the surface of the sterile filter paper. Blot briefly both the top and bottom of each explant; immediately transfer the explant to the surface of the culture medium (one explant per tube). Remember to hold the culture tubes at a slight angle so that the hand grasping the forceps is not directly over the sterile surface of the medium.

9. After capping the culture tubes with foil, place them in the Pyrex storage jars and wrap the jars with foil. Transfer the cultured explants to an incubator adjusted to 25-27℃.

RESULTS

After a few days in culture, the explants become slightly rough in texture, and their surface may glisten (闪闪发光) in reflected light. This is a sign of the beginning of callus formation. Culture for a single incubation period (passage) may last from a few weeks to three months, depending on the rapidity of growth. For most potato cultivars, within four to six weeks after initiating a culture with picloram, there should be more than 1 g fresh weight of callus available for subculture. Subcultured callus typically increases tenfold in fresh weight within four or five weeks on the MS-picloram medium solidified with Gelrite. Depending on the friability of the callus, use either a spatula or a scalpel for transferring inocula(接种体, inoculum 的复数) from the callus mass to the fresh medium. Only gray or cream-colored tissue can be used; brownish tissue is a sign that localized necrosis has occurred. The instruments must be flamed and aseptic techniques used throughout the subculture procedure.

Examine the surface of the callus with a dissecting microscope or a hand lens, and notice the external appearance of the callus. With a dissecting needle scrape some of the cells onto a microscope slide. Add a drop of distilled water and a cover slip, and examine the cells with the light microscope (100 × magnification). The contrast can be enhanced by lightly staining the cells with an aqueous solution of toluidine blue O (甲苯胺蓝 O, 0.05% w/v). With a Pasteur pipette add a droplet of the stain solution to one edge of the cover slip. On the opposite side of the cover slip moisten a piece of lens paper with the aqueous mounting medium. The blotting action of the paper will draw the stain beneath the cover slip and into the field of vision.

QUESTIONS FOR DISCUSSION

1. Why are explants containing cambial cells excellent choices for the initiation of callus cultures?

2. Under the cultural conditions employed in your experiment, what is the optimal interval of time between subcultures (passage time)? How is this time determined?

3. Give some reasons for subculturing.

4. What are vascular nodules?

5. What is the significance of the three stages of callus development?

知识要点

愈伤组织是外植体中部分细胞脱分化，不断分裂所形成的一团无定形的松散的薄壁细胞团。愈伤组织具有分化产生出根、芽和胚状体的能力，是器官形成和再生植株的一个重要来源。愈伤组织很容易分散成单个细胞，因而也是植物单细胞悬浮培养的一个重要组织来源。愈伤组织的产生可分为诱导、细胞分裂和细胞分化3个阶段。除柠檬类水果的果肉和含形成层细胞的外植体外，愈伤组织的诱导一般需要植物生长调节类物质的作用，需要分别添加：①生长素；②细胞分裂素；③生长素和细胞分裂素；④成分复杂的天然提取物。愈伤组织中的细胞很少是均一的，一般都可观察到细胞分化的发生，如导管成分、筛管成分、栓细胞、分泌细胞和毛状体等。愈伤组织长到一定大小后，可用镊子将其转移到新鲜的培养基中继续培养，称为传代培养。愈伤组织的及时传代，既促进了愈伤组织的生长，又避免了次生代谢产物积累所产生的毒性。褐化的愈伤组织要丢掉，不能传代培养。

CHAPTER 6　ORGANOGENESIS(器官发生)

The capability to induce the formation of adventitious roots and shoots in vitro is of the utmost importance in plant tissue culture methodology. Studies involving the transformation of protoplasts would be of little value unless the genetically altered plant material could be regenerated into a plantlet. Plant regeneration by tissue culture techniques can be achieved either by zygotic embryo culture, somatic embryogenesis, or organogenesis. The latter approach is employed in micropropagation(微繁,快繁) from bud and shoot material and in organ production from callus and suspension cultures. This chapter is devoted entirely to organogenesis. Roots, shoots, and flowers are the organs that may be initiated from tissue cultures. Embryos are not classified as organs because these structures have an independent existence; that is, embryos do not have vascular connections with the parent plant body.

The underlying basis for organogenesis is poorly understood and involves the interplay of a host of factors: donor plant growth, source of the explant, culture medium, supplements of growth regulators, and environmental conditions. The first major breakthrough came with the discovery that in vitro organogenesis in tobacco cultures could be chemically regulated. The addition of auxin to the medium served to initiate root formation, whereas shoot initiation was inhibited. The latter effect on shoot formation could be partially reversed by increasing the concentration of both sucrose and inorganic phosphate. Later it was found that adenine sulfate was active in promoting shoot initiation, and this chemical reversed the inhibitory effect of auxin. The studies of Skoog's group led to the hypothesis that organogenesis is regulated by a balance between cytokinin and auxin. A relatively high auxin: cytokinin ratio induced root formation in tobacco callus, whereas a low ratio of the same compounds favored shoot production (Fig. 6.1).

Probably the most precise regulation of organ formation has been achieved with epidermal and subepidermal explants consisting of a few cell layers in thickness. The formation of floral buds, vegetative buds, and roots has been demonstrated in thin cell-layer explants of several species by regulating the auxin: cytokinin ratio, carbohydrate supply, and environmental conditions. Certain isolated tissue layers in species that readily regenerate organs in vivo showed a remarkable potential to form organs during culture. Root initiation occurred in the presence of (-NAA plus zeatin, and shoot formation required the addition of either zeatin or benzylaminopurine(苯甲酸嘌呤) in the absence of auxin.

For plantlet regeneration in many dicot callus cultures, the callus is removed from the maintenance medium and subcultured on a shoot-induction medium. The latter medium usually has a cytokinin: auxin ratio in the range of 10:1 to 100:1, in many cases by supplementing the medium with cytokinin as the sole growth regulator. In comparison to dicots, monocot cultures are more difficult to regenerate. With monocot cultures exogenous cytokinin may be unnecessary for the initiation of shoots. The omission of auxin from the maintenance medium may suffice (足够) to induce shoot formation in these cultures, and two successive transfers on auxin-free media have been recommended. Root initiation frequently occurs spontaneously after the culture has initiated buds, and shoot development undoubtedly alters the endogenous hormones within the culture. Regenerated shoots are transferred to a root-inducing medium. In many cases, auxin alone or in combination with a low level of cytokinin will enhance root primordia formation. The appropriate cultural conditions for rhizogenesis(生根) in some species may be completely ineffective in a closely related plant. There is some evidence that phenolic compounds may act with auxin to promote rooting. For example, the combination of phloroglucinol(间苯三酚) with indolebutyric acid (吲哚丁酸, IBA) was more effective in stimulating rooting than auxin alone.

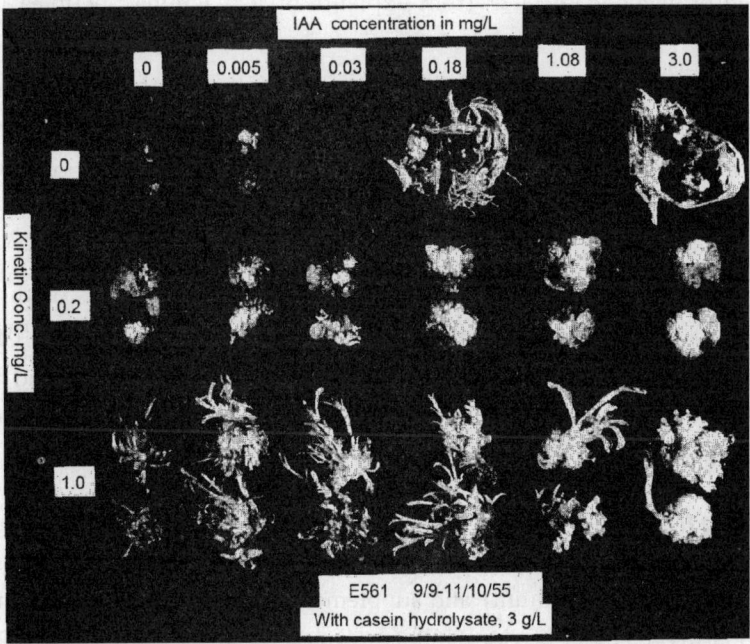

Figure 6.1 The regulation of organ formation in explants of tobacco (*Nicotiana tabacum*) pith by varying the auxin: cytokinin ratio. Note the occurrence of shoots induced by high levels of kinetin. Moderate levels of kinetin, in the presence of auxin, stimulate the production of callus. Auxin induced root formation in the absence of kinetin. (By F. Skoog)

In addition to auxin and cytokinin, there are reports involving the possible roles of other growth regulators in the induction of organogenesis. Although there are a few exceptions, gibberellins (赤霉素, GB) tend to suppress both root and shoot initiation in cultures. Endogenous ethylene may be a factor in shoot initiation. Ethylene may block the early stages of organogenesis, but enhances the further development of primordia. Endogenous ethylene was identified as a factor in bud induction arising from cultured tobacco cotyledons (子叶). Indirect evidence suggests a similar role for ethylene in cultured *Lilium* (百合) bulb (鳞茎) tissues.

Cultured explants are typically incubated in the dark for the initiation and subsequent development of callus, although low-level illumination may be beneficial. A light requirement has been reported for adventitious bud formation in hairy roots of horseradish (山葵). The roots had been inoculated with *Agrobacterium rhizogenes* (发根农杆菌, strain 15834; Ri plasmid). The hairy roots produced buds on a hormone-free medium in the presence of red light, but not far-red light. Thus, phytochrome appeared to be involved in this phenomenon. Excised roots from nontransformed plants did not exhibit this response.

In the present experiment the student will attempt to induce the formation of plantlets from explants of *Saintpaulia ionantha* Wendl. (非洲紫罗兰, African violet). This plant has been propagated in vitro from explants of leaf lamina, petioles (叶柄), and floral organs.

Obtain from the florist a healthy African violet plant with an abundance of dark-green foliage. Explants will be prepared from the youngest leaves by transversely slicing the petioles into segments approximately 10 mm in length. In addition, prepare some explants from the base of the lamina, that is, at the point of attachment of the petiole to the blade of the leaf. Because of the fuzzy texture of the epidermal layer, a wetting agent must be used during the sterilization procedure. The objectives of the experiment are two-fold: (a) three different concentrations of benzylaminopurine (苯甲酸嘌呤) will be tested for shoot initiation (cytokinin : auxin ratios of 5 : 1, 10 : 1, and 50 : 1), and (b) regeneration of African violet plants will be accomplished by rooting the newly formed shoots.

PROCEDURE

Prepare 1 liter of MS medium, and supplement with myo-inositol (100 mg/L), nicotinic acid (0.5 mg/L), pyridoxine-HCl (0.5 mg/L), and thiamine-HCl (0.4 mg/L). Prepare three 50-cm^3 aliquots. Each aliquot will contain sucrose (2.0% w/v), Gelrite (0.2% w/v), (-NAA (0.1 mg/L), and a supplement of benzylaminopurine (苯甲酸嘌呤, BA). In aliquot 1 add 0.5 mg/L BA; in aliquot 2 add 1.0 mg/L BA; and in aliquot 3 add 5.0 mg/L BA. Using a magnetic stirrer adjust the pH of each aliquot to 5.7. The autoclaved medium is dispensed into 15 culture tubes (five tubes per treatment; 10 cm medium in each tube).

Culture Procedure:

1. Excise 10 small-to-medium-sized leaves, including the petioles, and wash the leaves briefly in cool, soapy water. Rinse in running tap water, and prepare for aseptic procedures.

2. Immerse the leaves in a 20% (v/v) aqueous solution of Clorox or other commercial bleach for 10 min. The bleach solution should contain a few drops of Tween 20. Rinse the leaves in three successive baths of DDH_2O (200 cm^3 each). Each rinse should last about 30 sec to 1 min.

3. Each leaf is transferred to a sterile Petri dish containing filter paper, and explants are prepared with the aid of forceps(镊子) and scalpel(解剖刀). The filter paper will remove the excess moisture from the rinse water. Petiole segments about 1 cm in length make excellent explants. Also prepare 1-cm^2 explants from the lamina tissues located near the point of attachment of the petiole and from the center of the blade. Position the petiole explants flat, that is, parallel to the surface of the medium. The laminar explants should be positioned with the lower epidermis touching the surface of the medium.

4. Place the cultures in a growth chamber maintained at 25-29℃ with 16-hr photoperiods furnished by a combination of GroLux and cool-white fluorescent tubes. The light intensity should be about 1,000-1,500 lux.

RESULTS

Shoots will appear within about four weeks, and after about eight weeks of culture the regenerated shoots will be ready to be aseptically subdivided and subcultured for the initiation and development of root systems. Rooting is promoted by transferring the shoots to a fresh medium that is devoid of plant growth regulators and has a sucrose concentration of about 1.6% (w/v). This subculture step may be unnecessary since the subdivided shoots apparently can establish a root system in a sterile potting soil mixture. The miniature pots(小型花盆)should be maintained under a relatively high humidity with adequate lighting. Direct sunlight, however, can be harmful. The requirements for hardening off the plantlets are similar to those for plantlets regenerated by the micropropagation of the shoot apex.

QUESTIONS FOR DISCUSSION

1. What are the main tissue culture methods of plant regeneration?
2. Is embryo a plant organ? Explain.

3. What was Skoog's hypothesis on organogenesis? Is it used today? Can you think of some reasons why application of Skoog's hypothesis may fail to give the expected results?

4. Why can a petiole explant from an immature African violet leaf give a better organogenetic response than an explant taken from the tip of a mature leaf?

知识要点

植株的再生有3条途径,即合子胚发育、器官形成和胚状体形成。因此,不定根和不定芽的诱导(即器官形成)是植物组织培养的重要内容之一。器官发生的机制还不是很清楚,但可以肯定的是,与供体植物的生长、外植体的来源、培养基中添加的生长调节类物质和环境条件有关。Skoog的激素调节器官发生理论是该领域中最重要的发现。细胞分裂素和生长素的比例决定了愈伤组织的器官发生:当细胞分裂素/生长素的比例大于1时,诱导愈伤组织分化成芽;当细胞分裂素/生长素的比例小于1时,诱导愈伤组织分化成根;当细胞分裂素/生长素的比例等于1时,愈伤组织只生长而不分化。

CHAPTER 7　CELL SUSPENSIONS(细胞悬浮培养)

The culture of cell suspensions is as close as cell culture comes to the fermentation biology of microbial systems. According to King (1980) the term "suspension culture" has no clear-cut biological definition, and such tissue culture systems are evidently more than simply aggregates of cells suspended in a liquid medium. A suspension culture originates with a "random critical event" occurring during the early exposure of the plant cells to the liquid medium. Cells undergoing this transition in metabolism and growth rate produce a "cell line." Some of the characteristics of cell lines include the following: (1) a high degree of cell separation, (2) homogeneous cell morphology, (3) distinct nuclei and dense cytoplasm, (4) starch granules, (5) relatively few tracheary elements, (6) doubling times of 24-72 h, (7) loss of totipotency, (8) hormone habituation, and (9) increased ploidy levels.

Cell suspension cultures are generally initiated by transferring fragments of undifferentiated callus to a liquid medium, which is then agitated during the culture period (Fig. 7.1). Although a longer time is required, suspension cultures can be started by inoculating the liquid medium with an explant of differentiated plant material (e. g., a fragment of hypocotyl(胚轴) or cotyledon(子叶)). The dividing cells will gradually free themselves from the inoculum because of the swirling action of the liquid. It should be kept in mind, however, that no suspension culture has been shown to be composed entirely of single cells. After a short time the culture will be composed of single cells, cellular aggregates of various sizes, residual pieces of the inoculum, and the remains of dead cells. The term "friability(易碎)" is used to describe the separation of cells following cell division. Formation of a "good suspension" (i. e., a culture consisting of a high percentage of single cells and small clusters of cells) is much more complex than finding the optimum environmental conditions for cell separation. The degree of cell separation of established cultures already having the characteristics of high friability can be modified by changing the composition of the nutrient medium. Increasing the auxin: cytokinin ratio will, in some cases, produce a more friable culture. On the other hand, some cultures exhibit low friability regardless of cultural conditions. There is no standard procedure that can be recommended for starting cell suspension cultures from callus; the choice of suitable conditions is largely determined by trial and error.

The initiation of a cell suspension culture requires a relatively large amount of callus to serve as the inoculum, for example, approximately 2-3 g for 100 cm^3. When the plant material is first placed in the medium there is an initial lag period prior to any sign of cell division. This is followed by an exponential rise in cell number and a linear increase in the cell population. There is a gradual deceleration (减速) in the division rate. Finally, the cells enter a stationary or nondividing stage. In order to maintain the viability of the culture, the cells should be subcultured early during this stationary phase.

Because cells from different plant material vary in the length of time they remain viable during the stationary phase, it may be prudent to subculture during the period of progressive deceleration. Passage time can be learned only from experience, and a given suspension culture should be subcultured at a time approximating the maximum cell density. For many suspension cultures the maximum cell density is reached within about 18-25 days, although the passage time for some extremely active cultures may be as short as 6-9 days. At the time of the first subculture it will be necessary to filter the culture through a nylon net or stainless steel filter to remove the larger cell aggregates and residual inoculum that would clog the orifice(口) of a pipette. A small sample should be withdrawn and the cell density determined before subculturing. There is a critical cell density below which the culture will not grow; for example, this value is 9×10^3-15×10^3 cells/cm^3 for a clone of sycamore (小无花果树) cells.

Cell suspension cultures must be agitated or subjected to forced aeration, and a platform (orbital) shaker is used for this purpose in most laboratories. The best speed range for cultures in 250-cm^3 flasks is 100-120 rpm. The volume of liquid in relation to the size of the flask is important for adequate aeration (i.e., the liquid medium should occupy about 20% of the total volume of the flask). Other devices for aeration include magnetic stirrers, roller cultures, and Steward's auxophyton (a kind of device). The latter apparatus slowly rotates the cultures in nipple (奶头) flasks and tumble (转鼓) tubes. Microcultures do not need any device for oxygenation of the nurtured cells.

There is some terminology associated with cell suspension cultures. The present experiment involves the preparation of a "batch culture," defined as a culture grown in a fixed volume of culture medium. Our experiment is also a "closed culture" because all cells are retained, and a continual increase in cell density will occur until the stationary phase is reached. A "closed continuous system" involves a continuous influx of fresh medium and a withdrawal of spent medium. An "open continuous system" is similar to the closed in the replenishment of the nutrient medium; in addition, however, the cells are harvested. Examples of open continuous systems are "chemostats" (恒化器) and "turbidostats(恒浊器)". In a chemostat the continuous flow of fresh medium into the system is set at a predetermined rate; this influx of nutrients will largely determine the growth rate of the culture. In the turbidostat, cell density is set at some predetermined level, and fresh medium is added periodically to maintain that density within the preset limits. Cell density in the turbidostat is determined with a photocell control device.

The callus culture could be used as the inoculum for the cell suspension culture. Carrot callus is an excellent inoculum for a suspension culture.

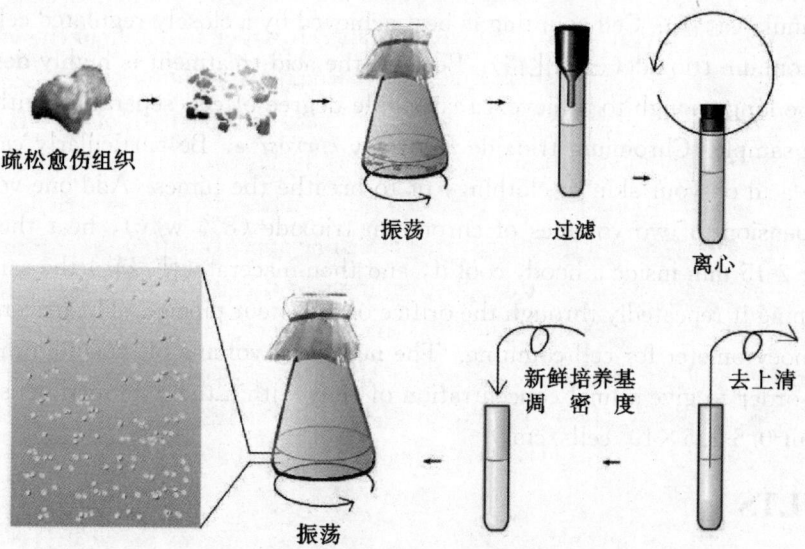

Figure 7.1 Initiation of cell suspension culture from callus

PROCEDURE

Carrot Cell Suspension Culture. Carrot cell suspension cultures can be initiated on a basal MS medium supplemented with 2,4-D (1.0 mg/L) and sucrose. A suspension culture of carrot does not require an exogenous cytokinin, and the presence of auxin as the sole plant growth regulator in the medium may improve the friability of the culture. For the present experiment each 125-cm^3 flask will contain 25 cm^3 of basal MS medium, plus 2, 4-D (1 g/cm^3) and sucrose (750 mg).

The callus obtained should be removed from the culture tubes with forceps and transferred to a Petri dish containing Whatman No. 1 filter paper. Trim the callus blocks with the scalpel and use only the young, actively growing callus for the inoculum. Each flask should receive an inoculum of about 500-750 mg of callus in order to ensure the initiation of the culture. Brownish callus may be indicative of senescence and should be discarded. Place the inoculated flasks on the shaker and set the speed at 100 rpm. The shaker with the flasks should be placed in an air-conditioned enclosure maintained at 25-27℃. If, during the first few days, the medium appears "milky," this is a sign that contamination occurred during the inoculation. The initial subculture can be performed after 710 days, although it is first necessary to filter the culture through an industrial nylon mesh filter in order to remove the residual inoculum and larger clumps of cells. The next step is to determine the cell density of the culture. It is impossible to subculture and maintain the culture unless the cell density is within a given range. According to Street (1977), most suspension cultures contain $0.5\text{-}2.5\times10^5$ cells/cm^3 after dilution with the fresh medium. A sample is taken with a syringe equipped with a wide-bore cannula (套管). Cell counting is best achieved by a closely regulated cell separation with chromium trioxide(三氧化铬). Because the acid treatment is highly destructive, it should be long enough to achieve a reasonable degree of cell separation without destroying the sample. Chromium trioxide *is highly corrosive*. Be particularly careful not to spill the acid on your skin or clothing, or to breathe the fumes. Add one volume of the cell suspension to two volumes of chromium trioxide (8% w/v), heat the mixture to 70℃ for 2-15 min inside a hood, cool it, and then macerate(使浸软) the sample further by pumping it repeatedly through the orifice of a Pasteur pipette. The macerate is placed in a hemocytometer for cell counting. The necessary volume of inoculum may be calculated in order to give a final concentration of cells within the minimum density level (i. e., about $0.5\text{-}2.5\times10^5$ cells/cm^3).

RESULTS

Remove a 1-cm^3 sample of the cell suspension culture with a sterile Pasteur pipette and discharge the contents into a small beaker. Place about 0.1 cm^3 of the suspension in a depression slide and examine the preparation with a dissection microscope. Then examine a droplet of the suspension with the light microscope (100×). In what obvious ways do these cultured cells differ from the cells of the primary explant excised from the cambial zone of the carrot root? With the aid of an ocular micrometer (目微尺), the approximate range of sizes of the cells can be estimated. Some of the cellular details may be enhanced by employing a biological stain to increase the contrast.

QUESTIONS FOR DISCUSSION

1. What is suspension culture? List some of the characteristics of plant cell suspension culture.

2. In addition to providing a source of oxygen, what are other possible effects of agitating a cultured plant cell or organ?

3. From the time of inoculation, what growth stages may be exhibited by a cell suspension culture? Why does the rate of growth resemble an S-shaped curve (i. e., what are the reasons for each of these fluctuations in the growth curve)?

4. Explain the following terms: batch culture, continuous culture.

知识要点

将疏松愈伤组织压碎,接种到液体培养基中,通过震荡分散成单细胞进行培养,是常见的植物细胞悬浮培养方法。与微生物和动物的悬浮细胞培养不同,即使接种的是单细胞悬液,植物细胞的悬浮培养物也一般是由单细胞和细胞团组成的混合物。这是由于植物细胞分裂后形成胞间连丝,细胞不易彼此分开造成的。植物细胞悬浮培养时的接种密度不能太低,一般为 $10^4 \sim 10^5$ 个细胞/毫升。细胞密度太低的话,细胞不易分裂生长;太高的话,易形成嵌合体。

CHAPTER 8　SOMATIC EMBRYOGENESIS(体细胞胚胎发生)

The capacity of flowering plants to produce embryos is not restricted to the development of the fertilized egg; embryos ("embryoids") can be induced to form in cultured plant tissues. This phenomenon was first observed in suspension cultures of carrot by Steward, Mapes, and Mears (1958) and in carrot callus grown on an agar medium by Reinert (1959). This is a general phenomenon in higher plants, and experimental somatic embryogenesis has been reported in tissues cultured from more than 30 plant families.

Somatic embryoids(体细胞胚)may arise in vitro from three sources of cultured diploid cells: (1) vegetative cells of mature plants, (2) reproductive tissues other than the zygote, and (3) hypocotyls(胚轴)and cotyledons(子叶)of embryos and young plantlets without any intervening callus development.

Somatic embryogenesis may be initiated in two different ways. In some cultures embryogenesis occurs directly in the absence of any callus production from "preembryonic determined cells" that are programmed for embryonic differentiation. The second type of development requires some prior callus proliferation, and embryos originate from "induced embryogenic cells" within the callus. Carrot cells are an example of the latter case. Although individual carrot cells are totipotent and carry all the genetic templates necessary for the development of the whole plant, isolated single cells do not generally become transformed into embryos by repeated divisions. Embryoids are initiated in callus from superficial clumps of cells associated with highly vacuolated cells that do not take part in embryogenesis. The embryoid-forming cells are characterized by dense cytoplasmic contents, large starch grains, and a relatively large nucleus with a darkly stained nucleolus. Staining reagents indicated that these embryogenic cells have high concentrations of protein and RNA. These cells also exhibited high dehydrogenase activity with tetrazolium(四唑)staining. Each developing embryoid passes through the sequential stages of embryo formation (i. e. , globular, heart shape, and torpedo shape) (Fig. 8.1). Two critical events are involved in the early programming of this process: (1) the induction of cytodifferentiation of the proembryoid cells, and (2) the unfolding of the developmental sequence by these proembryoid cells.

Although a given culture may differentiate these embryogenic cells, their further development may be blocked by an imbalance of chemicals in the culture medium. Abnormalities, known as "embryonal budding" and "embryogenic clump formation," may occur if relatively high levels of auxin are present in the medium after the embryogenic cells have been differentiated. In other words, two distinctly different types of media may be required: one medium for the initiation of the embryonic cells and another for the subsequent development of these cells into embryoids. The first (induction) medium must contain auxin. The second generally consists of a mixture either lacking auxin, with a lower concentration of the same auxin, or with reduced levels of a different auxin. With some plants, however, both embryo initiation and subsequent maturation occur on the first medium, and a second medium is employed for plantlet development.

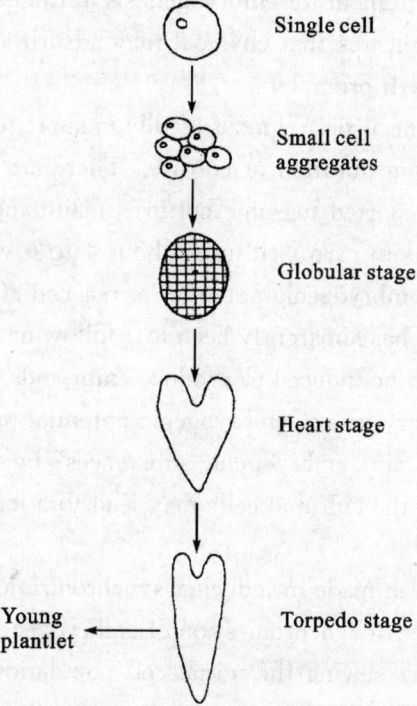

Figure 8.1 Stages of somatic embryogenesis. Following repeated cell divisions, cell aggregates progressively develop and pass through globular, heart, and torpedo stages before ultimately forming plantlets.

The most important chemical factors involved in the induction medium are auxin and reduced nitrogen. Substantial amounts of reduced nitrogen are required in both the first and second media. In wild carrot cultures the addition of 10 mM NH_4Cl to an embryogenic medium already containing KNO_3 (1240 mM) produced near-optimal numbers of embryoids. Glutamine(谷氨酰胺), glutamic acid(谷氨酸), urea, and alanine, respectively, were found partially to replace NH_4Cl as a supplement to KNO_3. These various nitrogen sources are not specific for the induction of embryogenesis, although at low concentrations organic forms are much more effective than inorganic nitrogen compounds.

The role of cytokinins in embryogenesis is somewhat obscure because of conflicting results. Although zeatin (0.1 μM) stimulates embryogenesis in carrot cell suspensions during the auxin-free subculture (secondary culture), the process is inhibited by the addition of either kinetin or benzylaminopurine(苯甲酸嘌呤)to the medium. The inhibitory effect of exogenous cytokinins may result from the increase in endogenous cytokinins in the developing embryoids.

Supplementing the medium with activated charcoal has facilitated embryogenesis in several cultures. The induction of embryogenesis was successful in *Daucus carota*(胡萝卜) cultures containing charcoal when auxin depletion failed to produce the desired results. Charcoal was a requirement for embryogenesis in English ivy (常春藤, *Hedera helix*) cultures. Evidence indicates that charcoal may adsorb a wide variety of inhibitory substances as well as growth promoters.

In general, embryogenesis occurs most readily in short-term cultures, and this ability decreases with increasing duration of culture. There are exceptions, however, and embryogenesis has been reported in some cultures maintained over a period of years. Embryoid formation begins in carrot cultures about 4 to 6 weeks after isolation of the tissues, and an optimum embryogenic potential is reached after about 15 weeks. After the embryogenic potential has apparently been lost following 36 weeks in vitro, the carrot cultures can once again be induced to produce embryoids by transfer to an appropriate medium. This temporary loss of embryogenic potential presumably results from the lack of biosynthesis of certain "embryogenic substances" by the cultured cells. In addition, changes in ploidy of the cultured cells may lead to a loss of morphogenetic potential.

Some progress has been made in inducing synchronization(同步化) of somatic embryogenesis. A high degree of synchronization of embryogenesis was achieved in a carrot suspension culture by: (a) sieving the initial cell populations, (b) employing density gradient-centrifugation in Ficoll solutions, and (c) using repeated low-speed centrifugation for 5-sec periods.

The resulting cell clusters, cultured in an auxin-free medium containing zeatin, gave a greater than 90% frequency of embryoid formation. Although the regeneration of whole plants by embryogenesis has been relatively rare in the Gramineae(禾本科), somatic embyros have been formed directly from leaf mesophyll cells of orchard grass (果园草) without an intervening callus.

Endogenous polyamines appear to be required for the induction of embryogenesis in cultures of wild carrot. Embryogenic cultures of *Daucus carota* treated with a specific inhibitor of arginine decarboxylase(精氨酸脱羧酶)showed a sharp reduction in embryo production compared to untreated controls. The cultures containing the inhibitor also had relatively low levels of the polyamines(多胺), putrescine(腐胺)and spermidine(亚精

胺). Supplementing the culture medium with either putrescine, spermidine, or spermine (精胺)restored embryogenesis to the inhibitor blocked cultures. In recent years the multiplication of somatic embryos in bioreactors has begun for commercial micropropagation.

PROCEDURE

The preparation for this experiment starts with an actively proliferating callus initiated from the taproot (直根) of carrot. Healthy fragments of the callus are transferred to a liquid culture medium, and a carrot cell suspension is initiated. This liquid medium consists of basal MS salts supplemented with 2,4-D (1.0 mg/L) and sucrose (3% w/v).

Embryoids are initiated in the present experiment in the following manner:

1. Ten Petri dishes are prepared as follows:

(a) five plates contain MS salts, zeatin (0.2 mg/L), 2,4-D (0.1 mg/L), sucrose (2% w/v), and agar (1% w/v); and

(b) five plates contain MS salts, zeatin (0.2 mg/L), sucrose (2% w/v), and agar (1% w/v).

The latter plates do not contain a source of exogenous auxin.

2. Aliquots (2 cm^3) of carrot suspension culture are added by pipette to the surface of the medium in the Petri dishes. The dishes are sealed with Parafilm and incubated at 25℃ in the dark for two to three weeks.

3. The test for embryogenic potential is based on a visual count of the embryoids. The "callus" from the agar surface is gently dispersed in DDH_2O, and the number of embryoids present is determined by placing the Petri dish over a black card marked with 1-cm^2 grid lines.

4. Small aliquots of this dispersed sample can be placed on a microscope slide and examined with a compound or dissecting microscope for the various stages of somatic embryogenesis.

RESULTS

In the Petri dish containing the auxin medium the carrot cells develop into a callus and grow into small compact clumps. Embryoids, however, are not formed in these dishes. The carrot cells grown on the auxin-free medium produce large numbers of embryoids (Fig. 8.2). The embryoids are not formed in a synchronous manner: When an inoculum of this material is examined under the microscope, a wide range of developmental stages similar to those shown in Figure 8.2a-c can be seen.

After the carrot cultures have reached the late torpedo stage of development, they can be transferred to filter-paper bridges (Fig. 8.2d). (A filter-paper bridge can be made by folding a strip of filter paper [Whatman No. 1; 9×90 mm] in the shape of the letter "M." Its arms are immersed in the liquid medium, thus acting as a wick. The carrot emryoids are nurtured in the central "V" of the bridge.) To the culture tubes add 3 cm³ of liquid medium composed of MS salts, kinetin (0.2 mg/L), and sucrose (2% w/v). The plantlets that are formed can be potted in sterile soil and grown to maturity (Fig. 8.2e).

The plantlets must be maintained under a high relative humidity to prevent excessive water loss.

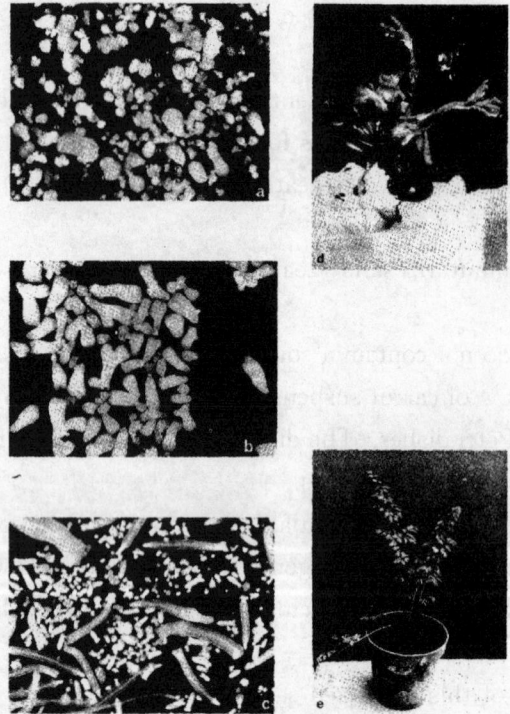

Figure 8.2 Stages of development of carrot (Daucus carota) embryoids. a. young globular stage; b. heart stage; c. torpedo stage; d. carrot plantlet growing on filter-paper bridge; e. mature carrot plant derived from cultured embryoid. (By L. A. Withers)

QUESTIONS FOR DISCUSSION

1. What is somatic embryogenesis? Describe the features of embryoid-forming cells

2. What are the sequential stages of somatic embryogenesis?

3. Compare the induction medium for the initiation of the embryonic cells with the second medium for the subsequent development of these cells into embryoids.

4. What difficulties may be encountered with clonal propagation of plants by the means of somatic embryogenesis?

知识要点

体细胞胚是指在植物组织培养中起源于非合子细胞,经过胚胎发育所形成的胚状结构,又称胚状体。与芽的发生不同,胚状体与母体植物的维管组织没有直接联系,具有根芽两极,可直接再生成植株。胚状体的发生途径多种多样,但其起始细胞一般具有致密的细胞质、液泡小、淀粉颗粒大、核质比大、核仁染色深等特点。胚状体的诱导培养基中必须添加生长素,但胚状体的发育则受到生长素的抑制。还原性氮如 KNO_3 是胚状体的诱导和发育都必不可少的。

CHAPTER 9　CULTURE OF ISOLATED ROOTS(离体根的培养)

Special attention has been given recently to the potential of hairy-root cultures for the production of secondary metabolites. Liquid media have been employed in the culture of excised roots and cell suspensions. A liquid medium has certain advantages over the use of nutrients in a gel-solidified matrix. When a callus is grown on a semisolid medium, diffusion gradients of nutrients and gases within the callus will lead to moderate growth and metabolism. The matrix itself may release contaminants to the culture medium, and metabolites secreted by the growing callus will accumulate in the gel matrix.

The first successful organ culture, that is, potentially unlimited growth of the isolated organ, was reported by White (1934) with excised tomato roots. Since then the excised roots of numerous herbaceous species have been cultured. Less success has been achieved in starting root cultures from woody plants.

Some general comments can be made about the nutrition of cultured roots. In some cases, the minimum growth requirements were met with the essential mineral elements, a carbon source, a vitamin supplement, and a few amino acids. Some responded favorably to the addition of auxin and other growth regulators. Another effective growth stimulant for some isolated roots is *myo*inositol(肌醇), as it plays a role in secondary vascular tissue formation in excised roots of radish (*Raphanus*). A marked improvement over White's medium is the substitution of a chelated form of iron for $Fe_2(PO_4)_3$. NaFeEDTA, the sodium salt of ferric EDTA, is generally used as a chelated form of iron. Sucrose is the carbon source of choice, although some monocot roots grow equally well with D-glucose. A sugar level of 1.5%-2.0% (w/v) is sufficient, and higher concentrations may alter root metabolism (Figure 9.1). Although several amino acids have been tested, most of them inhibit the growth of cultured roots. The vitamin requirements vary for different species, although all species require the addition of thiamine(维生素B1). The effect of light on the growth of cultured roots appears to vary with the species and the cultural conditions.

A question has been raised about the relationship between cultured roots and similar roots produced by an intact plant. Although isolated roots and "intact" roots are alike in many anatomical and metabolic ways, certain differences have been reported. Excised roots gradually lose the capability of forming secondary vascular tissues during culture. Cultured tomato roots, in contrast to seedling roots of the same plant, fail to show the normal geotropic response to gravity. In addition, the biochemical composition of cultured roots may differ from that of seedling roots.

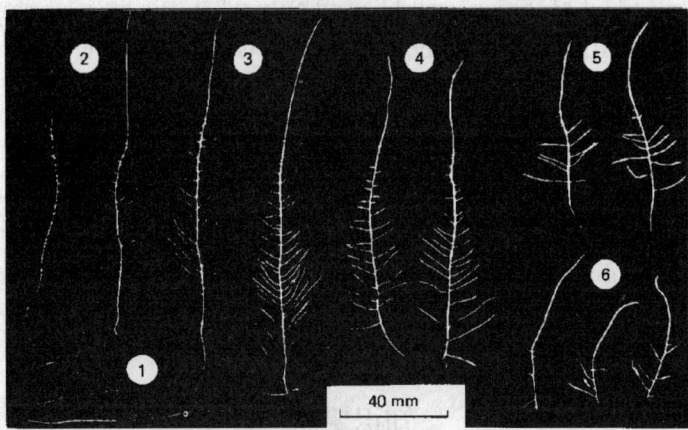

Figure 9.1 Excised tomato (*Lycopersicon esculentum*) roots cultured 7 days at 27℃ in White's medium containing sucrose at a concentration of ①0.5%, ②1.0%, ③1.5%, ④2.0%, ⑤3.0%, and ⑥4.0%. Note the formation of lateral roots from the main root axis. In order to subculture roots, the root is cut into sectors. Each sector, transferred to a fresh medium, contains a portion of the main root axis plus several lateral roots. (By H. E. Street)

Rapidly growing hairy-root cultures, obtained by the genetic transformation of plant cells *by Agrobacterium rhizogenes*（发根农杆菌）, may revolutionize（改革）the commercial production of rare chemicals. These roots are characterized by a high degree of lateral branching, a profusion（大量）of root hairs, and a stable and high-level biosynthesis of secondary metabolites. The transformed roots can be cultured indefinitely, with genetic stability, on a defined medium devoid of any growth regulators. *A. rhizogenes* can be a vector to introduce genes into hairy-root cultures. Biosynthesis genes of secondary metabolite formation can be engineered to be expressed in the cultured root and flanked by T-DNA border sequences in an expression plasmid. Inserted genes in the plasmid are cotransported to hairy roots along with the wild-type Ri T-DNA genes. Large-scale cultures involving plant roots have been achieved in Japan. Cell aggregates with profuse proliferating roots of *Panax ginseng*（人参）have been cultured in a 20,000-liter bioreactor for saponin（皂角苷）production by the Nitto Electric Co.

The technique employed to initiate and subculture roots requires some explanation. Root tips about 10 mm in length are removed from young seedlings produced during the axenic（无菌）germination of seeds, and these apical tips are transferred to an aqueous culture medium. After about a week the primary root produces lateral roots. The main root axis is subdivided into "sectors," each containing a portion of the main axis plus several lateral branch roots. Each of these sectors is transferred separately to a fresh medium for an additional period of growth. The lateral roots subsequently produce laterals themselves. Each of these sector cultures provides the investigator with a constant supply of lateral root tips for experiments, as well as a source of material for the propagation of the clone. This technique can be used only when the excised root produces laterals in some sequential order; in some species this does not occur.

QUESTIONS FOR DISCUSSION

1. What are advantages of using an aqueous medium in comparison with a gel-solidified medium? Can you think of any advantages that were not mentioned in this chapter?
2. What are the nutritional requirements for the culture of tomato roots?
3. What can we get by culturing the hairy-roots?
4. How to initiate and subculture fairy-roots?

知识要点

发根培养物也是次生代谢产物生产的一种有效途经。我们可以通过一种土壤细菌：发根农杆菌的根诱导质粒(Ri-T DNA)将次生代谢产物相关基因整合到发根细胞中，通过表达，从而实现高效生产靶基因产物的目的。发根的培养采用液体培养基，以利于发根的生长和目的产物的分离纯化。培养的发根在结构和代谢方式上与植物来源的发根有所不同。

CHAPTER 10   MICROPROPAGATION BY BUD PROLIFERATION(芽的快繁)

Micropropagation of ornamental and crop plants by the proliferation of shoot tips has resulted in the commercial production of plants on a worldwide scale. The entire method of micropropagation with shoot tips is based on the cytokinin-induced outgrowth of bud primordia, each of which produces a miniature shoot. Once a cluster of shoots has been formed, it can be subdivided into smaller clumps of offshoots, transferred to a fresh medium, and the process repeated. The rates of micropropagation vary greatly from species to species, but it is often possible to produce several million plants in the period of a year starting from a single isolated shoot tip.

The application of shoot-apex cultures for the clonal multiplication of plants was first realized by Morel (1960) during his studies on the propagation of the orchid *Cymbidium*(兰属), and modifications of his technique are currently used for the commercial production of orchids. The rapid progress in micropropagation techniques using multiple-shoot formation was due largely to the efforts of Murashige. Murashige (1974) subdivided the procedure into three stages:

Ⅰ. Establishment of the aseptic culture;

Ⅱ. Multiplication of propagula（繁植体）by repeated subcultures on a multiplication medium; and

Ⅲ. Preparation of the plantlets for establishment in the soil.

Since Murashige's terminology, workers now recognize stages 0 and Ⅳ:

0. Preparation of the mother plant prior to explant removal;

Ⅳ. Careful nurture of the plantlets in a potting mix under in vivo conditions.

During stage 0 there are several practices for preparing the mother plant. For example, subirrigation（下方灌溉）to provide water and nutrition ensures lower humidity and cleaner explants. The regulation of temperature, daylength, and light intensity is important to ensure that the donor plant is in the proper physiological condition for explant removal. Treatment with either low temperature or GA_3 may be used to facilitate breaking dormancy for certain bulbs, tubers, or other dormant organs.

Factors influencing stage I include the choice of a suitable explant, the composition of the medium, and the appropriate environmental conditions. Shoot tips and buds excised from healthy and actively growing herbaceous plants are generally ideal material for multiple-shoot production. The larger the tip explant, the more rapid the growth and greater the ability to survive. A procedure that does not involve a callus phase is preferable, because the genetic instability of callus leads to a high degree of genetically aberrant plants. Although some groups of plants have unique nutritional needs, the MS formulation is satisfactory in most cases. The basal medium is supplemented with vitamins, sucrose, and the appropriate growth regulators. The cultures can be grown on agar or on filter-paper bridges over a liquid medium. Light, necessary for photomorphogenesis(光形态建成) and chlorophyll biosynthesis(叶绿素生物合成), is provided with Gro-Lux(一种灯管品牌) and white fluorescent tubes. Murashige (1974) found that many cultures grew best with 1,000 lux for stages Ⅰ and Ⅱ, with the light intensity increased to 3,000-10,000 lux for stage Ⅲ. Photoperiods of 16 h were optimum for several species, although relatively little research has been done on the effects of light and temperature on micropropagation.

Typically, the same medium and environmental conditions are used for both stages Ⅰ and Ⅱ. The choice and concentration of growth regulators are the most important consideration in preparation of the medium. Cytokinins may be added in the form of kinetin, benzylaminopurine (BAP), 2iP, or zeatin. Some of Murashige's multiplication media that are commercially available contain combinations of adenine sulfate plus either kinetin or IPA (N^6-[Δ^2-isopentyl]adenine). The source of exogenous auxin is usually IAA, α-NAA, or IBA (indole-3-butyric acid). The auxin 2,4-D is unsatisfactory, since it stimulates callus formation and suppresses organogenesis. Gibberellic acid (GA_3) may be required for the culture of some shoot apices.

Stage Ⅲ involves the development of a root system, hardening the young plants to moisture stress, increasing resistance to certain pathogens, and conversion of the plants to an autotropic state. Root initiation may be facilitated by adding a low concentration of either α-NAA or IBA to the medium. The auxin treatment must be limited to a brief period of time. Auxin at this stage of the process may have undesirable side effects-that is, may stimulate callus production and inhibit root elongation. The formation of roots often occurs after transfer to a medium lacking hormones. In fact, the shoots of some species can be rooted by conventional root procedures after removal from the in vitro environment.

During stage Ⅳ the plantlets are transferred from the culture tubes to a soil mixture. The plantlets must undergo a period of acclimation to in vivo environmental conditions. They must be protected from direct sunlight, and the relative humidity should be gradually decreased over a period of time. During the acclimation period the rate of photosynthesis is initially low, and survival of the plantlets may depend on the accumulation of carbohydrates during the culture. Another difficulty is the relatively thin layer of cuticular(角质) wax present, which may lead to a serious dehydration of the aerial tissues.

Although the micropropagation of many tree crop species has been achieved using shoot-apex cultures, woody plants pose some unusual problems. Bud cultures must be taken either from shoots in the juvenile growth phase or selected from rejuvenated shoots. Buds taken from mature trees in the adult phase have little capacity for micropropagation since the cultured material is incapable of producing roots.

Shoot-apex cultures of woody plants require several consecutive treatments. The excised bud, with the formation of a rosette of leaves, requires exogenous gibberellin and cytokinin for growth. The explant must then be transferred to another medium lacking exogenous growth regulators in order to promote stem elongation. Finally, the culture must be transferred to a third medium containing exogenous auxin for the initiation of roots.

Tissues containing relatively high concentration of phenolic compounds are difficult to culture. Polyphenolase(多酚氧化酶) stimulated by tissue injury will oxidize these phenolics(酚类) to growth-inhibiting, dark-colored compounds. Techniques used to suppress this metabolic sequence include the following: (a) adding antioxidants to the medium, (b) presoaking the explants in antioxidant solutions prior to culture, (c) subculturing to a fresh medium on signs of enzymatic browning, and (d) providing little or no light during the initial period of culture.

The time required for the excised shoot tips to initiate growth and begin micropropagation, that is, with the outgrowth of axillary(腋生) buds, may vary from a few weeks to several months depending on the species and the cultural conditions.

Potato is one of the fastest-growing plants in micropropagation. Within five or six weeks a dense cluster of proliferating shoots should be evident (Fig. 10.1). After the shoots are well developed they are excised and transferred to a root-induction medium, which is the same as that used in this experiment except for the omission of BAP. Root initiation normally takes four to six weeks.

(a)   (b)   (c)   (d)

Figure 10.1 Micropropagation of potato (*Solanum tuberosum*) (a) Excised shoot tip grows to produce a plantlet. (b) In micropropagating culture multiple shoots are produced by outgrowth of axillary buds. (c) In vitro plantlet transferred to "jiffy" pot. (d) Plantlet derived from shoot tip ready for transfer to field. (By John H. Dodds)

Once an adequate root system has developed, the young plantlets are ready to be transferred to nonsterile condition. The plantlets are first placed in a sterile soil mixture and maintained under humid conditions by mist irrigation. This is preferable to sealing the pots with plastic, which tends to encourage the growth of fungi. Gradually the young plants are hardened off by reducing the mist irrigation, and eventually they are transferred to a cool greenhouse. Finally, the plants regenerated from the shoot-apex cultures are planted in the field.

QUESTIONS FOR DISCUSSION

1. What are the problems associated with the growth and development of cultured shoot tips excised from woody plants?

2. Why must the micropropogated plants gradually become accustomed to in vitro conditions? List some physiological basis for acclimatization.

知识要点

通过茎尖培养和次生芽诱导可快速繁殖花卉和农作物。芽快繁技术可分为 5 个阶段:母本植株的准备、无菌培养体系的建立、快繁、驯化和移栽。芽的诱导和生根培养所使用的培养基对植物激素的要求是不同的。驯化要循序渐进,给组培苗足够的时间长出新叶,以应付体外环境的干燥、强烈光照和微生物污染,否则的话,移栽的组培苗很容易干枯死亡。

CHAPTER 11　ANTHER AND POLLEN CULTURES
（花药和花粉的培养）

The cells of haploid plants contain a single complete set of chromosomes, and these plants are useful in plant-breeding programs for the selection of desirable characteristics. The phenotype is the expression of single-copy genetic information, there being no masking of a trait through gene dominance. The purpose of anther and pollen culture is to produce haploid plants by the induction of embryogenesis from repeated divisions of monoploid spores, either microspores（小孢子）or immature pollen grains. "Microspores" represent the beginning of the male gametophyte generation; "pollen grains" are mature microspores, especially following their release from tetrads（四分孢子）. The chromosome complement of these haploids can be doubled by colchicine（秋水仙碱）or by regeneration techniques to yield fertile homozygous（纯合）diploids. Although the number of successful pollen culture systems is still relatively small, this technique has resulted in several improved varieties of crop plants in China.

Tulecke (1953) first observed that mature pollen grains of the gymnosperm（裸子植物）*Ginkgo biloba*（银杏）could be induced to form a haploid callus following culture on a suitable medium. Repeated divisions of cultured pollen grains of angiosperms（被子植物）were described much later by Guha and Maheshwari (1966), who made a remarkable discovery by accident. These investigators were conducting experiments with cultured pollen grains of *Datura innoxia*（毛曼陀罗）in order to determine the feasibility of this system for the study of factors regulating meiosis. The growth response of the pollen grains, enclosed in mature anthers, was of three types (Fig. 11.1) and reflected the nature of the medium. Although the pollen grains were unresponsive in the presence of IAA, callus was initiated on media containing either yeast extract or casein hydrolysate（酪蛋白水解物）. Torpedo-shaped embryoids, which later developed into plantlets, were produced following culture of the anthers on media containing either kinetin or coconut water. Acetocarmine（醋酸洋红）staining revealed that these newly formed plantlets contained only a single set of chromosomes.

The particular stage of development of the anthers at the time of culture is the most important factor in achieving success in the formation of embryoids. In angiosperms with an indeterminate number of anthers in each flower bud, buds can be selected that will contain several anthers in various stages of pollen development. In species with a determinate number of anthers per flower, a series of buds must be examined in order to give all the stages of development. Two basic methods are used: (a) excised anthers are cultured on an agar or liquid medium, and embryogenesis occurs within the anther; or (b) the pollen is removed from the anther, either by mechanical means or by natural dehiscence (裂开) of the anther, and the isolated pollen is cultured on a liquid medium.

It may take three to eight weeks for haploid plantlets to emerge from the cultured anthers.

Sunderland (1979) reported that in flowers of many plants anthers fall into one of three categories: premitotic, mitotic, or postmitotic. In the premitotic category the best response is obtained by using anthers in which the microspores have completed meiosis but have not yet started the first pollen division (e. g., *Hyoscyamus* (大麦), *Hordeum vulgare* (莨菪)). Anthers of plants belonging to the mitotic group respond optimally about the time of the first pollen division (e. g., *Nicotiana tabacum* (烟草), *Datura innoxia* (毛曼陀罗), *Paeonia* (芍药)). The early bicellular stage of pollen development is best in the postmitotic plants (e. g., *Atropa belladonna* (颠茄), *Nicotiana* spp. (烟草)). In the case of *N. tabacum*, floral buds with the corolla(花冠) barely visible beyond the calyx(花萼) will probably contain anthers at the appropriate stage of development, although there may be slight differences among different cultivars.

As discussed previously, activated charcoal has a stimulatory effect on somatic embryogenesis as well as on the initiation of embryos from haploid anther tissue. This charcoal effect has been demonstrated for anther cultures of tobacco, rye (裸麦), potato, and other plants. The removal of inhibitory substances from the agar is considered to be a factor since a similar response was obtained by dialyzing(透析)agar against activated charcoal, or by employing highly purified agar. Another possibility is adsorption (吸附) by charcoal of 5(hydroxymethyl)2-furfural(5-羟甲基-2-糠醛), a degradation product of autoclaved sucrose. Although the precise role of activated charcoal in this developmental process remains unknown, the use of charcoal for the enhancement of haploid plantlet production should be encouraged.

Another factor in anther culture is the physiological status of the parent plant (e. g., photoperiod, light intensity, temperature, and mineral nutrition). Anthers should be taken from flowers produced during the beginning of the flowering period of the plant. Higher yields of embryos have been reported from donor plants grown under short days and high light intensities.

Various types of anther pretreatment have been found to improve embryo production in some plants. Low-temperature pretreatment of anthers for periods of 2-30 days at temperatures of 3-10℃ may stimulate embryogenesis. Other types of pretreatment include soaking the detached inflorescence in water for several days and centrifugation of the anthers at 3-5℃ for approximately 30 min.

The presence of anther tissue in the culture introduces several ill-defined factors that influence embryoid production. Raghavan (1978) suggested that a gradient of endogenous auxin within the anther may play a role in pollen grain development. Embryogenic pollen grains were observed to be confined to the periphery of the anther locule（小室）in close proximity to the tapetum（珠被绒毡层）, and possibly substances released from the tapetum initiate embryogenic divisions in pollen grains within cultured anther segments of *Hyoscyamus niger*（天仙子）. Pollen is also sensitive to toxic substances released following injury to anther wall tissue.

It is important to determine the chromosome number of the newly formed plantlets because there may be considerable variation in ploidy levels, depending on the developmental events that led to embryoid formation. Diploid heterozygous plants may arise from anther tissue or from growth of the microspore mother cells and unreduced microspores. The further development of dyads（二分体）and incomplete tetrads often produces plants that are heterozygous（杂合）at certain loci because of crossing-over prior to the first reduction division. Chromosome doubling and fusion of nuclei can produce homozygotes with varying ploidy levels. Plantlets formed from the callus tissue arising from haploid microspores can exhibit mutations and chimeras（嵌合体）. It was found that the ploidy levels of 2,496 rice plants derived from pollen cultures were 35.3% haploid, 53.4% diploid, 5.2% polyploid, and 6.0% mixoploid. Some developmental pathways exhibited by microspores are shown in Figure 11.1.

There are several techniques for doubling the chromosome number of the haploid plants, and two approaches can be taken: (a) regeneration by tissue culture methods, and (b) chemically induced doubling with colchicine（秋水仙素）. The ploidy level of the plant involved must first be confirmed with standard cytological procedures before additional experiments are undertaken. One method of chromosome doubling employs aged leaf tissue from haploid plants because older leaves have the potential to regenerate both haploid and diploid plants. The diploids result from chromosome endoreduplication, which frequently occurs in cultured plant tissues. In addition, the chromosome number can be doubled by the application of colchicine to either the embryos or the haploid plants. A simple procedure is to immerse the anthers containing the newly formed plantlets in an aqueous solution of coichicine (0.5% w/v) for 24-48 h. Another approach is to apply a preparation of colchicine in lanolin paste（羊毛脂膏,0.4% w/v）to the axillary buds of decapitated（去顶芽）mature haploid plants.

The potential for using haploid plants and homozygous lines in plant-breeding programs has been recognized. One important area of research concerns the development of homozygous lines for the production of hybrids in self-incompatible species (e.g., rye（黑麦）and rape). Microspore culture is important in mutagenic studies: Mutations are not masked in haploids because there cannot be a dominant gene. For these studies to be successful, however, large numbers of microspores must be induced to undergo embryogenesis and develop on nutrient media. The haploids must remain genetically stable, and it must be possible to regenerate diploid plants from the haploids. This has been achieved in the case of *Nicotiana tabacum*(烟草).

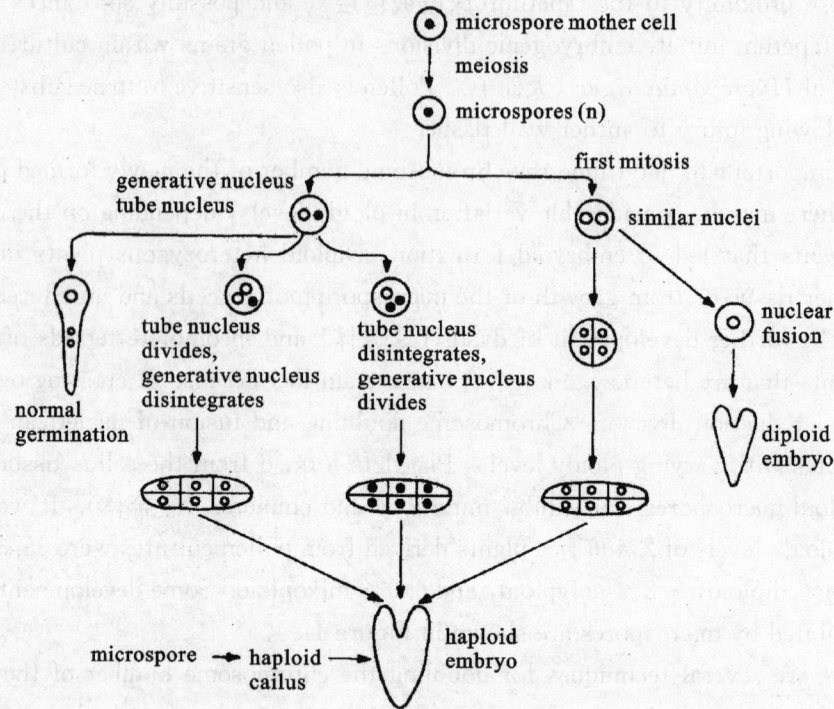

Figure 11.1 Some possible developmental pathways of microspores under in vivo and in vitro conditions
Normal development (in vivo) results in the production of two sperm, a tube nucleus (粉管核,营养核) or cell, and pollen tube formation (*far left*). Several possibilities exist following the in vitro culture of isolated microspores or anthers. Either the tube or generative nucleus (生殖核) degenerates, and the surviving nucleus divides repeatedly and ultimately produces a haploid embryoid. The first mitosis of the microspore may produce two similar nuclei, and repeated division of these nuclei can produce a haploid embryoid. The similar nuclei, however, may fuse and produce a diploid embryoid (*far right*). Haploid callus of microspore origin can form embryoids de novo. (By John H. Dodds)

Main procedures for anther culture are listed in Fig. 11.2 and 11.3. Formation of haploid plantlets can also be induced via the culture of isolated pollen removed from excised anthers. Final success in the culture of anther or pollen is to obtain chromosome doubling and the production of diploid plants from the haploids.

PROTOCOL FOR ANTHER CULTURE

1. Plants selected for experimentation are cultivated until they reach the flower bud stage. In the case of tobacco plants, the buds should have a corolla length of 21-23 mm, and at this stage the pollen will have completed the first mitosis.

PART Ⅰ   CYTOTECHNOLOGY IN PLANTS（植物细胞工程）

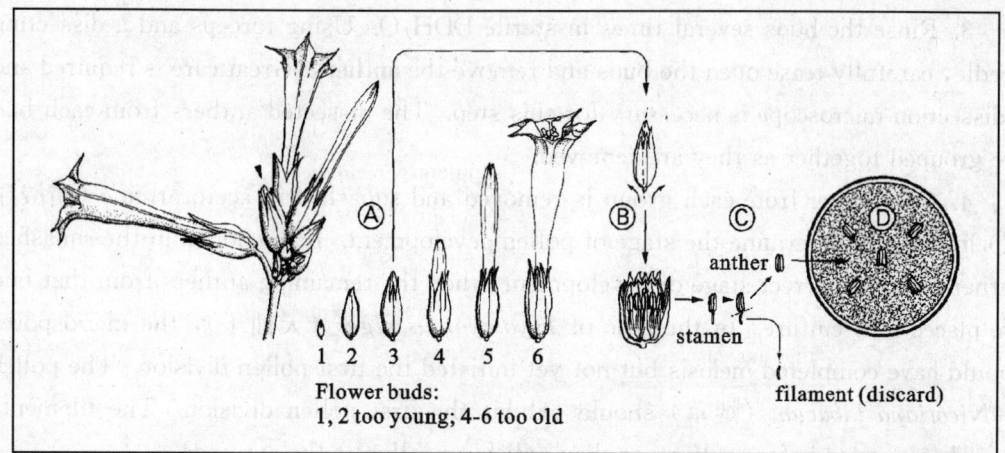

Figure 11.2   Isolation of anthers from the flowers of *Nicotiana tabacum* （By J. Reinert and M. M. Yeoman, 1982）

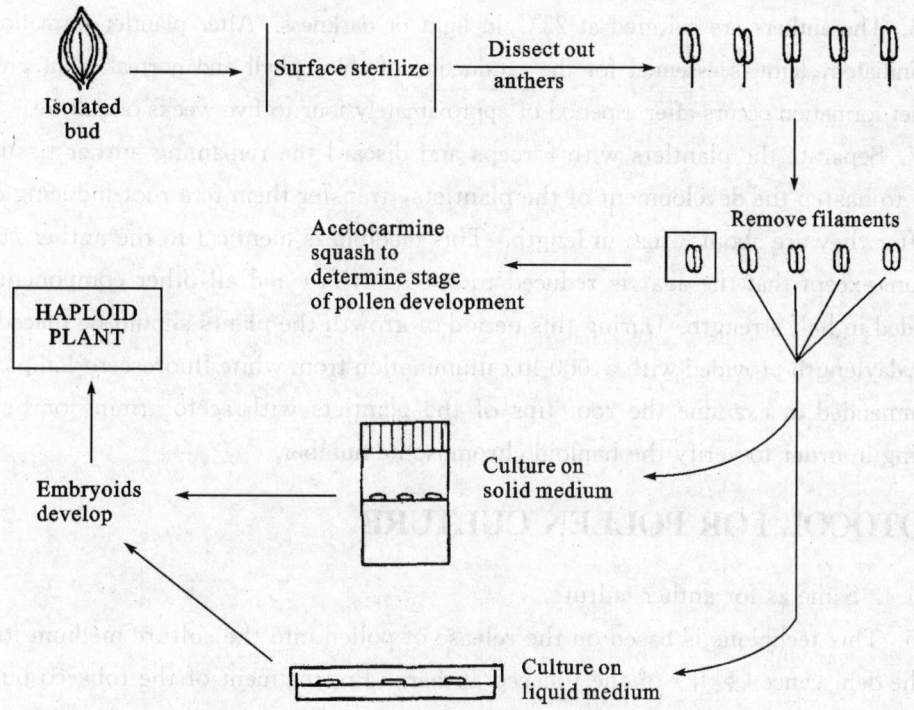

Figure 11.3   Basic procedure for the production of haploid plants from anther culture. Isolated buds are surface sterilized and the anthers aseptically removed. Individual anthers are screened by an acetocarmine staining method for the selection of the proper stage of pollen development. Subsequent culture results in embryoid and haploid plantlet formation. （By John H. Dodds）

2. If tobacco is employed, best results will be obtained by chilling the buds approximately 12 days (7-8℃) prior to culture. For surface sterilization the buds are transferred to a Petri dish containing hypochlorite solution plus a wetting agent (10 min).

· 69 ·

3. Rinse the buds several times in sterile DDH_2O. Using forceps and a dissecting needle, carefully tease open the buds and remove the anthers. Great care is required and a dissection microscope is necessary for this step. The dissected anthers from each bud are grouped together as they are removed.

4. One anther from each group is removed and squashed in acetocarmine（醋酸洋红）in order to determine the stage of pollen development. If the pollen in the squashed anther is in the correct stage of development, then the remaining anthers from that bud are placed into culture. In the case of *Hyoscyamus niger*（天仙子）, the microspores should have completed meiosis but not yet initiated the first pollen division. The pollen of *Nicotiana tabacum*（烟草）should exhibit the first pollen division. The filaments must be removed before culture or they will form callus at the cut ends.

5. Anthers can be cultured either on agar-solidified culture media or by floating them on the surface of a liquid medium.

6. The anthers are cultured at 25℃ in light or darkness. After plantlet formation has been initiated, light is essential for the production of chlorophyll and normal plant growth. Plantlet formation occurs after a period of approximately four to five weeks of culture.

7. Separate the plantlets with forceps and discard the remaining anther tissue. In order to hasten the development of the plantlets, transfer them to a root-inducing medium after they are about 3 mm in length. This medium is identical to the anther culture medium except that the agar is reduced to 0.5% (w/v) and all other components are provided in half strength. During this period of growth the plants should be placed on a 12-hr daylength provided with 5,000-lux illumination from white fluorescent lamps. It is recommended to examine the root tips of the plantlets with acetocarmine or Feulgen staining in order to verify the haploid chromosome number.

PROTOCOL FOR POLLEN CULTURE

1-4. Same as for anther culture.

5. This technique is based on the release of pollen into the culture medium following the dehiscence（裂开）of the tobacco anthers. Pretreatment of the tobacco buds by chilling (step 2) apparently facilitates the dehiscence. For each culture place the anthers from three tobacco buds in 5 cm^3 of liquid medium in a Petri dish. Remove and discard the anthers from some cultures after 6, 10, and 14 days. For anthers other than those of tobacco, it may be advisable to make a slight incision in the anther tissue with a fine scalpel blade and gently squeeze the contents of the anthers into the medium with forceps. Seal the dishes with Parafilm and incubate them at 28℃ in the dark for the first 14 days of culture. After 14 days transfer the cultures to an illuminated growth chamber (GroLux fluorescent lamps, 500 lux, 12-hr daylength, 25℃). The pollen released from

the anthers into the medium at 6-, 10-, and 14-day intervals will develop into haploid embryos.

REGENERATION OF DIPLOID PLANTS

In this experiment an attempt will be made to promote chromosome doubling of the plantlets formed from the anther culture experiment. When plantlets are beginning to emerge from the cultured anthers, immerse the anthers in an aqueous solution of colchicine (0.5% w/v) for 24-48 h. Colchicine is a powerful poison, the utmost care should be took in handling this alkaloid. Following the colchicine treatment, rinse the plantlets in DDH_2O and culture them as described in the experiment of anther culture, step 7. It is important to ascertain the chromosome number of the plantlets with acetocarmine or Feulgen staining. If diploidy is not achieved, the immersion time may be increased to 96 h, and it may be necessary to repeat the treatment several times in order to achieve success.

The medium recommended for the culture of anthers is the MS medium supplemented with 2% (w/v) sucrose. If desired, the medium may be solidified with agar (0.6%-0.8% w/v).

After two or three days in culture certain developmental changes can be detected (Fig. 11.4-a,b). Many of the pollen grains accumulate starch granules, others degenerate, and a small proportion divide and enter into the developmental pathway leading to the formation of haploid embryoids. The stages of embryoid development arising from cultured pollen grains are similar to the stages of zygotic embryo development normally found in diploid plants. The embryogenic pollen grains eventually rupture, and the developing system undergoes repeated cell divisions. This cellular proliferation gives rise to a globular-stage embryo (Fig. 11.4-c,d), which undergoes morphological changes to produce heart-shaped (Fig. 11.4-e) and torpedo forms. Eventually the haploid embryoids break through the anther wall (Fig. 11.4-f). This entire process takes about 14 days in *Hyoscyamus*, and 21-28 days in *Nicotiana*.

The plantlets can be dissected from the anthers and grown to maturity. It is possible to treat these haploid plants in such a way that they develop into homozygous and fertile diploid plants. Haploid plants are sterile because they are unable to undergo reduction division in meiosis.

QUESTIONS FOR DISCUSSION

1. What is haploid plant? Of what importance are haploid plants to the plant breeder?
2. Offer some explanations why cultured anthers will permit pollen to develop into embryos, whereas cultured isolated pollen grains may not form embryos.

3. Discuss the role of activated charcoal in the haploid plantlet production.
4. Discuss the ploidy variation in the newly formed plantlets during anther culture.
5. How to double the chromosome number of the haploid plants?

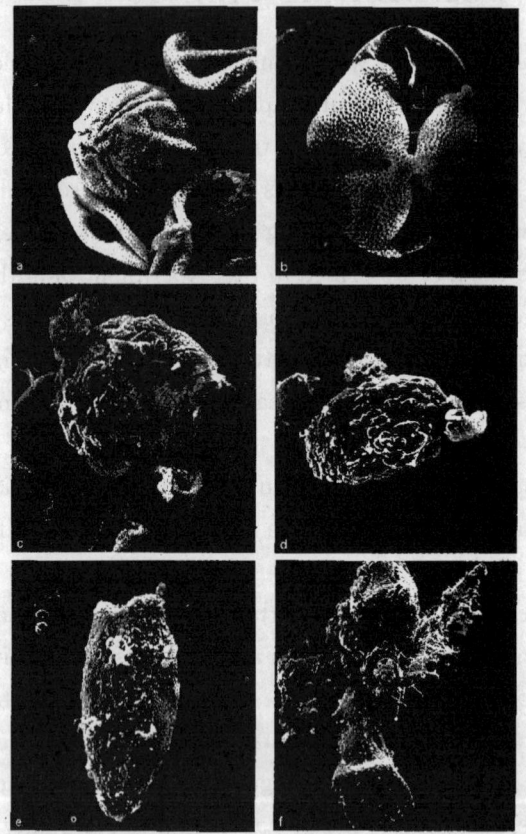

Figure 11. 4 (*facing*). **Observations on embryoid development in cultured anthers of *Hyoscyamus niger* (henbane) with scanning electron microscopy.** a. Developing pollen grain after culturing for 20 h. b. Swollen grain beginning to break open at raphe after 5 days of culture. c. Globular stage of embryoid after 7 days of culture showing association with the parental pollen grain. d. Another embryoid showing globular stage after 7 days of culture. e. Heart-shaped embryoid after 9 days of culture. f. Embryoid bursting through the anther wall after 14 days of culture; note the developing roots and shoots. (From Dodds & Reynolds, 1980)

知识要点

花药和花粉培养是获得单倍体植株的一个重要途径。单倍体育种可大大缩短遗传性状的选择周期。对花粉发育时期的选择，即单核期花粉，是成功获得单倍体植株的关键。活性炭可吸附有害物质，因而可促进花粉的胚胎发生。母本植株的生理状态也可影响花粉的培养。低温处理花药可促进花粉的胚胎发生。花粉培养得到的植株的倍性必须进行鉴定，因为往往混有大量二倍体和多倍体。单倍体植株生活能力弱，必须进行染色体加倍以得到纯合二倍体植株。

CHAPTER 12 ISOLATION AND CULTURE OF PROTOPLASTS(原生质体的分离和培养)

Isolated protoplasts have been described as "naked" plant cells because the cell wall has been experimentally removed by either a mechanical or an enzymatic process. The isolated protoplast is unusual because the outer plasma membrane is fully exposed and is the only barrier between the external environment and the interior of the living cell. Despite technical difficulties that have limited their potential use in some investigations, protoplasts are currently utilized in several areas of study:

1. Two or more protoplasts can be induced to fuse and the fused product carefully nurtured to produce a hybrid plant. The regeneration of *Atropa belladonna* (颠茄) plants from single isolated protoplasts is shown in Figure 12.1.

2. After removal of the cell wall, the isolated protoplast is capable of ingesting "foreign" material into the cytoplasm by a process similar to endocytosis as described for

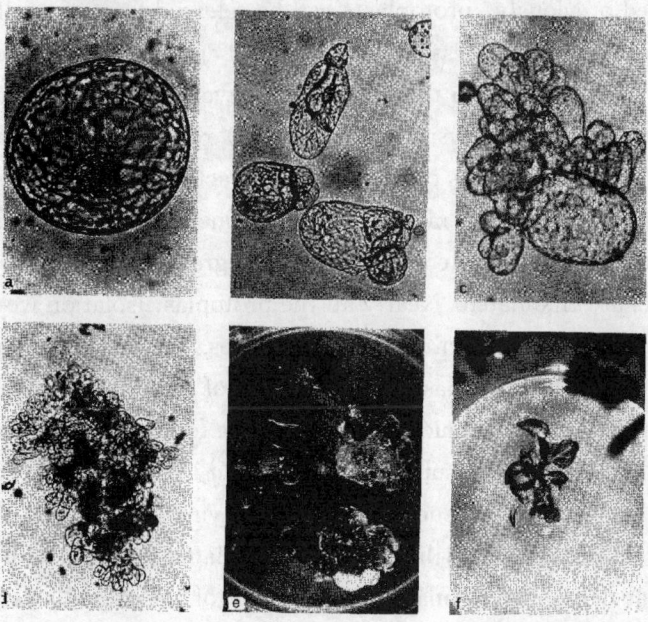

Figure 12.1 Sequence of development of a plantlet of *Atropa belladonna* (颠茄) from a single isolated protoplast. a. A single isolated protoplast. b. Cell wall regeneration and initiation of cell division. c, d. Development of cell aggregates. e. Appearance of embryoids on surface of callus. f. Formation of plantlet on agar medium. (By H. Lorz)

certain animal cells and protozoans. Experiments are in progress on the introduction of nuclei, chloroplasts, mitochondria, DNA, plasmids, bacteria, viruses into protoplasts.

3. The cultured protoplast rapidly regenerates a new cell wall and this developmental process offers a novel system for the study of wall biosynthesis and deposition.

4. Populations of protoplasts can be studied as a single cellular system. Microbiological methods have been developed for the selection of mutant cell lines and the cloning of cell populations.

One must remember that a chief function of the cell wall is to exert a wall pressure on the enclosed protoplast and thus prevent excessive water uptake leading to bursting of the cell. Before the cell wall is removed the cell must be bathed in an isotonic plasmolyticum(质壁分离剂), which is carefully regulated in relation to the osmotic potential of the cell. In general, mannitol(甘露糖) or sorbitol (山梨醇, 13% w/v) has given satisfactory results. A technique involving the use of a relatively low osmotic potential with sucrose (0.2 M) and polyvinylpyrrolidone (聚乙烯吡咯烷酮, 2% w/v) has been reported. In developing an original technique it may be advantageous to test a range of mannitol concentrations of 8%-15% (w/v). Some authors have pointed out that preparations bathed in a plasmolyticum of too low a concentration may lead to multinucleate protoplasts owing to the spontaneous fusion of two or more protoplasts during the isolation procedure.

As mentioned previously, protoplasts can be released from the cell wall by either a mechanical or an enzymatic process. The mechanical approach involves cutting a plasmolyzed (胞浆分离) tissue in which the protoplasts have shrunk and pulled away from the cell wall. Subsequent deplasmolysis(质壁分离复原) results in expansion and release of the protoplasts from the cut ends of the cells. In practice this technique is difficult, and the yield of viable protoplasts is meager(很少). One advantage, however, is that the complex and often deleterious effects of the wall-degrading enzymes on the metabolism of the protoplasts are eliminated. Nearly all the protoplast-isolation work since the early 1960s has been performed with enzymatic procedures. By using enzymes one obtains a high yield (2×10^6-5×10^6 protoplasts/g leaf tissue) of uniform protoplasts after removal of cellular debris. The basic technique consists of the following: (a) surface sterilization of leaf samples; (b) rinsing in a suitable osmoticum(渗透压调节剂); (c) peeling off the lower epidermis or slicing the tissue to facilitate enzyme penetration; (d) sequential or mixed-enzyme treatment; (e) purification of the isolated protoplasts by removal of enzymes and cellular debris; and, finally, (f) transfer of the protoplasts to a suitable medium with the appropriate cultural conditions (see Figs. 12.2 and 12.3).

The plant cell wall consists of a complex mixture of cellulose, hemicellulose(半纤维素), pectin(果胶质), and lesser amounts of protein and lipid. Because of the chemical bonding of these diverse constituents, a mixture of enzymes would appear necessary to degrade the system effectively. Cellulose is a polymer consisting of subunits of D-glucose. Xylans(木聚糖) form the bulk of the hemicellulose fraction in angiosperms(被子植物). These polymers consist, however, of several monosaccharides in addition to xylose(木糖). Pectins are polysaccharides containing the sugars galactose(半乳糖), arabinose(阿拉伯糖), and the galactose derivative galacturonic acid(半乳糖醛酸). Protoplast isolation is achieved by using cellulase(纤维素酶)in combination with pectinase(果胶酶) and hemicellulase(半纤维素酶). There have been two approaches to the use of wall-degrading enzymes. In the "mixed-enzyme method" both pectinase and cellulase-hemicellulase are applied simultaneously, whereas the "sequential method" involves treatment of the leaf material with pectinase to loosen the cells, followed by a cellulase-hemicellulase digestion.

In this chapter the standard procedure for obtaining isolated protoplasts from mature leaf mesophyll tissue is described. This technique can be employed with a reasonable degree of success on other tissues. In the following experiment we will isolate, purify, and culture mesophyll protoplasts and regenerate plantlets from the protoplast-seeded callus.

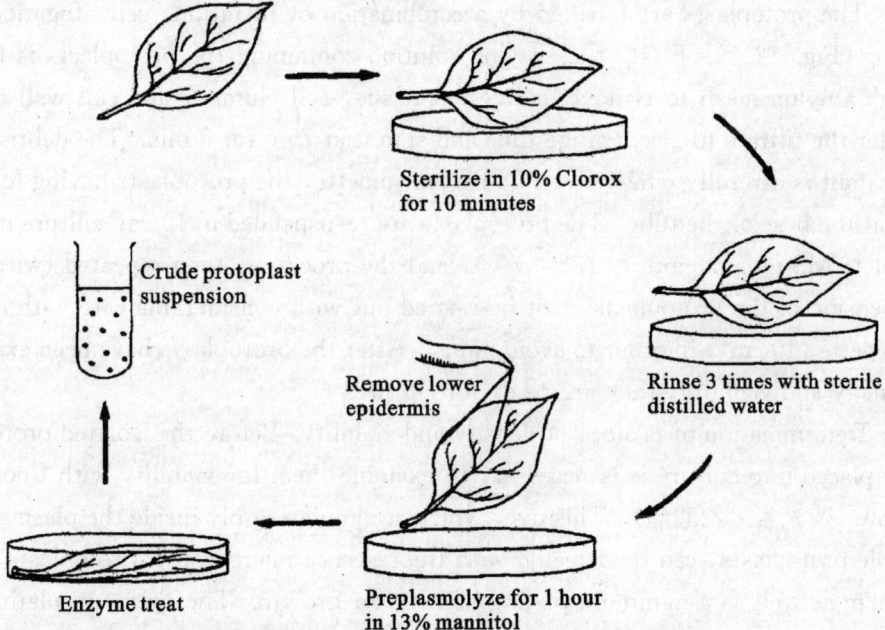

Figure 12.2 **Basic technique for the isolation of protoplasts from an excised leaf.** The leaf is surface sterilized, rinsed repeatedly in sterile distilled water, and the cells are plasmolyzed in a solution of mannitol. The lower epidermal layer is stripped from the leaf to enhance enzyme penetration into the mesophyll tissue. Following treatment with one or more wall-degrading enzymes, a crude suspension of mesophyll protoplasts is obtained. (By John H. Dodds)

PROCEDURE

1. Mature healthy leaves are removed from the plants and rinsed briefly in tap water. Immerse the leaves in the hypochlorite-detergent solution for 10 min. Rinse the leaves three times in order to remove all traces of the hypochlorite solution.

2. The waxy cuticle covering the leaves restricts access of the enzyme solution to the mesophyll cells. While the leaves are in the final rinse, the lower epidermis is peeled from the leaves with pointed forceps. Cut the leaves into small and transfer approximately 1 g of peeled leaf strips to a Petri dish (100×15 mm) containing 10 cm^3 of enzyme solution that has been sterilized by membrane filtration (0.45 μm). The enzyme solution contains Macerozyme R-10 (0.5% w/v) plus cellulase Onozuka R-10 (2.0% w/v) dissolved in mannitol (13% w/v) at pH 5.4. Seal the Petri dishes with Parafilm and wrap them with aluminum foil. Usually the leaf material is incubated in the enzyme solution overnight (12-18 h, 25℃), although the mesophyll cells should be in contact with the enzymes for as short a time as possible. The leaf strips are then teased gently with forceps to release the protoplasts.

3. The protoplasts are purified by a combination of filtration, centrifugation, and washing (Fig. 12.3). First, the enzyme solution containing the protoplasts is filtered through a nylon mesh to remove undigested tissue, cell clumps, and cell wall debris. Transfer the filtrate to a centrifuge tube and spin it at 75 g for 5 min. The debris in the supernatant is carefully removed with a Pasteur pipette, the protoplasts having formed a pellet at the base of the tube. The protoplasts are resuspended in 10 cm^3 culture medium (complete MS plus mannitol, 13% w/v), and the process is then repeated twice. The resuspension of the protoplasts must be carried out with considerable care with a wide-bore pipette (10 cm^3) in order to avoid injury. After the protoplasts have been examined for density and viability, they are ready for culture.

4. Determination of protoplast density and viability. Before the isolated protoplasts can be placed into culture it is necessary to examine them for viability with fluorescein diacetate (荧光素二乙酸酯). This dye, which accumulates only inside the plasmalemma of viable protoplasts, can be detected with fluorescence microscopy. Protoplasts have a maximum as well as a minimum plating density for growth. The optimum plating efficiency for tobacco protoplasts is about 5×10^4 protoplasts/cm^3; the protoplasts fail to divide when plated at one-tenth of this concentration. Because the protoplast preparation will be diluted by an equal quantity of agar-containing medium, the sample should be adjusted to a concentration of 10^5 protoplasts/cm^3.

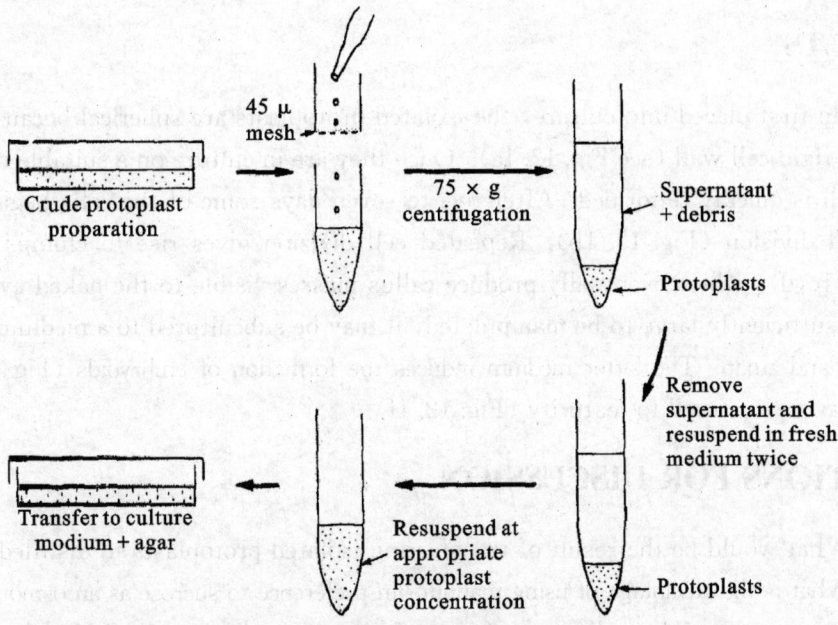

Figure 12.3 Purification procedure for isolated protoplasts. The crude protoplast suspension is filtered through a nylon mesh (45-m pore size), and the filtrate is centrifuged for 5 min at 75 g. The supernatant, carefully removed by Pasteur pipette, is discarded. The protoplasts, resuspended in 10 cm³ of fresh culture medium, are again centrifuged. Once again the supernatant is removed. The centrifugation-resuspension process is conducted three times. Before transfer of the protoplasts to a culture medium, the preparation is examined for protoplast density and viability. (By John H. Dodds)

5. Culture of protoplasts. Protoplasts have been cultured in several ways; for example, in hanging-drop cultures, in microculture chambers, and in a soft agar (0.75% w/v) matrix. The agar-embedding technique is one of the better methods as it ensures support for the protoplasts and permits observation of their development. The suspension of isolated protoplasts is adjusted with the hemocytometer by the addition of the culture medium plus mannitol (13% w/v) to yield a concentration of 10^5 protoplasts/cm³. The culture of embedded protoplasts is incubated at 25℃ in the presence of a dim white light.

6. Regeneration of plants from the protoplasts. Once the protoplasts have regenerated a cell wall, they undergo cell division and form a callus. This callus can be subcultured to plates or flasks containing a freshly prepared medium. If the callus of some species is transferred to a medium lacking both mannitol and auxin, embryogenesis begins on the callus after about three to four weeks. These embryoids, dissected from the callus, are nurtured in the same manner as those produced by somatic embryogenesis or by anther culture. With proper care and attention the embryoids will develop into seedlings and eventually grow into mature plants.

RESULTS

When first placed into culture, the isolated protoplasts are spherical because of the lack of a rigid cell wall (see Fig. 12.1a). Once they are in culture on a suitable medium, a cell wall is quickly re-formed. After five to seven days some of the cells begin to undergo cell division (Fig. 12.1b). Repeated cell division gives rise to clumps of cells (Fig. 12.1c,d), which eventually produce callus masses visible to the naked eye. Once callus is sufficiently large to be manipulated, it may be subcultured to a medium lacking mannitol and auxin. The latter medium induces the formation of embryoids (Fig. 12.1e), which may be nurtured to maturity (Fig. 12.1f).

QUESTIONS FOR DISCUSSION

1. What would be the result of transferring isolated protoplasts to distilled water?
2. What is the advantage of using mannitol in preference to sucrose as an osmoticum?
3. List some possible applications of isolated plant protoplasts to the field of agriculture.
4. What are advantages of using a mechanical technique over enzymatic digestion in the isolation of protoplasts?
5. What kinds of "foreign" material have been introduced into isolated protoplasts?

APPENDIX

Observations on Cell Wall Regeneration. Mesophyll protoplasts start to regenerate a new cell wall within a few hours following isolation, although it may take several days to complete wall biosynthesis. These initial events occurring on the surface of the plasmalemma can be observed microscopically by using Calcofluor White M2R（一种荧光染料）, purified (Polysciences Inc.). This white dye binds to wall material and exhibits fluorescence on irradiation with blue light. The regenerating cells are incubated in 0.1% (w/v) Calcofluor dissolved in the appropriate osmoticum for 5 min. After rinsing to remove excess dye, the protoplasts can be examined microscopically. Cellulose layers will fluoresce when irradiated with UV light at 366 nm.

知识要点

植物原生质体是脱去细胞壁的植物细胞。可通过机械切割经质壁分离处理的植物组织获得,但完整的原生质体的获得率很低。通过纤维素酶、半纤维素酶和果胶酶消化植物组织,可获得大量完整的原生质体,但消化酶对原生质体的损伤较大。原生质体的分离和培养一定要在轻微高渗的培养基中进行,以防止低渗导致的细胞破裂;原生质体在体外培养几天后即可再生出新的细胞壁,之后,培养基的渗透压则需逐渐恢复等渗状态。植物原生质体培养的成功,为植物细胞杂交育种和遗传工程育种提供了新的途径。

CHAPTER 13 PROTOPLAST FUSION AND SOMATIC HYBRIDIZATION(原生质体的融合和体细胞杂交)

The preceding chapter describes the methods used for the isolation, purification, and culture of isolated protoplasts, and offers some insight into the way in which whole plants may be regenerated from single isolated protoplasts. The interest in protoplast fusion techniques is related to the prospect that wider crosses that are not possible by sexual means may be achieved with protoplast fusion. For example, some plants that show physical or chemical incompatibility in normal sexual crosses may be produced by the fusion of protoplasts obtained from two cultures of different species. It should be emphasized, however, that hybrid whole plants have been regenerated in a relatively small number of fusion systems, and there are no instances of the successful use of somatic hybridization in a plant-breeding program. In the early research work, protoplasts could be isolated from a small number of plants; the number of successful protoplast isolations increased following improvements in technique. Doubtless, the number of successful fusion experiments will rapidly increase after the techniques have been perfected.

The fusion of plant protoplasts is not a particularly new phenomenon; Kuster in 1909 described the process of random fusion in mechanically isolated protoplasts. When two or more isolated protoplasts are fused together, there is always a coalescence(合并) of the cytoplasms of the various protoplasts (Fig. 13.1). The nuclei of the fused protoplasts may fuse together or remain separate. Cells containing nonidentical nuclei are referred to as "heterokaryons(异核体)" or "heterokaryocytes(异核细胞)". The fusion of nuclei in a binucleate heterokaryon results in the formation of a true hybrid protoplast or "synkaryocyte(合核细胞)". The fusion of two protoplasts from the same culture results in a "homokaryon(同核体)". Frequently genetic information is lost from one of the nuclei. If one nucleus completely disappears, the cytoplasms of the two parental protoplasts are still hybridized (Fig. 13.1), and the fusion product is known as a "cybrid" (胞质杂交体, cytoplasmic hybrid) or "heteroplast(异质体)". Certain genetic factors are carried in the cytoplasmic inheritance system instead of in the nuclear genes. The formation of cybrids, therefore, has application in a plant-breeding program.

Spontaneous fusion of protoplasts may occur, or they may be induced to fuse in the presence of "fusigenic agents(融合剂)". During enzymatic digestion of the cell wall, the protoplasts of contiguous cells may fuse together through their adjoining plasmodesmata (胞间连丝).

CYTOTECHNOLOGY (细胞工程技术)

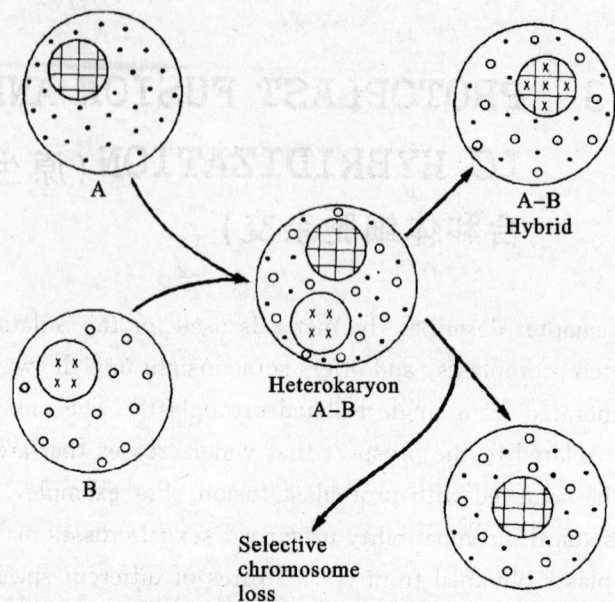

Figure 13.1 Some fusion products resulting from protoplast culture. The fusion of protoplasts A and B results in a binucleate heterokaryon containing the cytoplasmic contents of the two original protoplasts. Fusion of the two nuclei results in a tetraploid hybrid cell or synkaryocyte. If one of the nuclei degenerates, a cybrid or heteroplast is produced. (By John H. Dodds)

   The dissolution of the wall allows the plasmodesmatal(胞间连丝)strand interconnecting the cells to enlarge; the cytoplasm and organelles from two or more cells then flow together. These spontaneous fusions are always intraspecific (i. e., originate from plant tissue of the same species). This phenomenon rarely occurs because the negative charges on the surface of the protoplasts cause them to repel each other. Although the fusigenic agent lowers the surface charge, which permits the protoplast membranes to come into proximity, the adhesion of the protoplasts is insufficient to bring about fusion without molecular alterations in the bilayer structure of the plasma membranes.

   Several compounds have been shown to have a fusigenic effect on protoplasts. The addition of sodium nitrate to the culture medium induced fusion of root protoplasts from oat (燕麦) and maize. Fusion is also promoted by a combination of high pH (10.5), a high concentration of calcium ions (50 mM $CaCl_2 \cdot 2H_2O$), and high temperature (37℃). At present polyethylene glycol(聚乙二醇,PEG) is the most widely used fusigenic agent. The molecular weight and concentration of PEG, the density of protoplasts, the incubation temperature, and the presence of divalent cations are all factors that play a role in the fusion process. Polyethylene glycol had been used to induce the fusion of animal cells; therefore, it was not surprising that a heterokaryon was produced between an animal cell and a plant cell. The latter heterokaryon involved the fusion of a hen erythrocyte and a yeast protoplast. The fusion of cultured amphibian cells with protoplasts of a higher plant has also been reported.

The experiment outlined in this chapter involves an attempt to induce fusion between isolated protoplasts from two different plant sources. The inductive treatment, which involves low-speed centrifugation of the mixed culture in the presence of PEG, will result in a range of fusion products. The protoplast population will consist of unfused parental cells from the two tissues, homokaryons, heterokaryons, and multiple fusion products. If this mixture is plated on a culture medium, some of the various cell types will divide and develop callus. The next problem for the investigator is the recognition of callus formed by somatic hybrid and cybrid cells. The selection procedures are generally of two types: visual and biochemical.

Visual selection has been restricted to the fusion of colorless protoplasts with those containing chloroplasts. Protoplasts that demonstrate the completely integrated structural characteristics of both parental types are heterokaryons and potentially hybrid cells. An example of the fusion of plant protoplasts is shown in Figure 13.2. It is possible to attach a fluorescent label to the outer membranes of two parent protoplast populations and then separate the fusion products from the parental mixture by a method of fluorescent cell sorting.

Somatic hybrids can be selected by using a medium to encourage selective growth. A method was developed for the selection of hybrids resulting from the fusion of protoplasts of Nicotiana glauca and N. langsdorffii (Figure 13.3). Neither of the parental protoplasts is capable of growth on a medium deprived of auxin. The protoplast fusion products of the two species are auxin autotropic and are capable of callus formation on an auxin-free medium. The callus can then be subcultured and induced to regenerate hybrid plants.

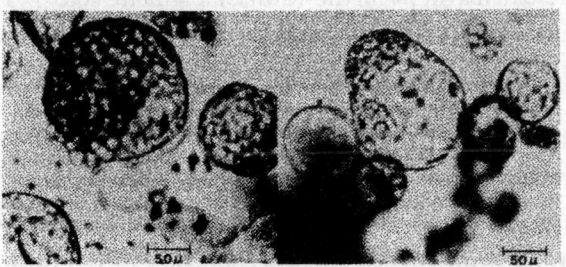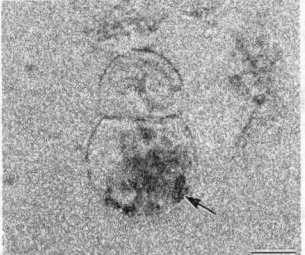

Figure 13.2 Fusion of colorless and chloroplast-containing protoplasts (By G. Melchers)

Another method involves the use of biochemical mutants and a selection of somatic hybrids by a form of complementation(互补). The antibiotic actinomycin D(放线菌素D) was used in the detection of fusion products of two species of *Petunia*（矮牵牛花）. Cultured cells of *P. hybrida* cannot grow in the presence of actinomycin D, whereas cells of *P. parodii* are capable of growth in the presence of this antibiotic. The cells of the latter species, however, are unable to regenerate *Petunia* plants from callus cultures. The only cells capable of growth in the presence of actinomycin D and capable of regeneration of whole plants are the fusion products of the two parental protoplast lines. Two mutant strains of Nicotiana tabacum, which have the characteristics of light sensitivity and chlorophyll deficiency, were used in a complementation selection procedure by Melchers and Labib (1974) and Bottcher, Aviv, & Galun (1989).

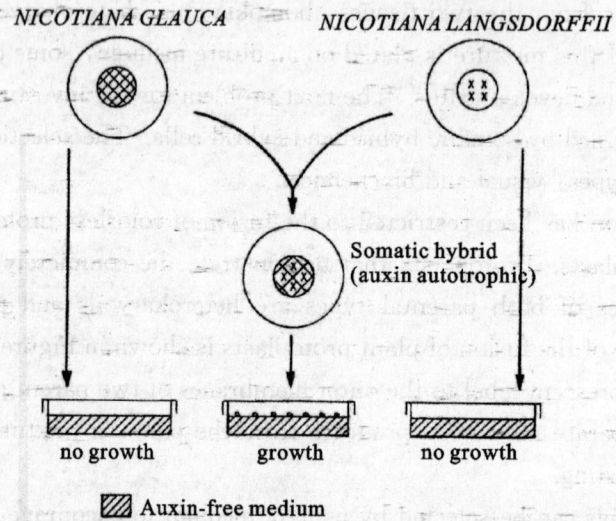

Figure 13.3 Screening method for the detection of somatic hybrids that are auxin autotropic in nutrition. The isolated protoplasts of both Nicotiana glauca and N. langsdorffii are unable to grow in the absence of exogenous auxin. The fusion product of the two parental types is auxin autotropic and grows on an auxin-free medium. (By John H. Dodds)

The fusion experiment should not be attempted until protoplasts have been successfully isolated and cultured. The present experiment involves the induction of fusion of two protoplasts carrying distinct visual markers in the form of pigment color. The fusion products can be identified by microscopic examination of the cells.

PROCEDURE

1. After selection of the appropriate plants, isolate two sets of protoplasts by employing the technique outlined in the previous chapter. This should result in one tube containing green protoplasts of mature leaf material and a second tube of red protoplasts from petal, tuber, or taproot tissue. Thus the markers for fusion are chloroplasts and vacuoles containing anthocyanin（花青素）pigment.

2. The basic principle of fusion is shown in Figure 13.1. Similar numbers of protoplasts A and B are mixed in a centrifuge tube containing PEG (20% w/v) as the fusigenic agent. The tube is centrifuged at 75-100 g for 10 min. This relatively slight pressure forces the protoplasts into close contact and allows fusion to take place.

3. The pellet of fused and unfused protoplasts is carefully resuspended and assayed

for viability and density, and the mixture is plated out as described in the previous Chapter.

RESULTS

After the protoplast mixture has been plated and the agar has solidified, the plate may be viewed microscopically for the identification of the fusion products. Heterokaryons can be identified by the presence of chloroplasts and an anthocyanin-containing vacuole. If any of the fusion products initiates a callus, plant regeneration can be attempted as previously described.

QUESTIONS FOR DISCUSSION

1. What chemical compounds have been employed as fusigenic agents?
2. What are the advantages of protoplast fusion over traditional methods of sexual hybridization?
3. What types of procedures can be used for the selection of hybrid cells?
4. What is a cybrid? How does this phenomenon occur? Does it have any significance in the breeding of plants?
5. List some interesting protoplast fusions that may result in unusual hybrids (e.g., potato and tomato).

知识要点

酶解法制备原生质体时,由于胞间连丝的存在,很易发生原生质体的自发融合。聚乙二醇是最常用的人工诱导原生质体融合的化学诱融剂,高 pH 值、高钙和高温处理以及 $NaNO_3$ 处理等也可以成功诱导原生质体的融合。杂交细胞的筛选可通过肉眼观察和选择性培养基培养来实现。原生质体融合在植物细胞的远缘杂交育种上有重要意义。

CHAPTER 14 CRYOPRESERVATION OF GERMPLASM
(种质资源的低温保存)

Considerable interest has been shown in recent years in the application of tissue culture technology to the storage of plant germplasm. The conventional methods of germplasm preservation are prone to possible catastrophic(灾难性的)losses because of: (1) attack by pests and pathogens, (2) climatic disorders, (3) natural disasters, and (4) political and economic causes. In addition, the seeds of many important crop plants lose their viability in a short time under conventional storage systems.

CRYOPRESERVATION

It will be evident from the previous discussion that an in vitro system with a high multiplication rate, although ideal for purposes of clonal propagation, is entirely unsuitable as a means of germplasm conservation. Such systems require frequent attention and maintenance, and in some cases, carry the risk of genetic instability. This instability is related to growth, particularly disorganized growth such as in callus cultures. Consequently, an ideal system for germplasm storage would be to store material in such a manner as to achieve complete cessation of cell division and growth. This can be accomplished by storing the plant material at the temperature of liquid nitrogen ($-196°C$). Although such techniques have been applied to a range of tissue cultures, the success rates have been variable.

Protocol for Freeze Preservation of Potato Shoot Tips

1. Excise some sterile potato shoot apices. Wrap them carefully in sterile squares of aluminum foil.

2. While wearing protective gloves and a face mask, plunge a foil packet into a Dewar flask of liquid nitrogen ($-196°C$). Permit the packet to remain in the liquid nitrogen for several minutes. The sample can be left for longer periods if a liquid nitrogen storage tank is available.

3. Remove the foil packet and allow it to warm to room temperature.

4. Open the packet and carefully transfer aseptically the frozen and thawed apices to culture tubes containing the potato shoot micropropagation medium (10 tubes).

Results

The time required for excised shoot apices to initiate growth varies greatly, but growth should be easily visible after four to six weeks with potato. The success rate or percentage of survivors will be relatively low, and results will vary from one species to another.

MINIMAL-GROWTH STORAGE

Techniques of germplasm conservation based on the storage of shoot-tip cultures or meristem-derived plantlets under conditions that permit only minimal rates of growth will have widespread application in the near future. Such systems already have important uses in several international germplasm resource centers, mainly because the stored material is readily available for use, it can easily be seen to be alive, and the cultures may be readily replenished when necessary.

There have been several approaches to growth suppression in plant tissue cultures. Three principal methods are used: (1) The physical conditions of culture can be altered (e.g., temperature or the gas composition within the culture vessels); (2) The basal medium can be altered, for example, using sub-(在下)or supra-(在上)optimal concentrations of nutrients. Some factor essential for normal growth may either be omitted or be employed at a reduced level; (3) The medium can be supplemented with growth retardants(延缓剂)(e.g., abscisic acid (脱落酸) or osmoregulatory compounds such as mannitol(甘露醇)and sorbitol(山梨醇).

Protocol for Storage of Shoots in the Presence of Growth Retardant

1. Actively propagating potato shoots are removed from the culture vessel and transferred to sterile Petri dishes for dissection of the individual shoots.

2. Using sterile instruments 10 shoots are transferred to culture tubes containing a fresh micropropagation medium as a control, and 10 shoots are transferred to tubes containing a similar medium supplemented with mannitol 6%.

3. The cultures are incubated in illuminated plant growth chambers (25℃). Growth measurements are made of the cultures at monthly intervals.

Results

The control shoots transferred to the normal micropropagation medium grow rapidly under these culture conditions. The growth retardant has a severe effect on shoot growth. Growth retardant to control cultures can be observed after a six-week period.

QUESTIONS FOR DISCUSSION

1. Why are tissue cultures used to conserve germplasm of plants?
2. What techniques are available for in vitro conservation? Discuss the advantages and disadvantages of them.

CYTOTECHNOLOGY（细胞工程技术）

知识要点

目前，组织培养技术已发展成为一种新的种质保存方法。与传统的种植保存和贮藏保存相比，组织培养技术可不受病虫害和气候变化等自然灾害的影响。低温冷冻保存和生长抑制保存是常见的组织培养技术种质保存方法。

CHAPTER 15 PRODUCTION OF SECONDARY METABOLITES(次生代谢产物的生产)

Aside from the primary metabolic pathways common to all life forms, some reactions lead to the formation of compounds unique to a few species or even to a single cultivar. These reactions are classified under the term "secondary metabolism," and their products are known as "secondary metabolites". These substances include alkaloids(生物碱), antibiotics, volatile oils(精油), resins(树脂), tannins(丹宁酸), cardiac glycosides(强心苷), sterols (甾醇), and saponins(皂角苷). In addition to their economic importance, many secondary metabolites play ecological and physiological roles in higher plants. Investigations in the area of biochemical ecology indicate that some secondary compounds produced by plants are important either to protect these plants against microorganisms and animals, or to enhance the ability of one plant species to compete with other plants in a particular habitat.

Despite advances in the field of organic chemistry, plants are still an important commercial source of chemical and medicinal compounds. The chief industrial applications of secondary metabolites have been as pharmaceuticals (e. g., sterols and alkaloids), and as agents in food flavoring and perfumery. In some cases, these plants have not been subjected to intensive genetic programs for the optimum production of the compound. In addition, there have been technical and economic problems in the cultivation of these plants. Unfortunately, many Third World countries producing medicinal plants are politically unstable, and the supply of crude plant material for processing cannot be guaranteed.

It has been proposed that many of these secondary metabolites produced by intact plants could be synthesized by cell cultures. The basic technology involved in submerged cell cultures on a large scale was described by Nickell (1962). Patents have been obtained for production from cell cultures of such metabolites as allergens(过敏原), diosgenin(薯蓣皂苷配基), L-dopa, ginseng saponin glycosides(人参皂苷), and glycyrrhizin(甘草酸)(Staba, 1977). Tissue cultures have produced compounds previously undescribed, and cultures of higher plant cells may provide an important source of new, economically important compounds.

Although the production of secondary metabolites by cell cultures may be impractical, in some cases the techniques of plant tissue culture can be used to improve the cultivation of these plants. These culture procedures include vegetative propagation, the isolation of virus-free stock, mutation studies with haploid plants, protoplast fusion, and the screening of disease-resistant lines.

There are numerous reasons why progress has been slow in the industrial application of cell cultures for the production of secondary metabolites. The cultures exhibit relatively slow rates of growth, and the biosynthesis of the desired compounds is often at a much lower level than in the intact plant. In order for cell cultures to be used as commercial sources of these compounds, the in vitro production must be comparable to or exceed the amount produced by the intact plant. Several reports have been published indicating yields approaching or exceeding yields from the whole plant. In some cases, the production of secondary metabolites does not show a positive correlation with the maximal growth rate of the culture. This observation may reflect a competition for metabolites utilized in primary metabolism with those pathways leading to the formation of secondary products; for example, competition could exist for amino acids in the formation of proteins, alkaloids, and phenylpropanoids.

The relationship between the degree of tissue organization and the biosynthesis of secondary products is obscure. The spatial orientation of enzymes, compartmentalization of enzymes and substrates, and reservoir sites for product accumulation may be some of the factors involved in the biosynthesis of secondary products by specialized tissues. The metabolic requirements for some of these biosynthetic pathways, however, do not depend on the level of cytodifferentiation.

In a review on secondary product formation in cell cultures, Butcher (1977) has subdivided these compounds into four general groups: (1) Some compounds occur throughout the plant kingdom and are not associated with any level of cytodifferentiation (e.g., phytosterols (植物甾醇) and certain flavonoids (类黄酮)); (2) Some widely distributed compounds are restricted to certain types (e.g., lignin(木质素) and tannins (丹宁酸)); (3) Some compounds are restricted to certain plant families and species, although the biosynthesis is not associated with any form of cytodifferentiation (e.g., specific flavonoids and anthraquinones (蒽醌)); (4) The biosynthesis of some compounds is restricted to highly specialized cells or tissues (e.g., essential oils, resins, and latex (橡胶)). Within this group the level of differentiation is directly related to the biosynthesis of the compound. In the last category, we can assume that progress toward inducing certain levels of cytodifferentiation in cell cultures must be made before success will be achieved in the in vitro biosynthesis of these secondary metabolites.

QUESTIONS FOR DISCUSSION

1. Name some important pharmaceutical chemicals produced by plants.
2. What are the advantages of producing pharmaceutical compounds from cell and tissue cultures?

PART I　CYTOTECHNOLOGY IN PLANTS（植物细胞工程）

知识要点

植物通过次级代谢可产生多种有用的次级代谢产物，如生物碱、精油、树脂、丹宁酸和强心苷等，尤其是甾醇和生物碱类化合物，结构复杂，很难通过化学方法人工合成。通过体外培养可合成和分泌次级代谢产物的植物细胞，既可以提供大量有用的药物，又不必受种植植物时病虫害和自然灾害的限制。

CHAPTER 16　TRANSGENIC PLANTS(转基因植物)

Transgenic plants are plants possessing a single or multiple genes, transferred from a different species. Though DNA from another species can be integrated into a plants' genome via natural processes, the term "transgenic plants" refers to plants created in a laboratory using recombinant DNA technology, frequently called transformation. The aim of creating transgenic plants is to design plants with specific characteristics through artificial insertion of genes from other species. The first modern recombinant crop approved for sale in the U.S. was the *Flavr Savr* tomato in 1994, which had a longer shelf life. The first conventional transgenic cereal created by scientific breeders was actually a hybrid between wheat and rye（黑麦）in 1876. As of 2006 there were around 250 million acres of genetically engineered crops being grown commercially in 22 countries. The U.S. has adopted the technology most widely whereas Europe has almost no genetically engineered crops. The EU had a formal ban on genetically modified crops, until it was overturned in 2006.

Transformation is usually achieved using gold particle bombardment or through the process of horizontal gene transfer using a soil bacterium, *Agrobacterium tumefaciens* （根癌农杆菌）, carrying an engineered plasmid.

DIRECT GENE TRANSFER

Several different methods or strategies for direct gene transfer have been developed over the years. Some of these, particularly particle bombardment (biolistics), have become widely adopted by plant biotechnologists. Direct gene transfer methods, which have found particularly widespread use in the transformation of cereal crops (that initially proved difficult to transform with *Agrobacterium*), have some advantages and disadvantages (Table 16.1) when compared with *Agrobacterium*-mediated transformation. One of the major disadvantages with direct gene transfer methods is that they tend to lead to a higher frequency of transgene rearrangement and a higher transgene copy number. This can lead to high frequencies of gene silencing.

Particle bombardment, Electroporation and Silicon carbide fibres will be considered in some detail below. Other, less-reproducible methods, which will not be considered further, such as laser-mediated uptake of DNA, microinjection, ultrasound and in planta exogenous application, have mainly been used for the analysis of transient gene

Table 16.1 Direct gene transfer methods

Direct gene transfer method	Comments
Particle bombardment	Very successful method. Risk of gene rearrangements and high copy number. Useful for transient expression assays Transgenic plants obtained from a range of cereal crops. Low efficiency. Requires careful optimisation
Electroporation	DNA uptake into protoplasts Used for all major cereal crops. Requires optimisation with a regenerable cell suspension that may not be available
Silicon carbide fibres	Requires regenerable cell suspensions. Transgenic plants obtained from a number of species

expression, although stable transformation has been reported for some of these techniques on rare occasions.

Particle Bombardment

Particle bombardment is the most important and most effective direct gene transfer method in regular use. In this technique, tungsten(钨) or gold particles are coated with the DNA that is to be used to transform the plant tissue. The particles are propelled at high speed into the target plant material, where the DNA is released within the cell and can integrate into the genome. The delivery of DNA using this technology has allowed transient gene expression to be widely studied, but integration of the transgene occurs only infrequently. In order to generate transgenic plants, the plant material, the tissue culture regime and the transformation conditions have to be optimised quite carefully.

Practical bombardment systems can be used in the transformation of both dicotyledonous(双子叶)and monocotyledonous (单子叶)plants. All the major cereals were able to be transformed, and the first commercial genetically modified crops, such as maize containing the Bt-toxin gene, were produced by this method. Developments to this technology are based on the impelling force for the particles, from gunpowder exposure to electrostatic discharge and helium(氦)-driven. Attempts to optimise the system have focused on three aspects of the process: particle type and preparation; particle acceleration; and choice of target material. A balance has to be reached between the number and size of particles fired into the target cells, the damage they do and the amount of DNA they deliver. Too little DNA may lead to low transformation frequencies, but too much DNA may lead to a high copy number and rearrangements of the transgene constructs. Plant tissues are first bombarded and then induced to regenerate. To protect the plant tissue from the damage sustained during the bombardment procedure, treatments to induce limited plasmolysis（质壁分离）or culture on high-osmoticum media have been used.

Electroporation

The electroporation of cells can be used to deliver DNA into plant cells and protoplasts. The vectors used can be simple plasmids; the genes of interest require high-voltage-induced pores in the plasma membrane and integrate into the genome. Electroporation has been successfully used to transform all the major cereals, particularly rice, wheat and maize. Initially, protoplasts were used for transformation, but one of the advantages of the system is that both intact cells and tissues (such as callus cultures, immature embryos and inflorescence material) can be used. This reduces some, but not all, of the tissue culture problems. However, the plant material used for electroporation may require specific treatments, such as pre-and post electroporation incubations in high osmotic buffers. The efficiency of electroporation is also questionable. It is very dependent on the condition of the plant material used and the electroporation and tissue treatment conditions chosen.

Silicon Carbide Fibres（碳化硅纤维）

This is a simple technique for which no specialised equipment is required. Plant material (such as cells in suspension culture, embryos and embryoderived calluses) is introduced into a buffer containing DNA and the silicon carbide fibres, which is then vortexed. The fibres, which are about 0.3-0.6 μm in diameter and 10-100 μm long, penetrate the cell wall and plasma membrane, allowing the DNA to gain access to the inside of the cell. The drawbacks of this technique relate to the availability of suitable plant material and the inherent dangers of the fibres, which require careful handling.

AGROBACTERIUM-MEDIATED GENE TRANSFER

Agrobacterium tumefaciens（根癌农杆菌）is a soil-borne, Gram-negative bacterium. It is the causative agent of 'crown gall'（冠瘿瘤） disease, an economically important disease of many plants, particularly grapes（彩页 Fig. 16.1）. The ability to cause crown galls (tumorous tissue growths) depends on the ability of *Agrobacterium* spp. to transfer bacterial genes into the plant genome. This startling feature is, to date, a unique example of inter-kingdom gene transfer, which has been used by biotechnologist to establish an ideal plant transformation method, called Agrobacterium-mediated gene transfer.

Crown-gall formation depends on the presence of a plasmid in *A. tumefaciens* known as the 'Ti (tumour-inducing) plasmid'. Part of this plasmid (the T-DNA region) is actually transferred from the bacterium and into the plant cell, where it becomes integrated into the genome of the host plant. The T-DNA carries genes that encode proteins involved in both hormone (auxin and cytokinin) biosynthesis and the biosynthesis of novel plant metabolites called 'opines'（冠瘿碱）and 'agropines'（农杆碱）. The production of auxin and cytokinin causes the plant cells to proliferate and so form the gall.

These proliferating cells also produce opines (which are amino acid derivatives) and agropines (sugar derivatives) which are used by A. tumefaciens as its sole carbon and energy source. Different strains of A. tumefaciens contain different Ti plasmids that code for the production of different opines. Opines and agropines are not normally part of plant metabolism and are very stable chemicals, which the Ti plasmid provide a carbon and energy source that only A. tumefaciens can use. The genes in Ti plasmid that are not transferred to the plant encode proteins involved in opine uptake and catabolism. A. tumefaciens has therefore developed the ability to genetically transform plant cells in order to usurp(侵占) the plant's biosynthetic machinery and produce nutrients that only it can utilise.

Structure of the Ti Plasmid

The ability of A. tumefaciens to cause crown-gall disease was found to depend on the presence in the bacteria of a large (-200 kb) plasmid termed the Ti plasmid. Analysis of the nuclear DNA from plant tumours showed that a portion of the Ti plasmid was integrated into the genome of the host plant. This portion was termed 'transfer DNA' (T-DNA) and was found to be responsible for the tumorous phenotype. Analysis showed that one or more copies of the T-DNA could be integrated into the genome, but, in general, the T-DNA insertions in the plant genome were bordered by small (24 bp), nearly perfect, direct repeats, which also border the T-DNA in the Ti plasmid (Fig. 16. 2).

Nopaline strains(胭脂碱)
TG(G/A)CAGGATATAT(-/T)G(T/G)(G/C)G(T/G)GTAAAC

Octopine strains(章鱼碱)
(C/T)GGCAGGATATA(T/A)C(A/C)(A/G)TTGTAA(A/T)T
TAAGTCGCTGTGTATGTTTGTTTG (Enhancer or 'overdrive' sequence)

Figure 16. 2. **The consensus sequence of the border sequences from nopaline-and octopine-strain Ti plasmids.** The nucleotide sequence of the enhancer sequence is also shown.

Ti plasmids from different strains of A. tumefaciens generally have several features in common. They usually contain one (or more) T-DNA regions, a *vir* region, an origin of replication, a region enabling conjugative transfer and some genes for the catabolism of opines.

The T-DNA. The T-DNA region of any Ti plasmid is defined by the presence of the right-and left-border sequences (Fig. 16. 2). These border sequences are 24-bp imperfect repeats. Any DNA between the borders will be transferred into the genome of the host plant. Octopine-strain Ti plasmids contain an 'overdrive' or enhancer sequence associated with the right-border sequence, which is required for optimal T-DNA transfer.

The oncogenes. Two genes *auxA* (or *tmsl* or *iaaM*) and *auxB* (or *tms2* or *iaaH*) encode proteins involved in the production of the auxin indole acetic acid (IAA). *auxA* encodes tryptophan monooxygenase(色氨酸单加氧酶) and *auxB* encodes indole acetamide hydrolase(吲哚乙酰胺水解酶). Another gene (*cyt* (or *tmr* or *ipt*)) encodes an isopentenyl transferase(异戊烯转移酶) that catalyses the most important step in cytokinin production. These genes are the prime determinants of tumour phenotype and are therefore often referred to as 'oncogenes'.

The vir region. The genes responsible for the transfer of the T-DNA region into the host plant are also situated on the Ti plasmid, in an —40 kb region outside the T-DNA known as the *vir* (virulence) region. There are at least nine *vir*-gene operons.

Other genes present on the T-DNA. Genes for the production of opines (either octopine or nopaline) are present on the T-DNA. The *tml* gene, which is involved in determining tumour size in some species, is also found in the T-DNA (彩页 Fig. 16.3).

The Process of T-DNA Transfer and Integration

T-DNA transfer and integration into the plant genome can be divided into the following steps.

(1) Signal recognition and attachment of the Agrobacterium to the host cells (Fig. 16.3①). The *Agrobacterium* perceives signals, such as phenolics and sugars, which are released from wounded plant cells. Normally these substances are probably part of the plant's defense mechanism, being involved in phytoallexin(植物抗毒素) and lignin(木质素)synthesis. The substances released from wounded cells effectively signal the presence of plant cells that are competent for transformation. Attachment of *Agrobacterium* to plant cells is a two-step process, involving an initial attachment via a polysaccharide. Subsequently, a mesh of cellulose fibres is produced by the bacterium. Several chromosomal virulence genes (*chv* genes) are involved in the attachment of the bacterial cells to the plant cells.

(2) Sensing of specific signals by the Agrobacterium VirA/VirG two-component-signal-transduction system and activation of the *vir* gene region (Fig. 16.3②,③). VirA (a membrane-linked sensor kinase) senses phenolics and autophosphorylates(自磷酸化), subsequently phosphorylating and thereby activating VirG. VirG then induces expression of all the *vir* genes (including *virA* and *virG*). Many sugars, but in particular glucose, galactose(半乳糖) and xylose(木糖), enhance *vir* gene induction. This enhancement requires another chromosomal *vir* gene termed '*chvE*' which encodes a glucose/galactose transporter that interacts with VirA.

(3) T-strand production (Fig. 16.3④). The left and right borders are recognised by a VirDl/VirD2 complex and VirD2 produces single-stranded nicks in the DNA. After nicking, VirD2 becomes covalently attached to the 5' end of the displaced single-stranded T-DNA strand. Repair synthesis replaces the displaced strand.

(4) Transfer of T-DNA out of the bacterial cell (Fig. 16.3⑤). The T-DNA/VirD2

complex is exported from the bacterial cell by a 'T-pilus' (effectively a membrane-channel secretory system) composed of proteins encoded by the *virB* operon and VirD4.

(5) Transfer of the T-DNA and Vir proteins into the plant cell and nuclear localization (Fig. 16.3⑥~⑩). The T DNA/VirD2 complex and other Vir proteins cross the plant plasma membrane, possibly through channels formed from VirE2. Once inside the plant cytoplasm the T-DNA strand becomes covered with VirE2 proteins, which have been postulated to protect the T-DNA from nucleases, facilitate nuclear localisation and confer the correct conformation to the T-DNA/VirD2 complex for passage through the nuclear-pore complex (NPC). VirD2 contains a nuclear localisation signal (NLS) that facilitates its interaction with a plant protein.

VirE2 possess two NLSs, but its nuclear localisation is mediated by another plant protein, termed 'VIP1', which functions to facilitate VirE2 NLS recognition by importins. This contribution of VirE2 to the nuclear localisation of T-DNA complexes may be particularly important for large T-DNAs.

The T-DNA and associated proteins pass through the nuclear pore, with the bound VirE2 proteins also giving the correct conformation to the T-DNA strand. VIP1 and another VirE2 interacting protein, VIP2, are then thought to direct the T-DNA strand to chromatin and possibly promote integration. The T-DNA strand is integrated into the host plant genome by a process referred to as 'illegitimate recombination（非法重组）'. This process, unlike homologous recombination, does not depend on extensive regions of sequence similarity.

The Ri (root inducing) Plasmids

A. rhizogenes(发根农杆菌), another species of the genus *Agrobacterium*, is also capable of transferring genes to plants, and has been developed into a plant transformation system that is used in some specialised circumstances.

Although *Agrobacterium rhizogenes* also infects plants, it differs from *A. tumefaciens* in that the resulting pathology is not crown galls but a phenomenon known as 'hairy roots'. At the site of infection there is a proliferation of roots. Plasmids in *A. rhizogenes* (Ri plasmids) strains have been characterised, and it has been shown that there are of a number of different types, which can be classified based upon opine usage.

Hairy roots are important in some areas of plant biotechnology as they can be cultured *in vitro*. For many years they have been used as a source of secondary metabolites, but more recently they have been used as a system for the production of pharmaceutical proteins. *A. rhizogenes* transformation was, at one stage, considered an alternative strategy to *A. tumefaciens* for gene transfer as it led to the production of defined tissues (hairy roots) that could be regenerated into whole plants. This strategy seems to have been discarded, however, as more efficient *A. tumefaciens* systems have been developed.

In summary, a variety of techniques for plant transformation are available to the plant biotechnologist. These techniques can be split into two groups: *Agrobacterium*-mediated transformation; and direct gene transfer methods, of which the biolistics approach is probably the most widely used. These two groups of techniques are fundamentally different in mechanism, and are, in general, applied to different crops. *Agrobacterium*-mediated transformation is most widely used with dicotyledonous crops, which reflects the natural host range of members of the genus *Agrobacterium*. Direct gene transfer methods are most commonly used to transform monocotyledonous crops, such as cereals. In part, this reflects the initial difficulties with using *Agrobacterium* to transform monocotyledonous plants. Direct gene transfer methods and *Agrobacterium*-mediated methods have their own advantages and disadvantages. However, all plant transformation methods can suffer from a problem known as 'gene silencing', where transgene (and homologous endogenous genes) expression is actually repressed. Gene silencing is impossible to predict precisely, although some precautions in vector design and transformation protocol can be taken to reduce its frequency. Gene silencing can prove a major hurdle to the commercialisation of plant transformation products.

Despite any problems, improvements in plant transformation technologies, especially when coupled to an efficient plant regeneration protocol, have seen the list of crop species that can be routinely transformed grow. Crops that were once considered impossible to transform (cereals may well fall into this category) are now routinely transformed in many laboratories around the world.

VECTORS FOR AGROBACTERIUM-MEDIATED TRANSFORMATION

Development of efficient transformation vectors was very important in the Agrobacterium-mediated plant transformation. Now useful information about T-DNA can be considered in the design of plant expression vector: the only features of the T-DNA necessary for integration into the host plant genome were the short border sequences; the removal of the oncogene sequences enabled plants to be regenerated from transformed plant tissue by manipulating the plant hormone composition of the medium; and the *vir* genes function in *trans*.

A successful expression vector plasmid for plant transformation should have the following basic features: (a) To be replicated not only in *E. coli* (so that routine manipulations can be carried out) but also in *Agrobacterium*.; (b) Additional selectable markers need to be included so that the successfully transformed plants can be identified; (c) Border sequences need to be incorporated into the design of plasmid vectors for *Agrobacterium-mediated* transformation to ensure integration of the genes of interest into the host plant genome; (d) The genes (particularly if they are from prokaryotes or non-plant eukaryotes) that are to be integrated into the genome of the host plant may need to be made 'plant-like'. This includes the use of appropriate promoters and terminators to ensure that expression of the genes occurs. These features are often incorporated into the basic vector.

SELECTABLE MARKERS

Plant transformation is, in many cases, a very-low frequency event. It is therefore vital that some means for selecting the transformed plant tissue is provided by the plant transformation vector. In most cases this selection is based on the inclusion into the culture medium of a substance that is toxic to plants. The selectable marker on the vector confers resistance to the toxic substance when expressed in transformed plant tissue.

Antibiotic resistance genes for E. *coli* can be used as selectable markers in plants. Although plants are eukaryotic, antibiotics efficiently inhibit protein synthesis in the organelles, particularly the chloroplasts (叶绿体). Perhaps the most widely used selectable marker gene is the *nptll* gene that confers resistance to the antibiotic kanamycin. Other antibiotic resistance genes have also been used with some considerable success in plant transformation vectors. In part, this use of alternative selectable marker genes was driven by the observations that some plant species exhibited a very high degree of natural resistance to kanamycin (such as cereals), and that some species (some soft fruits for example) were too sensitive to kanamycin for it to be used successfully. This made the selection of transformed tissue very difficult, leading to a high number of false-positives or the inability to recover transformed plants. The development of alternative selectable marker genes also allows for the re-transformation of plant tissue that already expresses one or more different selectable markers. Thus genes conferring resistance to antibiotics such as bleomycin (博来霉素), spectinomycin (奇霉素) and hygromycin (潮霉素) are used quite widely. These selective agents can be used at lower concentrations than kanamycin, and therefore usually result in a cleaner selection of transformed tissue. Other resistance genes can also be used as selectable markers in plants. Amongst those widely used are genes that confer resistance to herbicides such as chlorsulphuron （氯磺隆）and bialaphos（双丙磷）.

Public concern has, in recent years, questioned the general desirability of growing transformed crops, and, more specifically, the merits of using antibiotic or herbicide resistance genes as selectable markers during plant transformation. There is obvious concern about the creation of so-called 'super weeds' and the transfer of antibiotic resistance genes. Although in the long term it is undoubtedly better to remove these selectable marker genes from the transformed plants once their job has been done, other, more acceptable, selectable marker genes are also being introduced. Some of the most interesting are based on the principal of facilitating alternative carbon-source utilisation. Thus genes from bacteria that allow the use of mannose or xylose as carbon sources have been successfully used as selectable markers in plant transformation. These genes have also generally proved to be superior to standard antibiotic selection genes in some transformation protocols.

Reporter genes are widely used in plant transformation vectors, both as a means of assessing gene expression by promoter analysis and as easily scored indicators of transformation (indeed, they have been used in place of selectable marker genes in some cases). Ideally, reporter genes should be easy to assay, preferably with a nondestructive assay system, and there should be little or no endogenous activity in the plant to be transformed. At present, only a small number of reporter genes are in widespread use in plant transformation vectors, these being β-glucuronidase (β-葡萄糖醛酸酶, *uidA or gus*), green fluorescent protein (绿色荧光蛋白, *gfp*), luciferase genes (荧光素酶基因, *lux and luc*) and, to a lesser degree (although it is widely used in animal systems), the chloramphenicol acetyltransferase gene (氯霉素乙酰转移酶, *cat*). Each of these reporter genes will be looked at in more detail.

β-Glucuronidase (GUS)

This is perhaps the most widely used reporter gene in plant transformation vectors. Its widespread acceptance is due to its many advantages over the previously existing marker genes. β-Glucuronidase can be assayed extremely sensitively using quick, easy and non-radioactive methods. It can be used to obtain both quantitative (i.e. the level of gene expression) and qualitative (i.e. localisation of gene expression) data. There is also little or no endogenous activity in most plant tissues (with the possible exception of reproductive tissues).

Quantitative data is obtained by assays utilising fluorogenic substrates such as 4-MUG (4-methylumbellifterryl-β-D-glucuronide, 4-甲基伞形基-β-D-吡喃半乳糖苷) which is hydrolysed to 4-MU (methylumbelliferone, 4-甲基伞形酮). Standard enzyme assay protocols can be used and results compared with a standard curve of 4-MU fluorescence.

Qualitative data can be obtained from histochemical assays that allow tissue-and cell-specific localisation of β-glucuronidase activity. These histochemical assays are conducted in situ with the chromogenic substrate X-gluc (5-bromo-4-chloro-3-indolyl β-D-glucuronide, 5-溴-4氯-3-吲哚葡萄糖苷). GUS activity results in the deposition of an insoluble blue precipitate, effectively identifying the precise location of the expression.

Green Fluorescent Protein (GFP)

Nowadays, *gfp* is rapidly becoming a very widely used reporter gene. GFP has the advantages that it is even easier to assay than GUS and the assay is non-destructive. This means that GFP can be used in situations where GUS cannot, for example in screening primary transformants, in time-course experiments, or analysing segregation in small seedlings. The gene was isolated from the jellyfish *Aequorea victoria*, which are brightly luminescent organisms. In order to work efficiently in plants it was found that the gfp gene had to be significantly modified in order to: (a) remove a cryptic intron (i. e. gfp mRNA is efficiently mis-spliced in some plants, resulting in the removal of 84 nucleotides); (b) make the codon usage more 'plant-like'; and (c) prevent accumulation in the nucleoplasm.

Luciferases

The firefly luciferase gene (luc) encodes an enzyme that catalyses the oxidation of D-luciferin (荧光素) in an ATP-dependent fashion. This oxidation results in the emission of light. Highly sensitive assays (based on photomultipliers (光电倍增管), luminometers (发光计) or film exposure) have been developed to detect the extremely rapid emission of light.

Chloramphenicol acteyltransferase

Although widely used as a reporter gene in mammalian cells the availability of the GUS and GFP reporter systems has generally limited the use of the CAT system in plants, although it was the first bacterial gene to be expressed in plants. Some plant transformation vectors do, however, carry this reporter gene as it can be assayed very sensitively, but it requires a radioactive assay procedure.

INTEGRATION AND EXPRESSION OF TRANSGENE

Multiple copies of the transgene can be incorporated into the target plant genome. Often high levels of transgene expression are associated with multiple copies of the transgene, but this is not always the case. Multiple T-DNA insertions into the genome may lead to erratic transgene expression.

Transgene integration into the plant genome is basically a random event, and it has been demonstrated that the position of the transgene in the genome can have a marked effect on transgene expression levels. Thus, independent transformants with single copies of the transgene can exhibit large differences in the level of transgene expression. This positional effect on transgene expression remains a problem in plant transformation, despite efforts to find ways of targeting gene integration. Unfortunately, higher plants do not possess an endogenous system enabling homologous recombination (genetic recombination involving the exchange of homologous loci) at a high efficiency that allows transgenes to be targeted to a particular region of the genome.

One approach taken to assuage (减轻) position effects is the inclusion of matrix-attachment regions (MARs). These AT-rich sequences are thought to be involved in maintaining the chromatin in an 'open' structure allowing for gene expression. The inclusion of MARs flanking transgenes has been shown to provide position-independent expression in animal systems, and some, albeit limited, success has been achieved with their use in plants. MARs may also help to stabilise transgene expression.

It is well documented that heterologous genes, particularly those from nonplant species, tend to express poorly in plants even when driven by a strong promoter. This can be due to a variety of factors, many of which are associated with the 'structure' of the transgene. Genes from different organisms tend to have different G+C contents, with bacterial genes having a particularly low G+C content. It has been found that a high A+T content in transgenes interferes with mRNA processing in plants, leading to little or no expression of the transgene. Differences in A+T content between the transgene and the isochore (long stretches of DNA with a homogeneous base composition) into which it is integrated may also contribute to the transgene being recognised as 'foreign'. A high A+T content also often results in the presence of sequences (AUUUA) that can destabilise mRNA and may also form potential plant polyadenylation signals. So-called 'cryptic' introns may also be found. These can be efficiently mis-spliced from transgenes (i.e. part of the coding sequence is recognised by the plant as an intron and spliced out) by the plants mRNA processing machinery, effectively resulting in a deletion mutation of the transgene.

Even if the transgene is efficiently transcribed it may not be translated efficiently, probably due to the architecture of the translation initiation region and the presence of codons that are used infrequently by plants. Consensus sequences flanking the ATG initiation codon are known to differ between plants and other species. The inclusion of plant-specific sequences upstream of the translation initiation codon is known to improve translatability. It has also been recently demonstrated that modification (by the insertion, after the translation initiation codon, of codons that are found in a number of highly expressed plant genes) of the region downstream of the translation initiation codon may also improve transgene translatability.

SAFETY OF TRANSGENIC PLANTS

Ecological Risks

The potential impact on nearby ecosystems is one of the greatest concerns associated with transgenic plants. Transgenes have the potential for significant ecological impact if the plants can increase in frequency and persist in natural populations. These concerns are similar to those surrounding conventionally bred plant breeds. Several risk factors should be considered: Is the transgenic plant capable of growing outside a cultivated area? Can the transgenic plant pass its genes to a local wild species, and are the offspring also fertile? Does the introduction of the transgene confer a selective advantage to the plant or to hybrids in the wild?

Many domesticated plants can mate and hybridize with wild relatives when they are grown in proximity, and whatever genes the cultivated plant had can then be passed to the hybrid. This applies equally to transgenic plants and conventionally bred plants, as in either case there are advantageous genes that may have negative consequences to an ecosystem upon release. This is normally not a significant concern, despite fears over 'mutant superweeds' overgrowing local wildlife. Although hybrid plants are far from uncommon, in most cases these hybrids are not fertile due to polyploidy, and will not multiply or persist long after the original domestic plant is removed from the environment. However, this does not negate the possibility of a negative impact.

In some cases, the pollen from a domestic plant may travel many miles on the wind before fertilising another plant. This can make it difficult to assess the potential harm of crossbreeding; many of the relevant hybrids are far away from the test site. Among the solutions under study for this concern are systems designed to prevent transfer of transgenes and the genetic transformation of the chloroplast, so that only the seed of the transgenic plant would bear the transgene.

There are at least three possible avenues of hybridization leading to escape of a transgene including hybridization with non-transgenic crop plants and wild plants of the same species, or closely related species, usually of the same genus. However, there are a number of factors which must be present for hybrids to be created. The transgenic plants must be close enough to the wild species for the pollen to reach the wild plants. The wild and transgenic plants must flower at the same time. The wild and transgenic plants must be genetically compatible. In order to persist, these hybrid offspring must carry the transgene and be viable, and fertile.

Agricultural Impact

Outcrossing of transgenic plants not only poses potential environmental risks, it may also trouble farmers and food producers. Many countries have different legislations for transgenic and conventional plants as well as the derived food and feed, and consumers demand the freedom of choice to buy genetical modification-derived or conventional products. The introduction of transgenic plants into agriculture has been vigorously opposed by some. There are a number of issues that worry the opponents. One of them is the potential risk of transgenes in commercial crops endangering native or nontarget species. For examples, A gene for herbicide resistance in transgenic crops escaping into a weed species could make control of the weed far more difficult. The gene for Bt toxin expressed in pollen might endanger pollinators like honeybees. Another worry is the inadvertent mixing of transgenic crops with nontransgenic food crops. Although this has occurred periodically, there is absolutely no evidence of a threat to human health. Despite the controversies, farmers are embracing transgenic crops. Worldwide, more than 100 million hectares (247 million acres) were planted to transgenic crops in 2006.

ACHIEVEMENTS IN TRANSGENIC PLANTS

Improved Nutritional Quality

Milled rice is the staple food for a large fraction of the world's human population. Milling rice removes the husk (壳) and any beta-carotene it contained. Beta-carotene is a precursor to vitamin A, so it is not surprising that vitamin A deficiency is widespread, especially in the countries of Southeast Asia. The synthesis of beta-carotene requires a number of enzyme-catalyzed steps. In January 2000, a group of European researchers reported that they had succeeded in incorporating three transgenes into rice that enabled the plants to manufacture beta-carotene in their endosperm (胚乳).

Insect Resistance

Bacillus thuringiensis (苏云金芽孢杆菌, Bt) is a bacterium that is pathogenic for a number of insect pests. Its lethal effect is mediated by a protein toxin it produces. Through recombinant DNA methods, the toxin gene can be introduced directly into the genome of the plant where it is expressed and provides protection against insect pests of the plant.

Disease Resistance

Genes that provide resistance against plant viruses have been successfully introduced into such crop plants as tobacco, tomatoes (Fig. 16.4), and potatoes.

Figure 16.4 Tomato plants infected with tobacco mosaic virus. The plants in the back row carry an introduced gene conferring resistance to the virus. The resistant plants produced three times as much fruit as the sensitive plants (front row). (By Monsanto Company)

Herbicide Resistance

Questions have been raised about the safety of some of the broad-leaved weed killers like 2,4-D, both to humans and to the environment. . Genes for resistance to some of the newer herbicides have been introduced into some crop plants and enable them to thrive even when exposed to the weed killer (彩页 Fig. 16.5).

Salt Tolerance

A large fraction of the world's irrigated crop land is so laden with salt that it cannot be used to grow most important crops. However, researchers at the University of California Davis campus have created transgenic tomatoes that grew well in saline soils. The transgene was a highly-expressed sodium/proton antiport pump（逆向运输泵）that sequestered excess sodium in the vacuole of leaf cells. There was no sodium buildup in the fruit.

"Terminator" Genes

This term is used for transgenes introduced into crop plants to make them produce sterile seeds (and thus force the farmer to buy fresh seeds for the following season rather than saving seeds from the current crop). The process involves introducing three transgenes into the plant: (1) A gene encoding a toxin which is lethal to developing seeds but not to mature seeds or the plant. This gene is normally inactive because of a stretch of DNA inserted between it and its promoter. (2) A gene encoding a recombinase-an enzyme that can remove the spacer in the toxin gene thus allowing to be expressed. (3) A repressor gene whose protein product binds to the promoter of the recombinase thus keeping it inactive.

Transgenes Encoding Antisense RNA

Messenger RNA (mRNA) is single-stranded. Its sequence of nucleotides is called "sense" because it results in a gene product (protein). The antisense strand is complement to mRNA. When mRNA forms a duplex with a complementary antisense RNA sequence, translation is blocked. This may occur because the ribosome cannot gain access to the nucleotides in the mRNA, or duplex RNA is quickly degraded by ribonucleases in the cell.

The Flavr Savr tomato is such an example. Most tomatoes that have to be shipped to market are harvested before they are ripe. Otherwise, ethylene synthesized by the tomato causes them to ripen and spoil before they reach the customer. Transgenic tomatoes (Flavr Savr tomato) have been constructed that carry in their genome an artificial gene (DNA) that is transcribed into an antisense RNA complementary to the mRNA for an enzyme involved in ethylene production. These tomatoes make only 10% of the normal amount of the enzyme. The goal of this work was to provide supermarket tomatoes with something closer to the appearance and taste of tomatoes harvested when ripe. However, these tomatoes often became damaged during shipment and handling and have been taken off the market.

Biopharmaceuticals

The genes for proteins to be used in human (and animal) medicine can be inserted into plants and expressed by them. Some of the proteins that are being produced by transgenic crop plants: human growth hormone, humanized antibodies against HIV and respiratory syncytial virus (RSV, 呼吸道合胞病毒), and lysozyme and trypsin, et al.

Bioremediation (生物修复) of contaminated soils

Mercury (水银), selenium (硒) and organic pollutants such as polychlorinated biphenyls (PCBs, 多氯联苯) have been removed from soils by transgenic plants containing genes for bacterial enzymes.

QUESTIONS FOR DISCUSSION

1. What is transgenic plant?
2. What is the principle for Agrobacterium-mediated gene transfer? Try to understand the gene structure and role of Ti plasmid in Agrobacterium-mediated gene transfer.
3. List some direct gene transfer methods for plants and understand their mechanism of gene transformation.
4. How to construct a plant transformation vectors? What kinds of selectable markers have been used in plant transformation vectors?
5. List some achievements in transgenic plants.
6. Talk about the safety of transgenic plants.

知识要点

基因组中稳定地整合有外源基因的植物称为转基因植物。外源基因的导入方法可分为两大类：直接基因转移技术和农杆菌介导的基因转移技术。基因枪法，又称散弹攻击法是最常用的直接基因转移技术，其主要原理是通过高速的金粉或钨粉携带外源 DNA 进入植物细胞中，优点是可以转化所有植物细胞。农杆菌是一种土壤细菌，所含有的 Ti 质粒或 Ri 质粒中有一段 T-DNA 序列，在农杆菌侵染植物细胞的过程中，可整合到植物细胞基因组中。因此，将外源基因插入这些质粒的 T-DNA 区，则可以通过农杆菌感染植物细胞而实现外源基因的转移。缺点是农杆菌不能感染大多数单子叶植物。为提高转基因植物的筛选效率，植物转化所用的载体质粒一般都携带有抗性基因或报告基因。基因的整合常常是多拷贝的，整合的基因不一定都能高效表达。不能将转基因植物释放到环境中，否则可能会破坏自然生态系统和危害农作物的生产。目前，人们已生产出很多种转基因植物，部分抗虫和抗除草剂农作物已被广泛种植。

PART II
CYTOTECHOLOGY IN ANIMALS
（动物细胞工程）

CHAPTER 17　INTRODUCTION TO ANIMAL CELL CULTURE(动物细胞培养简介)

BACKGROUND

Tissue culture was first devised at the beginning of this century [Harrison, 1907; Carrel, 1912] as a method for studying the behavior of animal cells free of systemic variations that might arise in the animal both during normal homeostasis(稳态)and under the stress of an experiment. As the name implies, the technique was elaborated first with undisaggregated(未解离)fragments of tissue, and growth was restricted to the migration of cells from the tissue fragment, with occasional mitoses in the outgrowth(生长晕). Properly, the term tissue culture(组织培养)refers to the maintenance of fragments of tissue *in vitro*, but now commonly applied as a generic(通用) term denoting tissue explant culture, organ culture, and dispersed-cell culture. The term organ culture(器官培养)will always imply a three-dimensional culture of undisaggregated tissue retaining some or all of the histological features of the tissue *in vivo*. Cell culture(细胞培养)refers to a culture derived from dispersed cells taken from original tissue, from a primary culture, or from a cell line or cell strain by enzymatic, mechanical, or chemical disaggregation.

The term histotypic culture(组织型培养)implies that cells have been reaggregated to recreate a three-dimensional tissue-like structure. Organotypic(器官型培养)implies the same procedures but recombining cells of different lineages, e. g. , epidermal keratinocytes(角质细胞)in combined culture with dermal fibroblasts.

Harrison (1907) chose the frog as his source of tissue, presumably because it was a cold-blooded animal, and consequently, incubation was not required. Furthermore, since tissue regeneration is more common in lower vertebrates, he perhaps felt that growth was more likely to occur than with mammalian tissue. The stimulus from medical science carried future interest into warm-blooded animals, in which both normal development and pathological development are closer to that found in humans. The accessibility of different tissues, many of which grew well in culture, made the embryonated (受孕)hen's egg a favorite choice; But the development of experimental animal husbandry, particularly with genetically pure strains of rodents, brought mammals to the forefront as the favorite material. While chick embryo tissue could provide a diversity of cell types in primary culture, rodent tissue had the advantage of producing continuous cell lines and a considerable repertoire of transplantable tumors. The development of transgenic mouse technology, together with the well-established genetic background of the mouse, has added further impetus(刺激)to the selection of this animal as a favorite species.

For many years, the lower vertebrates and the invertebrates were largely ignored, although unique aspects of their development (tissue regeneration in amphibian (两栖类), metamorphosis (变态) in insects) make them attractive systems for the study of the molecular basis of development. More recently, the needs of agriculture and pest control have encouraged toxicity and virological studies in insects, and developments in gene technology have suggested that insect cell lines with baculovirus(杆状病毒)and other vectors may be useful producer cell lines because of the possibility of inserting larger genomic sequences in the viral DNA and a reduced risk of propagating human pathogenic viruses. Furthermore, the economic importance of fish farming and the role of oceanic pollution have stimulated more studies of normal development and pathogenesis in fish. Procedures for handling nonmammalian cells have tended to follow those developed for mammalian cell culture, although a limited number of specialized media are now commercially available for fish and insect cells.

The development of cell culture owed much to the needs of two major branches of medical research: the production of antiviral vaccines and the understanding of neoplasia (肿瘤). The standardization of conditions and cell lines for the production and assay of viruses undoubtedly provided much impetus to the development of modern tissue culture technology, particularly the production of large numbers of cells suitable for biochemical analysis. This and other technical improvements, made possible by the commercial supply of reliable media and sera and by the greater control of contamination with antibiotics and clean-air equipments, have made tissue culture accessible to a wide range of interests.

In addition to cancer research and virology, other areas of research have come to depend heavily on tissue culture techniques. The introduction of cell fusion techniques and genetic manipulation established somatic cell genetics as a major component in the genetic analysis of higher animals, including humans. A wide range of techniques for genetic recombination now includes DNA transfer, monochromsomal transfer, and nuclear transfer, which have been added to somatic hybridization as tools for genetic analysis and gene manipulation. DNA transfer itself has spawned many techniques for the transfer of DNA into cultured cells, including calcium phosphate coprecipitation(磷酸钙共沉淀), lipofection(脂质体转染), electroporation(电穿孔), and retroviral infection(反转录病毒转染).

Tissue culture has contributed greatly, via the monoclonal antibody technique, to the study of immunology. Cell products such as human growth hormone, insulin, and interferon(干扰素) are now produced routinely by transfected prokaryotic and eukaryotic cells, although the absence of posttranscriptional modifications, such as glycosylation (糖基化), in bacteria suggests that mammalian cells may provide more suitable vehicles, particularly in light of developments in immortalization technology.

Other areas of major interest include the study of cell interactions and intracellular control mechanisms in cell differentiation and development and attempts to analyze nervous function. Tissue culture technology has also been adopted into many routine applications in medicine and industry. Chromosomal analysis of cells derived from the womb (子宫) by amniocentesis(羊膜穿刺术) can reveal genetic disorders in the unborn child, the quality of drinking water can be determined, and the toxic effects of pharmaceutical compounds and potential environmental pollutants can be measured in colony-forming and other *in vitro* assays.

Further developments in the application of tissue culture to medical problems have followed from the demonstration that cultures of epidermal cells form functionally differentiated sheets and endothelial cells may form capillaries, offering possibilities in homografting(自体移植) and reconstructive surgery using an individual's own cells, particularly for severe burns.

With the ability to transfect normal genes into genetically deficient cells, it has become possible to graft such "corrected" cells back into the patient. Transfected cultures of rat bronchial epithelium carrying the *β-gal* reporter gene have been shown to become incorporated into the rat's bronchial lining(支气管内壁) when they are introduced as an aerosol(气雾剂) into the respiratory tract.

The prospects for implanting normal cells from adult or fetal tissue-matched donors or implanting genetically reconstituted cells from the same patient are now very real. The technical barriers are steadily being overcome, bringing the ethical questions to the fore(前面). The technical feasibility of implanting normal fetal neurons into patients with Parkinson's disease has been demonstrated; society must now decide to what extent fetal material may be used for this purpose. Where a patient's own cells can be grown and subjected to genetic reconstitution by transfection of the normal gene-e. g. , transfecting the normal insulin gene into β-islet cells cultured from diabetics, or even transfecting other cell types, such as skeletal muscle progenitors-it would allow the cells to be incorporated into a low-turnover(低通量) compartment and, potentially, give a long-lasting physiological benefit.

ADVANTAGES OF TISSUE CULTURE

Control of the Environment

The two major advantages of tissue culture are control of the physiochemical environment(理化环境)(pH, temperature, osmotic pressure, and O_2 and CO_2 tension), which may be controlled very precisely, and the physiological conditions(生理条件), which may be kept relatively constant, but cannot always be defined.

Homogeneity(同质) of Sample

Tissue samples are invariably heterogeneous(异质). Replicates-even from one tissue-vary in their constituent cell types. After one or two passages, cultured cell lines assume a homogeneous (or at least uniform) constitution, as the cells are randomly mixed at each transfer and the selective pressure of the culture conditions tends to produce a homogeneous culture of the most vigorous cell type. Since experimental replicates are virtually identical, the need for statistical analysis of variance is reduced.

Economy, Scale, and Mechanization(机械化)

Cultures may be exposed directly to a reagent at a lower, and defined, concentration and with direct access to the cell. Consequently, less reagent is required than for injection *in vivo*, where 90% is lost by excretion and distribution to tissues other than those under study. Screening tests with many variables and replicates are cheaper, and the legal, moral, and ethical questions of animal experimentation are avoided.

In Vivo Modeling

Perfusion techniques allow the delivery of specific experimental compounds to be regulated in concentration, duration of exposure, and metabolic state. The development of histotypic and organotypic models also increases the accuracy of *in vivo* modeling.

LIMITATIONS OF TISSUE CULTURE

Expertise

Culture techniques must be carried out under strict aseptic conditions, because animal cells grow much less rapidly than many of the common contaminants, such as bacteria, molds(霉菌), and yeasts. Thus a level of skill and understanding on the part of the operator are needed in order to appreciate the requirements of the system and to diagnose problems as they arise. Also, care must be taken to avoid the recurrent problem of cross contamination and to authenticate(鉴别) stocks.

Quantity

A major limitation of cell culture is the expenditure(费用) of effort and materials that goes into the production of relatively little tissue. A realistic maximum per batch for most small laboratories might be 1-10 g of cells. With a little more effort and the facilities of a larger laboratory, 10-100 g is possible.

Dedifferentiation and Selection

When the first major advances in cell line propagation were achieved in the 1950s, many workers observed the loss of the phenotypic characteristics typical of the tissue from which the cells had been isolated. This effect was blamed on dedifferentiation, a process assumed to be the reversal of differentiation, but was later shown to be largely due to the overgrowth of undifferentiated cells of the same or a different lineage.

Origin of Cells

If differentiated properties are lost, for whatever reason, it is difficult to relate the cultured cells to functional cells in the tissue from which they were derived. Stable markers are required for characterization of the cells; in addition, the culture may need to be modified so that these markers are expressed.

Instability

Instability is a major problem with many continuous cell lines, resulting from their unstable aneuploid（非整倍体）chromosomal constitution. Even with short-term cultures of untransformed cells, heterogeneity in growth rate and the capacity to differentiate within the population can produce variability from one passage to the next.

TYPES OF TISSUE CULTURE

There are three main methods of initiating a culture (Fig. 17.1): (a) Organ culture implies that the architecture characteristic of the tissue *in vivo is* retained, at least in part, in the culture. Toward this end, the tissue is cultured at the liquid-gas interface (on a raft, grid, or gel), which favors the retention of a spherical or three-dimensional shape. (b) In primary explant culture, a fragment of tissue is placed at a glass (or plastic)-liquid interface, where, following attachment, migration is promoted in the plane of the solid substrate. (c) Cell culture implies that the tissue, or outgrowth from the primary explant, is dispersed (mechanically or enzymatically) into a cell suspension, which may then be cultured as an adherent monolayer on a solid substrate or as a suspension in the culture medium.

The formation of a cell line from a primary culture implies (a) an increase in the total number of cells over several generations and (b) the ultimate predominance of cells or cell lineages with the capacity of high growth, resulting in (c) a degree of uniformity in the cell population. The line may be characterized, and the characteristics will apply for most of its finite life span. The derivation of *continuous* or *established* cell lines usually implies a phenotypic change, or transformation.

When cells are selected from a culture, by cloning or by some other method, the subline is known as a cell strain（细胞株）. A detailed characterization is then implied. Cell lines or cell strains may be propagated as an adherent monolayer or in suspension. Monolayer（细胞单层）culture signifies that, given the opportunity, the cells will attach to the substrate and that normally the cells will be propagated in this mode. Anchorage dependence(贴壁依赖性)means that attachment to (and usually, some degree of spreading onto) the substrate is a prerequisite for cell proliferation. Monolayer culture is the mode of culture common to most normal cells, with the exception of hematopoietic cells. Suspension cultures(悬浮培养) are derived from cells that can survive and proliferate without attachment (anchorage-independent); this ability is restricted to hematopoietic cells, transformed cell lines, and cells from malignant tumors. It can be shown, however, that a small proportion of cells that are capable of proliferation in suspension exists in many normal tissues.

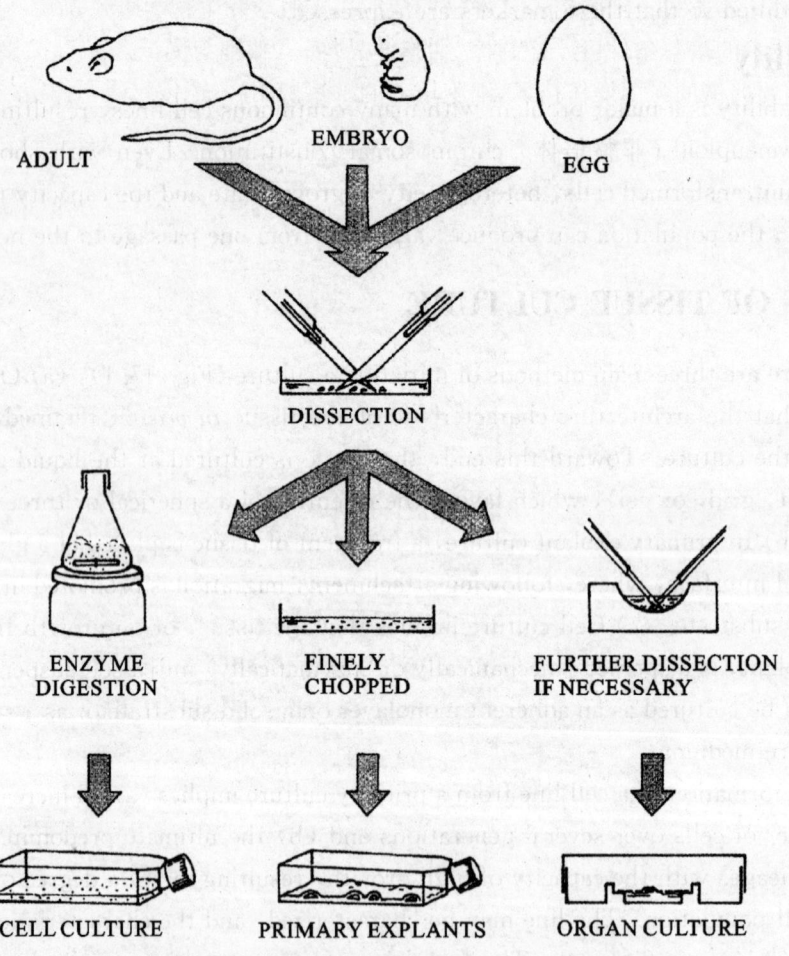

Fig. 17.1 Types of tissue culture (By R. Ian Freshney)

QUESTIONS FOR DISCUSSION

1. Explain the following concepts: cell culture, tissue culture, organ culture, cell strain, monolayer, histotypic culture, organotypic culture, anchorage dependence, outgrowth.

2. List the advantages and limitations of tissue culture.
3. There are three main methods for initiating a culture. What are they?
4. What does the formation of a cell line from a primary culture imply?

知识要点

依原始培养对象不同,动物细胞的体外培养可分为细胞培养、组织培养和器官培养三大类,但动物的组织培养常常作为一个通用的概念来使用,即泛指以上三种类型的动物细胞体外培养。组织型培养和器官型培养则是指分别将相同和不同谱系细胞重构成组织和器官的两种培养方式。细胞培养技术已广泛应用于医学和生物学研究领域,与活体相比,细胞培养技术的培养条件可严格调控,细胞样品均一,细胞与测试药品直接接触,药品用量少,但需要严格的无菌操作技术,细胞培养的维持费用高,遗传物质不稳定,突变率高。

# CHAPTER 18　BIOLOGY OF CULTURED CELLS(培养细胞的生物学特性)

CELL ADHESION AND RELATED MOLECULES

Most cells from solid tissues grow as adherent monolayers, and, unless they have transformed and become anchorage independent, following tissue disaggregation or subculture they will need to attach and spread out on the substrate before they will start to proliferate. Originally, it was found that cells would attach to, and spread on, glass that had a slight net positive charge. Subsequently, it was shown that cells would attach to some plastics as well, such as polystyrene(聚苯乙烯), if the plastic was appropriately treated with an electric ion discharge(电离子放电)or high-energy ionizing radiation(电离辐射). We now know that cell adhesion is mediated by specific cell surface receptors for molecules in the extracellular matrix, so it seems likely that spreading may be preceded by the secretion of extracellular matrix proteins and proteoglycans(蛋白多糖) by the cells. The matrix adheres to the charged substrate, and the cells then bind to the matrix via specific receptors. Hence, glass or plastic that has been conditioned by previous cell growth can often provide a better surface for attachment, and substrates pretreated with matrix constituents, such as fibronectin(纤连蛋白) or collagen, or derivatives, such as gelatin(明胶), will help the more fastidious(难贴附) cells to attach and proliferate. With fibroblast-like cells, the main requirement is for substrate attachment and spreading, because the cells migrate individually at low densities. Epithelial cells, however, appear to have to make the correct cell-cell contacts for optimum survival and growth; consequently, they tend to grow as patches(成片).

Three major classes of transmembrane proteins have been shown to be involved in cell-cell and cell-substrate adhesion. Cell-cell adhesion molecules, CAMs（钙调蛋白, Ca^{2+}-independent), and cadherins（钙黏蛋白, Ca^{2+}-dependent) are involved primarily in interactions between homologous cells. These proteins are self-interactive; that is, homologous molecules in opposing cells interact with each other. Cell-substrate interactions are mediated primarily by integrins(整合蛋白), which bind to the receptors in the matrix via a specific motif usually containing the arginine-glycine-aspartic acid（精氨酸-甘氨酸-天冬氨酸,RGD) sequence. Fibronectin, entactin(巢蛋白), laminin(层粘连蛋白), and collagen are such receptor molecules in the matrix. The third group of cell adhesion molecules is the transmembrane proteoglycans(蛋白多糖), also interacting with matrix constituents such as other proteoglycans or collagen, but not via the RGD motif. Cell adhesion molecules are attached to elements of the cytoskeleton. The attachment of integrins to actin microfilaments via linker proteins is associated with reciprocal signaling between the cell surface and the nucleus.

SELECTIVE GROWTH IN PRIMARY CULTURE

Primary culture is derived either by the outgrowth of migrating cells from a fragment of tissue or by enzymatic or mechanical dispersal of the tissue. Regardless of the method employed, primary culture is the first in a series of selective processes that may ultimately give rise to a relatively uniform cell line. In primary explantation, selection occurs by virtue of the cells' capacity to migrate from the explant. While with dispersed cells, only those cells that both survive the disaggregation technique and adhere to the substrate or survive in suspension will form the basis of a primary culture.

If the primary culture is maintained for more than a few hours, a further selection step will occur. Cells that are capable of proliferation will increase, some cell types will survive but not increase, and yet others will be unable to survive under the particular conditions of the culture. Hence, the relative proportion of each cell type will change and will continue to do so until, in the case of monolayer cultures, all the available culture substrate is occupied. After confluence is reached (i. e., all the available growth area is utilized and the cells make close contact with one another), cells whose growth is sensitive to density limitation will stop dividing, while any transformed cells, which are insensitive to density limitation, will tend to overgrow. Keeping the cell density low (e. g., by frequent subculture) helps to preserve the normal phenotype in cultures such as mouse fibroblasts, in which spontaneous transformants tend to overgrow at high cell densities.

EVOLUTION OF CELL LINES

After the first subculture, or passage, the primary culture becomes known as a cell line and may be propagated and subcultured several times. With each successive subculture, the component of the population with the ability to proliferate most rapidly will gradually predominate, and nonproliferating or slowly proliferating cells will be diluted out. This is most strikingly apparent after the first subculture, in which differences in proliferative capacity are compounded with varying abilities to withstand the trauma of trypsinization and transfer. Although some selection and phenotypic drift will continue, by the third passage the culture becomes more stable and is typified by a rather hardy, rapidly proliferated cell.

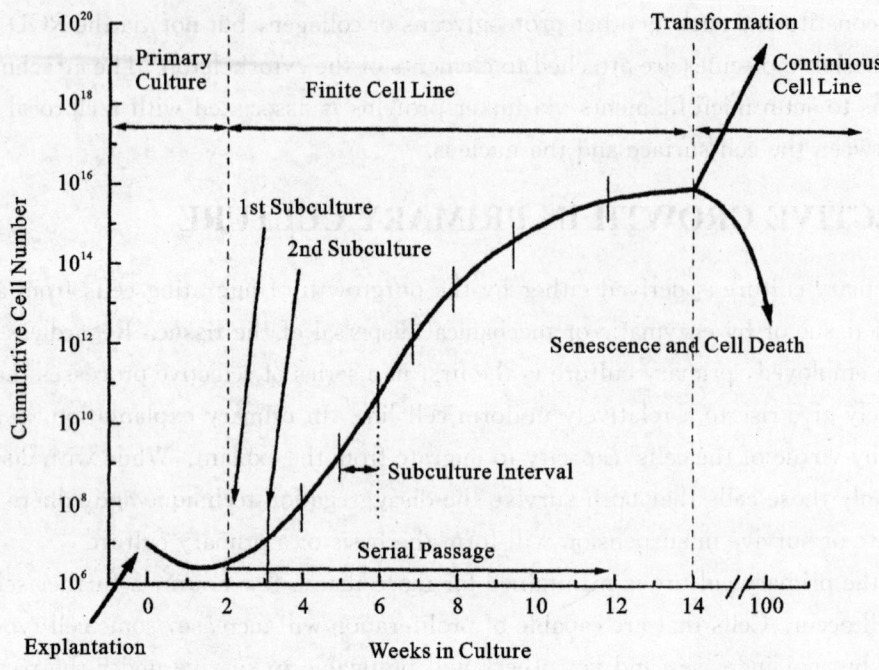

Fig. 18.1 Evolution of a cell line. The vertical (Y) axis represents total cell growth (assuming no reduction at passage) for a hypothetical cell culture. Total cell number (cell yield) is represented on this axis on a log scale, and the time in culture is shown on the X-axis on a linear scale. Although a continuous cell line is depicted as arising at 14 weeks, with different cells it could arise at any time. Likewise, senescence may occur at any time, but for human diploid fibroblasts it is most likely to occur between 30 and 60 cell doublings, or 10 to 20 weeks, depending on the doubling time. (By R. Ian Freshney)

Senescence(衰老,死亡)

Normal cells can divide only for a limited number of times; hence, cell lines derived from normal tissue will die out after a fixed number of population doublings, thus called finite cell line. This is a genetically determined event, known as senescence, and is thought to be determined by the inability of terminal sequences of the DNA in the telomeres(端粒) to replicate at each cell division. The result is a progressive shortening of the telomeres, until, finally, the cell is unable to divide further. Exceptions to this rule are germ cells, stem cells, and transformed cells, which often express the enzyme telomerase(端粒酶), which is capable of replicating the terminal sequences of DNA in the telomere and extending the life span of the cells, infinitely in the case of germ cells and some tumor cells.

The Development of Continuous Cell Lines

Some cell lines may give rise to continuous cell lines (Fig. 18.1.). The ability of a cell line to grow continuously probably reflects its capacity for genetic variation, allowing subsequent selection. Genetic variation often involves the deletion or mutation of the p53 gene, which would normally arrest cell cycle progression. Human fibroblasts remain predominantly euploid(整倍体) throughout their life span in culture and never give rise to continuous cell lines, while mouse fibroblasts and cell cultures from a variety of human and animal tumors often become aneuploid(非整倍体)in culture and frequently give rise to continuous cultures. Possibly the condition that predisposes to the development of a continuous cell line is inherent genetic variation, so it is not surprising to find genetic instability perpetuated in continuous cell lines. A common feature of many human continuous cell lines is the development of a subtetraploid chromosome number (Fig. 18.2). The alteration in a culture that gives rise to a continuous cell line is commonly called *in vitro* transformation and may occur spontaneously or be chemically or virally induced. The word transformation is used rather loosely and can mean different things to different people. In this chapter, immortalization means the acquisition of an infinite life span and transformation implies an alteration in growth characteristics (anchorage independence, loss of contact inhibition, and density limitation of growth) that will often, but not necessarily, correlate with tumorigenicity(致瘤性).

Continuous cell lines are usually aneuploid and often have a chromosome number between the diploid and tetraploid value (Fig. 18.2). There is also considerable variation in chromosome number and constitution among cells in the population (heteroploidy,异倍体). It is not clear whether the cells that give rise to continuous lines are present at explantation in very small numbers or arise later as a result of the transformation of one or more cells. The second alternative would seem to be more probable on cell kinetic grounds, as continuous cell lines can appear quite late in a culture's life history, long after the time it would have taken for even one preexisting cell to overgrow. The possibility remains, however, that there is a subpopulation in such cultures with a predisposition to transform that it is not shared by the rest of the cells.

The term transformation(转化)has been applied to the process of formation of a continuous cell line partly because the culture undergoes morphological and kinetic alterations, but also because the formation of a continuous cell line is often accompanied by an increase of tumorigenicity. A number of the properties of the continuous cell lines, such as a reduced serum requirement, reduced density limitation of growth, growth in semisolid media, aneuploidy, and more, are associated with *malignant* transformations (恶性转化). Similar morphological and behavioral changes can also be observed in cells that have undergone virally or chemically induced transformation.

Many (if not most) normal cells do not give rise to continuous cell lines. In the classic example, normal human fibroblasts remain euploid throughout their life span and at crisis (usually around 50 generations) will stop dividing, although they may remain viable for up to 18 months thereafter. Epidermal cells, on the other hand, have shown gradually increasing life spans with improvements in culture techniques and may yet be shown capable of giving rise to continuous growth. Such growth may be related to the self-renewal capacity of the tissue *in vivo*.

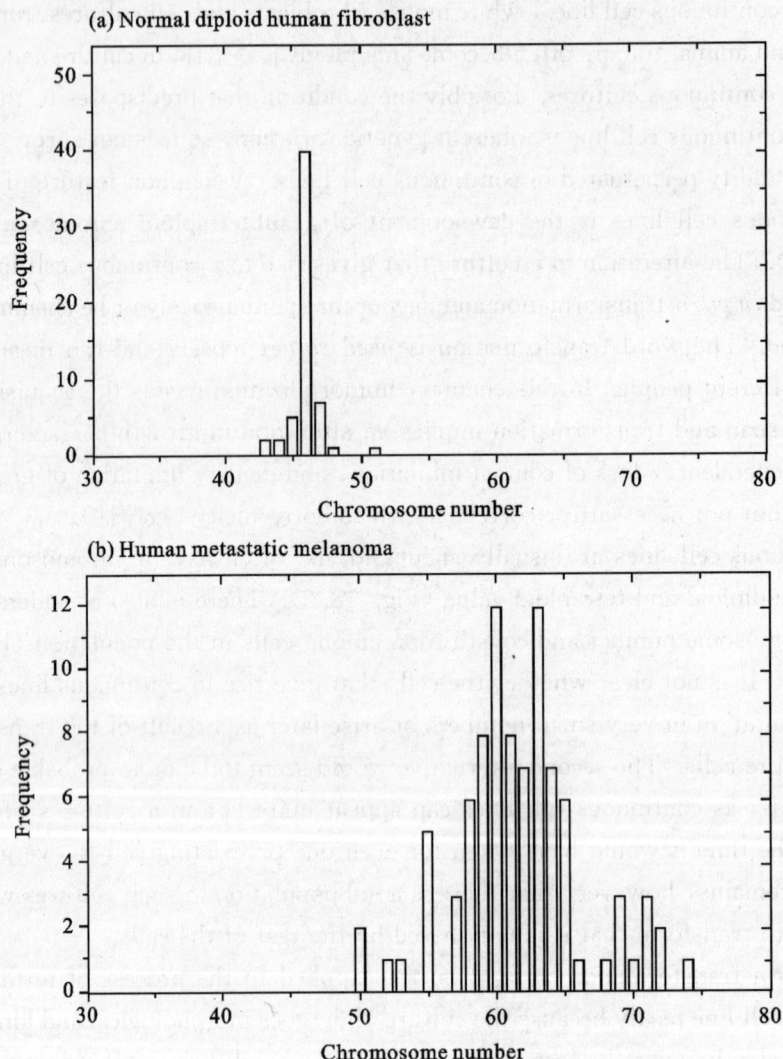

Fig. 18.2 Chromosome numbers of finite and continuous cell lines. (a) A normal human glial cell line. (b) A continuous cell line from human metastatic melanoma. (By R. Ian Freshney)

QUESTIONS FOR DISCUSSION

1. Explain the following concepts: confluence, senescence, immortalization, in vitro transformation, heteroploidy, aneuploidy.

2. Describe the mechanism of cell adhesion.

3. What may happen during the initiation of a primary cell culture?

4. The evolution of a cell line usually goes through three stages of primary culture,

finite cell line, continuous cell line and senescence. What may happen during these stages?

知识要点

体外培养动物细胞具有贴壁依赖性,只有血液细胞和某些转化细胞具有悬浮生长能力。细胞表面带净负电荷,因而易贴附于表面带净正电荷的玻璃或塑料表面。细胞在基质上的进一步铺展和贴附则与细胞的贴附分子有关。钙调蛋白和钙黏蛋白参与细胞和细胞之间的连接,整合蛋白和跨膜蛋白多糖参与细胞与基质之间的连接。细胞系的演化过程可分为3个阶段:原代培养、有限细胞系、连续性细胞系或衰老死亡。在原代培养细胞的迁移以及细胞系的早期传代过程中,都存在着选择性。不断选择的结果,导致传代3次后的培养物变得很均一。大部分细胞系在体外传代培养1年左右,就衰老死亡了,只有少数细胞系发生了转化,获得了永生性,才能发展成为连续性细胞系或永生性细胞系。

CHAPTER 19 EQUIPMENTS FOR ANIMAL CELL CULTURE(动物细胞培养器材)

Laboratory facilities required for the culture of animal cells are similar to those in a plant tissue culture laboratory. Sterile handling area equipped with laminar-flow hood and UV light, incubation area equipped with incubators, storage area with freezers, and the area for washup, sterilization and media preparation are essential layout and facilities in the laboratory (see chapter 2 and 3). But greenhouse and lighting equipments are unnecessary for the animal cell culture. In this chapter, only the specific needs for the culture of animal cells will be introduced.

CO_2 INCUBATOR

Although cultures can be incubated in sealed flasks in a regular dry incubator or a hot room, some vessels e.g., Petri dishes or multiwell plates, require a controlled atmosphere with high humidity and elevated CO_2 tension. The cheapest way of controlling the gas phase is to place the cultures in a plastic box, or chamber. Gas the container with the correct CO_2 mixture and then seal it. If the container is not completely filled with dishes, include an open dish of water to increase the humidity inside the chamber.

CO_2 incubators (Fig. 19.1) are more expensive, but their ease of use and superior control of CO_2 tension and temperature justify the expenditure. A controlled atmosphere is achieved by blowing air over a humidifying tray (Fig. 19.2) and controlling the CO_2 tension with a CO_2-monitoring device, which draws air from the incubator into a sample chamber, determines the concentration of CO_2 and injects pure CO_2 into the incubator to make up any deficiency. Air is circulated around the incubator by a fan to keep both the CO_2 level and the temperature uniform. Most inexpensive CO_2 controllers need to be calibrated every few months; in contrast, most "top-of-the-line" models reset the zero of the CO_2 detector automatically.

Frequent cleaning of incubators-particularly humidified ones-is essential so the interior should dismantle(拆除) readily without leaving inaccessible crevices(裂缝) or corners. Flasks or dishes, or boxes containing them, that are taken from the incubator to the laminar-flow or other sterile workstation, should be swabbed with alcohol before being opened.

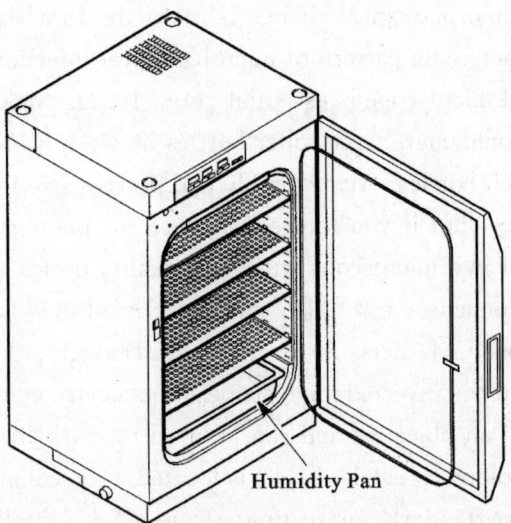

Fig. 19.1 CO_2 incubator. This incubator is designed to create a stable and reliable environment for cell culture. The desired temperature and CO_2 level can be set by the control panel. Distilled water is added into the humidity pan to maintain the moisture.

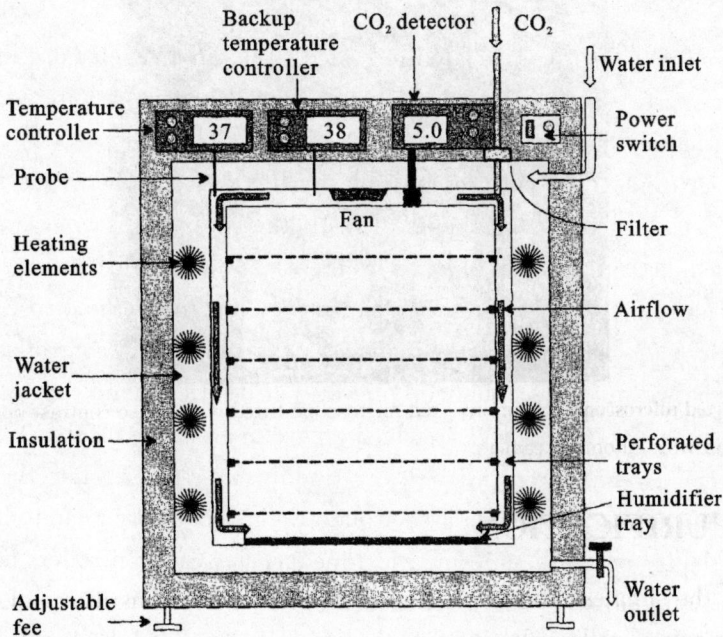

Fig. 19.2 CO_2 incubator design. Front view of control panel and section of chamber of a stylized humid CO_2 incubator. (By R. Ian Freshney)

INVERTED MICROSCOPE

It cannot be over-emphasized that, in spite of considerable progress in the quantitative analysis of cultured cells and remote sensing techniques, it is still vital to look at cultures regularly. A morphological change is often the first sign of deterioration in a culture, and the characteristic pattern of microbiological infection is easily recognized.

A simple inverted microscope is essential (Fig. 19.3). Make certain that the stage is large enough to accommodate large roller bottles between it and the condenser in case you should require such bottles. Many simple and inexpensive inverted microscopes are available on the market, but if you foresee the need for photographing living cultures, then you should invest in a microscope with high-quality optics and a long-working-distance phase-contrast condenser（聚光器）and objectives（物镜）, with provisions for a camera (e.g., Olympus CM, Zeiss Axiovert, Leitz Diavert). The Nikon marking ring is a useful accessory to the inverted microscope. This device is inserted in the nosepiece（物镜转盘）in place of an objective and can be used to mark the underside of a dish in which an interesting colony or patch of cells is located. The colony can then be picked or the development of a particularly interesting area in a culture followed.

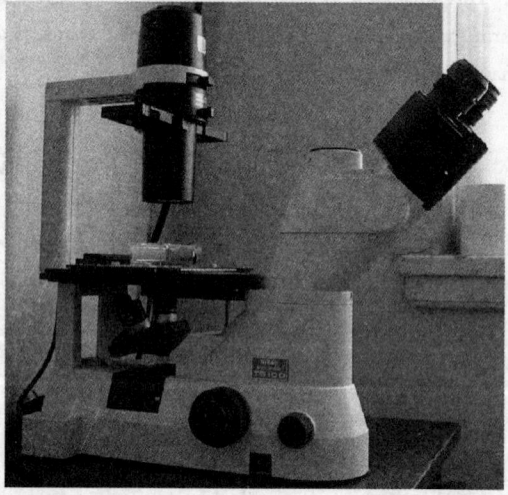

Fig. 19.3 Inverted microscope. Nikon inverted microscope fitted with phase-contrast optics and trinocular（三目镜）head with automatic camera.

WATER PURIFICATION

Water is the simplest, but probably the most critical, constituent of all media and reagents. For animal cells, ultrapure water (Fig. 19.4) or triple distilled water is required and should be used freshly or stored after sterilization.

CRYOSTORAGE CONTAINER

The freezing process can be carried out satisfactorily without sophisticated equipment, but storage requires a properly constructed liquid-nitrogen freezer. Two main types of freezer are available: narrow necked with slow evaporation, but more difficult access; and wide necked with easier access, but three times the evaporation rate. If the cost of liquid nitrogen and its supply present no problem, then a wide-necked freezer may be more convenient, although the holding time will be only about one week to 10 days. A narrow-necked freezer, on the other hand, will be more economical and will hold its contents up to two months.

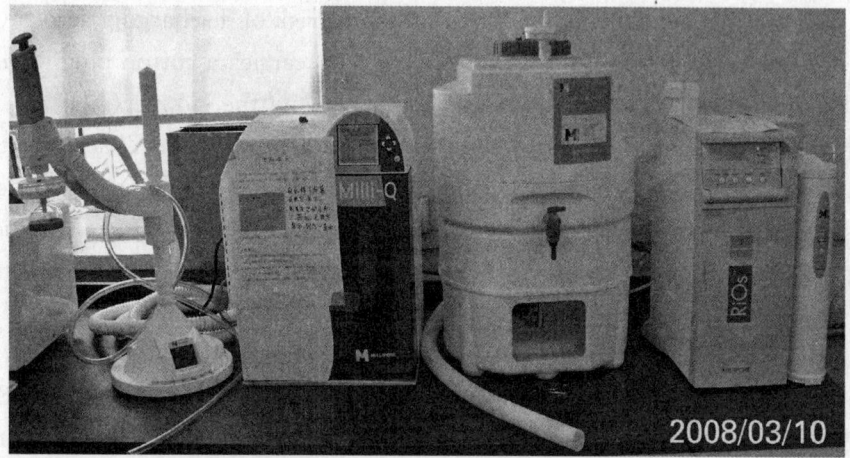

Fig. 19.4　Millipore ultrapure water system. Tap water first passes through a reverse-osmosis unit on the right and then goes to the storage tank on the left. It then passes through carbon filtration and deionization (center unit) before being collected via a micropore filter (Millipore MilliQ).

It is also possible to compromise by purchasing a narrow-necked freezer with a rack inventory system (Fig. 19.5). These freezers have the advantage of storing the tubes in rectangular drawers, rather than triangular, but still have a relatively narrow neck with consequent savings in evaporation of liquid nitrogen and a longer holding time. The total storage per unit volume is still not as great as that of a wide-necked freezer, and they can be a little awkward to use, but they do represent the best compromise between access and holding time.

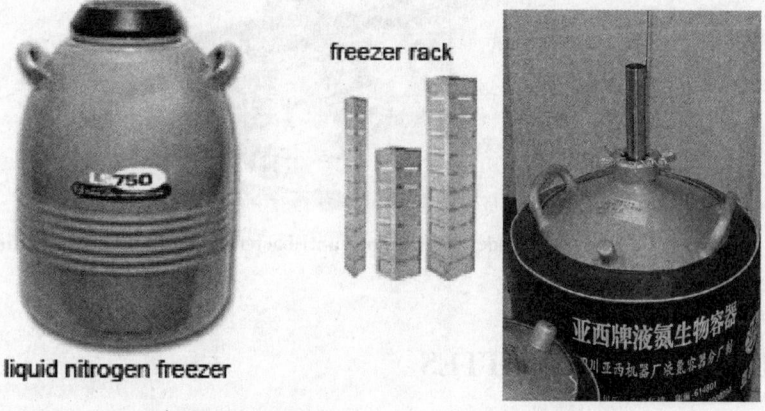

Fig. 19.5　Liquid nitrogen freezers and convenient box-type storage racks.

ASPIRATION PUMP(吸扬式泵)

A peristaltic pump(蠕动泵)may be used to remove spent medium or other reagents from a culture flask. Pumping is a rapid and efficient way to deal with several flasks at a time or with large volumes, and the effluent(流出物)can be collected directly into disinfectant(消毒剂) in a vented container, with minimal risk of discharging aerosol(气雾剂) into the atmosphere, particularly if the vent(排气口)carries a cotton plug or micropore filter. The inlet(输入) line should extend further below the stopper(制止器) than the outlet(输出), by at least 5 cm (2 in), so that waste does not splash back into the vent. The pump tubing should be checked regularly for wear, and it is more convenient if the pump is operated by a self-canceling foot switch (Figs. 19.6). To keep effluent from running back into a flask, always switch on the pump before inserting a pipette in the tubing.

A vacuum pump(真空泵), similar to that supplied for sterile filtration, may be used instead of a peristaltic pump. If necessary, the same pump could serve both purposes; however, a trap will be required to avoid the risk of waste entering the pump. The effluent should be collected in a reservoir(收集瓶) and a disinfectant such as hypochlorite added when work is finished and left for at least 2 h before the reservoir is emptied. A hydrophobic micropore filter and a second trap should be placed in the line to the pump to prevent fluid or aerosol from being carried over. Do not draw air through a pump from a reservoir containing hypochlorite, as the free chlorine will corrode the pump and could be toxic. Also, avoid vacuum lines; if they become contaminated with fluids, they can be very difficult to clean out.

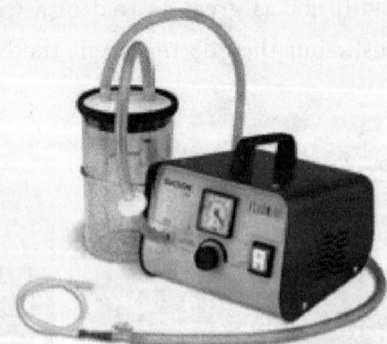

Fig. 19.6 Aspiration Pump Unit includes compressor, anti-bacterial filter and a suction line from the hood to a waste receiver.

PIPETTORS AND PIPETTES

Routine subculture, which should be rapid and secure from microbial and cross contamination, but need not be very accurate, is best performed with conventional graduated glass or disposable plastic pipettes (1-ml, 5-ml or 10-ml volume). Motorized or mechanical pipettors are required for this purpose (Figs. 19.7 and 19.8).

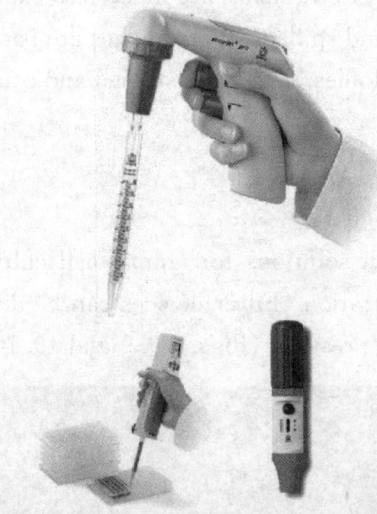

Fig. 19.7 Motorized pipetting device for use with conventional graduated pipettes.

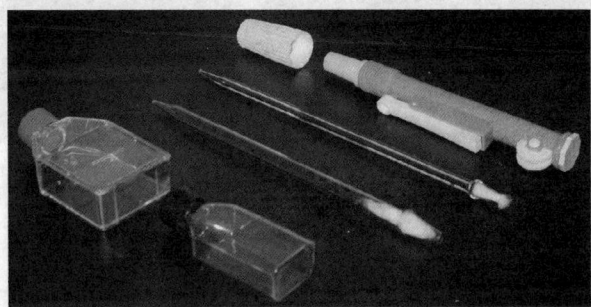

Fig. 19.8 Thumb controlled pipette pump, graduated glass pipettes (1-ml and 5-ml), plastic and glass flasks.

CULTURE VESSELS

The choice of culture vessels is determined by (a) the yield (number of cells) required; (b) whether the cell is grown in monolayer or suspension; and (c) the sampling regime (i.e., are the samples to be collected simultaneously or at intervals over a period of time?) "Shopping around" will often result in a cheaper price, but do not be tempted to change products too frequently, and always test a new supplier's product before committing yourself to it.

Care should be taken to label sterile, nonsterile, tissue-culture and non-tissue-culture grades of plastics clearly and to store them separately. Glass bottles with flat sides can be used instead of plastic, provided that a suitable washup and sterilization service is available. However, the lower cost tends to be overridden by the optical superiority, sterility, quality assurance, and general convenience of plastic flasks (Fig. 19.8). Nevertheless, disposable plastics can account for approximately 60% of the tissue culture budget-even more than serum. Petri dishes are much less expensive than flasks, though more prone to contamination and spillage. Petri dishes are particularly useful for colony-formation assays, in which colonies have to be stained and counted or isolated at the end of an experiment.

FILTER DEVICES

Media, serum and trypsin solutions for animal cell culture can not be autoclaved and should be sterilized by filtration. Filter devices can be divided into two categories: positive pressure and negative pressure (Figs. 19.9 and 19.10).

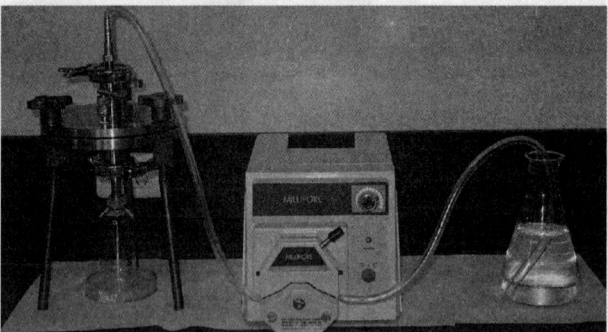

Fig. 19.9 Positive pressure filter device of peristaltic pump filtration. Sterile filtration with peristaltic pump between nonsterile reservoir and sterilizing filter.

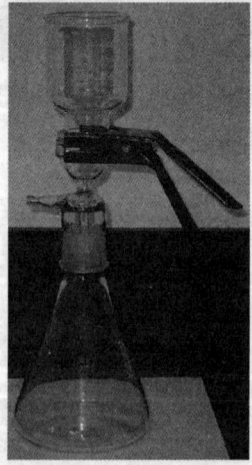

Fig. 19.10 Negative pressure filter device (Millipore).

SYRINGE FILTERS ADAPTERS

While permanent apparatus is available for sterile filtration, most laboratories now use disposable filters (Fig. 19.11). It is worthwhile to hold some of the more common sizes in stock, such as 25-mm syringe adapters and 47-mm bottle-top adapters or filter flasks. It is also wise to keep a small selection of larger sizes on hand.

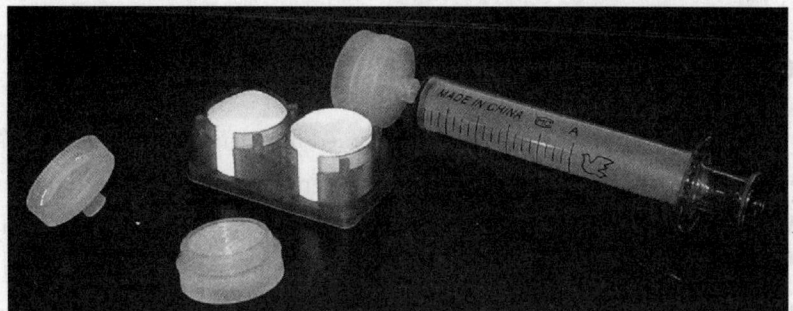

Fig. 19.11 Syringes, syringe filter adapters and micropore filter membranes.

QUESTIONS FOR DISCUSSION

Do you know how to use CO_2 incubator, inverted microscope and filter devices?

知识要点

与植物组织培养相比,动物细胞培养不需要配备温室和光照培养箱或培养架,但二氧化碳培养箱、倒置显微镜、抽滤装置、液氮罐以及动物细胞培养瓶等则是必不可少的装备。

CHAPTER 20 MEDIA FOR ANIMAL CELLS(动物细胞培养基)

DEVELOPMENT OF MEDIA

Initial attempts to culture cells were performed in natural media(天然培养基) based on tissue extracts and body fluids, such as chick embryo extract, serum, lymph(淋巴液), etc. With the propagation of cell lines, the demand for larger amounts of a medium with more consistent quality led to the introduction of chemically defined media based on analyses of body fluids and nutritional biochemistry. Eagle's Basal Medium and, subsequently, Eagle's Minimal Essential Medium [Eagle, 1959] became widely adopted, variously supplemented with calf, human, or horse serum, protein hydrolysates(蛋白水解产物), and embryo extract(胚胎抽提液). As more continuous cell lines(连续性细胞系) became available (L929 cells, HeLa, etc.), it was apparent that these media were perfectly adequate for the majority of those lines, and most of the succeeding developments were aimed at replacing serum, optimizing media for different cell types (e.g., RPMI 1640 for lymphoblastoid(淋巴母细胞)cell lines), or modifying for specific conditions (e.g., Leibovitz L15 to eliminate the need for adding CO_2 and $NaHCO_3$).

Currently, there is a divergence in media requirements: Some scientists wish to isolate and propagate cells of a specific lineage, while others simply require cells as substrates for the formation of products, as a host for viral propagation, or for non-cell-specific molecular studies. Production techniques make ever more use of selective media, usually serum-free, while viral and molecular work relies mainly on Eagle's MEM, Dulbecco's modification DMEM, or increasingly, RPMI1640. There is also a growing tendency to affect a compromise(折中) by mixing a complex medium, such as Ham's F12, with one with higher amino acid and vitamin concentrations, such as DMEM. This alternative will horrify(使震惊)the purist(力求纯正者), but it does generate a useful, all-purpose medium for primary culture as well as cell line propagation. Serum is still widely used, as most people do not want the added complexity and cost of serum-free media. However, those culturing specialized cells and those working in the production of biopharmaceuticals, in which serum is undesirable for a number of reasons, are now working without serum(血清). This chapter concentrates on the general principles of medium composition, using the widely used serum-supplemented media as examples.

PHYSICOCHEMICAL PROPERTIES

pH

Most cell lines grow well at pH 7.4. Although the optimum pH for cell growth varies relatively little among different cell strains, some normal fibroblast lines perform best at pH 7.4-7.7, and transformed cells may do better at pH 7.0-7.4. It was reported that epidermal cells could be maintained at pH 5.5, but this level has not been universally adopted. In special cases it may prove advantageous to do a brief growth experiment or a special function analysis to determine the optimum pH.

Phenol red(酚红) is commonly used as an indicator. It is red at pH 7.4 and becomes orange at pH 7.0, yellow at pH 6.5, lemon yellow below pH 6.5, more pink at pH 7.6, and purple at pH 7.8. Since the assessment of color is highly subjective, it is useful to make up a set of standards using a sterile balanced salt solution (BSS) and phenol red at the correct concentration and in the same type of bottle, with the same headspace for air that you normally use for preparing a medium.

CO_2 and Bicarbonate

Carbon dioxide in the gas phase appears in the medium as dissolved CO_2 in equilibrium with HCO_3^- and lowers the pH. The dissolved CO_2, HCO_3^- and pH are all interrelated, thus the atmospheric CO_2 tension will regulate the concentration of dissolved CO_2 directly. This regulation in turn produces H_2CO_3, which dissociates according to the reaction

$$H_2O + CO_2 \longleftrightarrow H_2CO_3 \longleftrightarrow H^+ + HCO_3^-$$

Since HCO_3^- has a fairly low dissociation constant with most of the available cations, it tends to reassociate, leaving the medium acid. The net result of increasing atmospheric CO_2 is to depress the pH, so the effect of elevated CO_2 tension is neutralized by increasing the bicarbonate concentration: $NaHCO_3 \longleftrightarrow Na^+ + HCO_3^-$

When preparing a new medium for the first time, add the specified amount of bicarbonate and then sufficient 1 N NaOH such that the medium equilibrates to the desired pH after incubation at 37℃ overnight. Each medium has a recommended bicarbonate concentration and CO_2 tension for achieving the correct pH and osmolality, but minor variations will occur in different methods of preparation.

In sum, cultures in open vessels need to be incubated in an atmosphere of CO_2 the concentration of which is in equilibrium with the sodium bicarbonate($NaHCO_3$) in the medium. Cells at moderately high concentrations ($\geqslant 1 \times 10^5$ cells/mL) and grown in sealed flasks need not have CO_2 added to the gas phase, provided that the bicarbonate concentration is kept low ($\sim$ 4 mM), particularly if the cells are high acid producers.

Buffering

Culture media must be buffered under two sets of conditions: (1) open dishes, wherein the evolution of CO_2 causes the pH to rise and (2) overproduction of CO_2 and lactic acid in transformed cell lines at high cell concentrations, when the pH will fall. A buffer may be incorporated into the medium to stabilize the pH, but in (1) exogenous CO_2 may still be required by some cell lines, particularly at low cell concentrations, to prevent the total loss of dissolved CO_2 and bicarbonate from the medium. In spite of its poor buffering capacity at physiological pH bicarbonate buffer is still used more frequently than any other buffer, because of its low toxicity, low cost, and nutritional benefit to the culture. HEPES is a much stronger buffer in the pH 7.2-7.6 range and is now used frequently at 10 or 20 mM. But HEPES is more toxic than bicarbonate.

Oxygen

The other major significant constituent of the gas phase is oxygen. While most cells require oxygen for respiration in vivo, cultured cells often rely on glycolysis(糖酵解), a high proportion of which, as in transformed cells, may be anaerobic(厌氧). Most dispersed cell cultures prefer lower oxygen tensions. Because the depth of the culture medium can influence the rate of oxygen diffusion to the cells, it is advisable to keep the depth of the medium within the range 2-5 mm in static culture.

Cultures vary in their oxygen requirement, the major distinction lying between organ and cell cultures. While atmospheric or lower oxygen tensions are preferable for most cell cultures, some organ cultures, particularly from late-stage embryos, newborns, or adults, require up to 95% O_2 in the gas phase. This requirement for a high level of O_2 may be a problem of diffusion related to the geometry and gaseous penetration of organ cultures, but may also reflect the difference between differentiated and rapidly proliferating cells. Oxygen diffusion may also become limiting in porous microcarriers(多孔微载体).

Osmolality(渗透压)

Most cultured cells have a fairly wide tolerance for osmotic pressure(渗透压). Since the osmolality of human plasma is about 290 mOsm/kg, it is reasonable to assume that this level is the optimum for human cells in vitro, although it may be different for other species (e.g., around 310 mOsm/kg for mice). In practice, osmolalities between 260 mOsm/kg and 320 mOsm/kg are quite acceptable for most cells, but, once selected, should be kept consistent at ± 10 mOsm/kg. Slightly hypotonic(低渗) medium may be better for Petri-dish or open-plate culture to compensate for evaporation during incubation.

Osmolality is usually measured by depression of the freezing point (Fig. 20.1), or elevation of the vapor pressure, of the medium. The measurement of osmolality is a useful quality-control step if you are making up the medium yourself, as it helps to guard against errors in weighing, dilution, etc. It is particularly important to monitor osmolality if alterations are made in the constitution of the medium. The addition of HEPES and drugs dissolved in strong acids and bases and their subsequent neutralization can all markedly affect osmolality.

Temperature

The optimal temperature for cell culture is dependent on (1) the body temperature of the animal from which the cells were obtained, (2) any anatomical(解剖学上的)variationin (e. g. , the temperature of the skin and testis(睾丸) may be lower than that of the rest of the body), and (3) the incorporation of a safety factor to allow for minor errors in regulating the incubator. Thus, the temperature recommended for most human and warm-blooded animal cell lines is 37℃, close to body heat, but set a little lower for safety, as overheating is a more serious problem than underheating.

Fig. 20.1　Osmometer(渗透压计)

Because of the higher body temperature in birds, avian cells should be maintained at 38.5℃ for maximum growth, but will grow quite satisfactorily, if more slowly, at 37℃. Cultured cells will tolerate considerable drops in temperature, can survive several days at 4℃, and can be frozen and cooled to −196℃, but they cannot tolerate more than about 2℃ above normal (39.5℃) for more than a few hours and will die quite rapidly at 40℃ and over.

Poikilotherms (变温动物)animals that do not regulate their blood heat within narrow limits tolerate a wide temperature range, between 15℃ and 26℃. Simulating in vivo conditions (e. g. , for cold-water fish) may require an incubator with cooling as well as heating, to keep the incubator temperature below ambient. If necessary, poikilothermic animal cells can be maintained at room temperature, but the variability of the ambient(周围) temperature in laboratories makes this undesirable, and a cooled incubator is generally preferred.

Apart from its direct effect on cell growth, the temperature will also influence pH due to the increased solubility of CO_2 at lower temperatures and, possibly, due to changes in ionization and the pKa of the buffer. The pH should be adjusted to 0.2 units lower at room temperature than at 37℃.

Viscosity(黏度)

The viscosity of a culture medium is influenced mainly by the serum content and in most cases will have little effect on cell growth. Viscosity becomes important, however, whenever a cell suspension is agitated or when cells are dissociated after trypsinization (胰蛋白酶消化). Any cell damage that occurs under these conditions may be reduced by increasing the viscosity of the medium with carboxymethylcellulose (羧甲基纤维素, CMC) or polyvinylpyrlidone (聚乙烯吡咯烷酮, PVP). This becomes particularly important in low-serum concentrations, in the absence of serum, and in stirred bioreactor cultures, in which Pluronic F68 (一种乳化剂) is often used, although its effect is probably pleiotropic(多效性).

Surface Tension and Foaming

The effects of foaming have not been clearly defined, but the rate of protein denaturation may increase, and it may increase the risk of contamination if the foam reaches the neck of the culture vessel. Foaming will also limit gaseous diffusion if a film from a foam or spillage gets into the capillary space between the cap and the bottle, or between the lid and the base of a Petri dish. Foaming can arise in suspension cultures when 5% CO_2 in air is bubbled through medium containing serum. The addition of a silicone antifoam (有机硅消泡剂) or Pluronic F68 (Sigma), 0.01%-0.1%, helps prevent foaming in this situation by reducing surface tension and may also protect cells against shear stress(剪切力) from bubbles.

BALANCED SALT SOLUTIONS

A balanced salt solution (BSS) is composed of inorganic salts and may include sodium bicarbonate and, in some cases, glucose. The compositions of some common BSS's are given in Table 20.1. HEPES buffer (5-20 mM) may be added to these solutions if necessary and the equivalent weight of NaCl omitted to maintain the correct osmolality. BSS forms the basis of many complete media. Hanks's would imply the use of sealed flasks with a gas phase of air, while Earle's salts would imply a higher bicarbonate concentration compatible with growth in 5% CO_2. BSS is also used as a diluent for concentrates of amino acids and vitamins to make complete media, as a washing or dissection medium, and for short incubations up to about 4 h (usually with glucose). These recipes are often modified-for instance, by omitting glucose or phenol red from Hanks's BSS or by leaving out Ca^{2+} or Mg^{2+} ions from Dulbecco's PBS. PBS without Ca^{2+} and Mg^{2+} is known as PBS Solution A, and the convention PBSA will be used throughout this book to indicate the absence of these divalent cations.

Table 20.1 Balanced salt solutions (By R. Ian Freshney)

component		Earle's BSS		Dulbecco's PBS A		Dulbecco's PBS B		Hank's BSS		Spinner salts (as in S-MEM)	
Inorganic salts	M.W.	g/L	mM	g/L	mM	g/L	mM	g/L	mM	g/L	mM
$CaCl_2$ (anhydrous)	111	0.02	0.18			0.2	1.80	0.14	1.3		
KCl	74.55	0.4	5.37	0.2	2.7	0.2	2.68	0.4	5.4	0.40	5.37
KH_2PO_4	136.1			0.2	1.5	0.2	1.47	0.06	0.4		
$MgCl_2 \cdot 6H_2O$	203.3							0.00	0.1	0.5	
$MgSO_4 \cdot 7H_2O$	246.5	0.2	0.81			0.98	3.98	0.1	0.4	0.20	0.81
NaCl	58.44	6.68	114.31	8	136.9	8	136.89	8	136.9	6.80	116.36
$NaHCO_3$	84.01	2.2	26.19					0.00	0.35	4.2	2.20	26.19
$Na_2HPO_4 \cdot 7H_2O$	268.1			2.2	8.1	2.1	68.06	0.09	0.3		
$NaH_2PO_4 \cdot H_2O$	138	0.14	1.01							1.40	10.14
Total salt			147.86		149.1		154.88		149.4		158.87
Other components											
D-glucose	180.2	1	5.55					1	5.5	1.00	5.55
Phenol red	354.4	0.01	0.03					0.01	0.0	0.01	0.03
Gas phase											
Buffer		5%CO_2				Air		Air		5%CO_2	
HEPES, Na salt	260.3	13.02	50.00	5.2	120.0	5.2	120	2.08	8.0	13.02	50.00

Note. If HEPES is used, the equivalent molarity of NaCl must be omitted and osmolality must be checked.

COMPLETE MEDIA

Complete media range in complexity from the relatively simple Eagle's MEM, which contains essential amino acids, vitamins, and salts, to complex media such as Medium 199 (M199), CMRL 1066, MB 752/1, RPMI 1640, and F12 (Table 20.2), and a wide range of serum-free formulations. The complex media contain a larger number of different amino acids, including nonessential amino acids and additional vitamins, and are often supplemented with extra metabolites (e.g., nucleosides(核苷), tricarboxylic acid cycle(三羧酸循环)intermediates, and lipids) and minerals.

Amino Acids

The essential amino acids (i.e., those which are not synthesized in the body) are required by cultured cells, together with cysteine(半胱氨酸) and tyrosine(酪氨酸) in addition, although individual requirements for amino acids will vary from one cell to another. Other nonessential amino acids are often added as well, to compensate either for a particular cell type's incapacity to make them or because they are made, but lost by leakage into the medium. The concentration of amino acids usually limits the maximum cell concentration attainable, and the balance may influence cell survival and growth rate. Glutamine(谷氨酰胺)is required by most cells, although some cell lines will utilize glutamate(谷氨酸); evidence suggests that glutamine is also used by cultured cells as a source of energy and carbon.

Vitamins

Eagle's MEM contains only the water-soluble vitamins (the B-group), plus choline (胆碱), folic acid (叶酸), inositol (肌醇), and nicotinamide (烟酰胺), but excluding biotin (生物素); see Table 20.2); other requirements presumably are derived from the serum. Biotin is present in most of the more complex media, including the serum-free recipes, and p-aminobenzoic acid (p-氨基苯甲酸, PABA) is present in M199, CMRL 1066 (which was derived from M199), and RPMI 1640. All the fat-soluble vitamins (A, D, E, K) are present only in M199, while vitamin A is present in LHC-9 and vitamin E in MCDB 110. Some vitamins (e.g., choline and nicotinamide) have increased concentrations in serum-free media.

Salts

The salts are chiefly those of Na^+, K^+, Mg^{2+}, Ca^{2+}, Cl^-, SO_4^{2-}, PO_4^{3-}, and HCO_3^- and are the major components contributing to the osmolality of the medium. Most media derived their salt concentrations originally from Earle's (high bicarbonate; gas phase, 5% CO_2) or Hanks's (low bicarbonate; gas phase, air) BSS. Divalent cations(阳离子), particularly Ca^{2+}, are required by some cell adhesion molecules, such as the cadherins(钙粘着蛋白). Ca^{2+} also acts as an intermediary in signal transduction, and the concentration of Ca^{2+} in the medium can influence whether cells will proliferate or differentiate. Na^+, K^+ and Cl^- regulate membrane potential, while SO_4^{2-}, PO_4^{3-} and HCO_3^- have roles as anions(阴离子)required by the matrix and nutritional precursors for macromolecules, as well as regulators of intracellular charge. Calcium is reduced in suspension cultures in order to minimize cell aggregation and attachment. The sodium bicarbonate concentration is determined by the concentration of CO_2 in the gas and has a significant nutritional role in addition to its buffering capability.

Glucose

Glucose is included in most media as a source of energy. It is metabolized principally by glycolysis(糖酵解)to form pyruvate(丙酮酸), which may be converted to lactate(乳酸)or acetoacetate(乙酰乙酸)and may enter the citric acid cycle(柠檬酸循环) to form CO_2. The accumulation of lactic acid in the medium, particularly evident in embryonic and transformed cells, implies that the citric acid cycle may not function entirely as it does in vivo, and recent data have shown that much of its carbon is derived from glutamine rather than glucose. This finding may explain the exceptionally high requirement of some cultured cells for glutamine or glutamate.

Organic Supplements

A variety of other compounds, including proteins, peptides, nucleosides, citric acid cycle intermediates, pyruvate, and lipids, appear in complex media. Again, these constituents have been found to be necessary when the serum concentration is reduced, and they may help in cloning and in maintaining certain specialized cells, even in the presence of serum.

Hormones and Growth Factors

Hormones and growth factors are not specified in the formulas of most regular media although they are frequently added to serum-free media (chapter 22).

Table 20.2 Frequently used media (By R. Ian Freshney)

Component	MEM	DMEM	F12	DMEM/F12	aMEM	CMRL 1066	RPMI 1640	M199	L15	McCoy's 5A	Fischer	MB 752/1
Amino acids												
L-alanine			1.0E-04	5.0E-05	2.8E-04	2.8E-04		2.8E-04	2.5E-03	1.5E-04		
L-arginine	6.0E-04	4.0E-04	1.0E-03	7.0E-04	6.0E-04	3.3E-04	1.1E-03	3.3E-04	2.9E-03	2.0E-04	7.1E-05	3.6E-04
L-asparagine			1.0E-04	5.0E-05	3.3E-04	3.8E-04	1.7E-03	3.0E-04	7.6E-05			
L-aspartic acid			1.0E-04	5.0E-05	2.3E-04	2.3E-04	1.5E-04	2.3E-04		1.5E-04		4.5E-04
L-cysteine			2.0E-04	1.0E-04	5.7E-04	1.5E-03		5.6E-07	9.9E-04	2.0E-04		5.0E-04
L-cystine	1.0E-04	2.0E-04		1.0E-04	1.0E-04	8.3E-05	2.1E-04	9.9E-05			9.9E-05	6.3E-05
L-glutamic acid			1.0E-04	5.0E-05	5.1E-04	5.1E-04	1.4E-04	4.5E-04		1.5E-04		1.0E-03
L-glutamine	2.0E-03	4.0E-03	1.0E-03	2.5E-03	2.0E-03	6.8E-04	2.1E-03	6.8E-04	2.1E-03	1.5E-03	1.4E-03	2.4E-03
Glycine		4.0E-04	1.0E-04	2.5E-04	6.7E-04	6.7E-04	1.3E-04	6.7E-04	2.7E-04	1.0E-04		6.7E-04
L-histidine	2.0E-04	2.0E-04	1.0E-04	1.5E-04	2.0E-04	9.5E-05	9.7E-05	1.0E-04	1.6E-03	1.0E-04	3.9E-04	8.3E-04
L-hydroxy-proline						7.6E-05	1.5E-04	7.6E-05		1.5E-04		
L-isoleucine	4.0E-04	8.0E-04	3.0E-05	4.2E-04	4.0E-04	1.5E-04	3.8E-04	1.5E-04	9.5E-04	3.0E-04	5.7E-04	1.9E-04
L-leucine	4.0E-04	8.0E-04	1.0E-04	4.5E-04	4.0E-04	4.6E-04	3.8E-04	4.6E-04	9.5E-04	3.0E-04	2.3E-04	3.8E-04
L-lysine HCl	4.0E-04	8.0E-04	2.0E-04	5.0E-04	4.0E-04	3.8E-04	2.2E-04	3.8E-04	5.1E-04	2.0E-04	2.7E-04	1.3E-03
L-methionine	1.0E-04	2.0E-04	3.0E-05	1.2E-04	1.0E-04	1.0E-04	1.0E-04	1.0E-04	5.0E-04	1.0E-04	6.7E-04	3.4E-04
L-phenylalanine	2.0E-04	4.0E-04	3.0E-05	2.2E-04	19E-04	1.5E-04	9.1E-05	1.5E-04	7.6E-04	1.0E-04	4.1E-04	3.0E-04
L-proline			3.0E-04	1.5E-04	3.5E-04	3.5E-04	1.7E-04	3.5E-04		1.5E-04		4.3E-04
L-serine		4.0E-04	1.0E-04	2.5E-04	2.4E-04	2.4E-04	2.9E-04	2.4E-04	1.9E-03	2.5E-04	1.4E-04	
L-threonine	4.0E-04	8.0E-04	1.0E-04	4.5E-04	4.0E-04	2.5E-04	1.7E-04	2.5E-04	2.5E-03	1.5E-04	3.4E-04	6.3E-04
L-tryptophan	4.9E-05	7.8E-05	1.0E-05	4.4E-05	4.9E-05	4.9E-05	2.5E-04	4.9E-05	9.8E-05	1.5E-05	4.9E-05	2.0E-04
L-tyrosine	2.0E-04	4.0E-04	3.0E-05	2.1E-04	2.3E-04	2.2E-04	1.1E-04	2.2E-04	1.7E-03	1.2E-04	3.3E-04	2.2E-04
L-valine	4.0E-04	8.0E-04	1.0E-04	4.5E-04	3.9E-04	2.1E-04	1.7E-04	2.1E-04	8.5E-04	1.5E-04	6.0E-04	5.6E-04

(续表)

Component	MEM	DMEM	F12	DMEM/F12	aMEM	CMRL 1066	RPMI 1640	M199	L15	McCoy's 5A	Fischer	MB 752/1
Vitamins												
p-Aminobenzoic acid						3.6E-07	7.3E-06	3.6E-07		7.3E-06		
L-Ascorbic acid					2.5E-04	2.8E-04		2.8E-04		3.2E-06		9.9E-05
Biotin			3.0E-08	1.5E-08	4.1E-07	4.1E-08	8.2E-07	4.1E-08		8.2E-07	4.1E-08	8.2E-08
Calciferol								2.5E-07				
Choline chloride	7.1E-06	2.9E-05	1.0E-04	6.4E-05	7.1E-06	3.6E-06	2.1E-05	3.6E-06	7.1E-06	3.6E-05	1.1E-05	1.8E-03
Folic acid	2.3E-06	9.1E-06	2.9E-06	6.0E-06	2.3E-06	2.3E-08	2.3E-06	2.3E-08	2.3E-06	2.3E-05	2.3E-05	9.1E-07
myo-inositol	1.1E-05	4.0E-05	1.0E-04	7.0E-05	1.1E-05	2.8E-07	1.9E-04	2.8E-07	1.1E-05	2.0E-04	8.3E-06	5.6E-06
Menadione								6.9E-08				
Nicotinamide	8.2E-06	3.3E-05	3.3E-07	1.7E-05	8.2E-06	2.0E-07	8.2E-06	2.0E-07	8.2E-06	4.1E-06	4.1E-06	8.2E-06
Nicotinic acid								2.0E-07		4.1E-06		
D-Ca pantothenate	4.2E-06	1.7E-05	2.0E-06	9.4E-06	4.2E-06	4.2E-08	1.1E-06	4.2E-08	4.2E-06	8.4E-07	2.1E-06	4.2E-06
Pyridoxal HCl	4.9E-06	2.0E-05	3.0E-07	1.0E-05	4.9E-06	1.2E-07		1.2E-07		2.5E-06	2.5E-06	
Pyridoxine HCl			3.0E-07	1.5E-07		1.2E-07	4.9E-06	1.2E-07		2.4E-06		4.9E-06
Riboflavin	2.7E-07	1.1E-06	1.0E-07	5.8E-07	2.7E-07	2.7E-08	5.3E-07	2.7E-08	1.9E-07	5.3E-07	1.3E-06	2.7E-06
Thiamin	3.0E-06	1.2E-05	1.0E-06	6.4E-06	3.0E-06	3.0E-06	3.0E-06	3.0E-08	2.4E-06	5.9E-07	3.0E-06	3.0E-05
Thiamin monoPO$_4$										4.8E-06		
a-tocopherol								2.3E-08				
Retinol acetate								3.5E-07				
Vitamin B12			1.0E-06	5.0E-07	1.0E-06		3.7E-09					1.5E-07
Antioxidants												
Glutathione						3.0E-05	3.0E-06	1.5E-07		1.5E-06		4.5E-05
Inorganic salts												
CaCl$_2$	1.8E-03	1.8E-03	3.0E-04	1.1E-03	1.8E-03	1.8E-03		1.3E-03	1.3E-03	9.0E-04	6.2E-04	8.2E-04
KCl	5.3E-03	5.3E-03	3.0E-03	4.2E-03	5.3E-04	5.3E-03	5.3E-03	5.3E-03	5.3E-03	5.3E-03	5.3E-03	2.0E-03
KH$_2$PO$_4$								4.4E-04	4.4E-04			5.9E-04
MgCl$_2$					1.2E-01							1.2E-03
MgSO$_4$	8.1E-04	8.1E-04		4.0E-04	8.1E-04	8.1E-04	4.0E-04	8.1E-04	1.6E-03	8.1E-04	4.9E-04	8.1E-04
NaCl	1.2E-01	1.1E-01	1.3E-01	1.2E-01		1.2E-01	1.0E-01	1.4E-01	1.4E-01	1.1E-01	1.4E-01	1.0E-01
NaHCO$_3$	2.6E-02	4.4E-02	1.4E-02	2.9E-02	2.6E-02	2.6E-02	2.6E-02	4.2E-03		2.6E-02	1.3E-02	2.7E-02
NaH$_2$PO$_4$	1.0E-03	9.1E-04		4.5E-04		1.0E-03				4.2E-03	5.7E-04	
Na$_2$HPO$_4$			1.0E-03	5.0E-04			5.6E-03	4.0E-04	1.6E-03		5.0E-04	2.1E-03
Trace elements												
CuSO$_4 \cdot 5H_2O$			1.6E-08	7.8E-09								
Fe(NO$_3$)$_3 \cdot 9H_2O$		2.5E-07		1.2E-07								
FeSO$_4 \cdot 7H_2O$			3.0E-06	1.5E-06								

(续表)

Component	MEM	DMEM	F12	DMEM/F12	aMEM	CMRL 1066	RPMI 1640	M199	L15	McCoy's 5A	Fischer	MB 752/1
$ZnSO_4 \cdot 7H_2O$			3.0E-06	1.5E-06								
Bases, nucleosides, etc.												
Adenine SO_4								5.4E-05				
Adenosine					3.7E-05							
AMP								5.8E-07				
ATP								1.8E-05				
Cytidine					4.1E-05							
Deoxyadenosine					4.0E-05	4.0E-05						
Deoxycytidine					4.2E-05	3.8E-05						
Deoxyguanosine					3.7E-05	3.7E-05						
2-Deoxyribose								3.7E-06				
DPN								9.5E-06				
FAD								1.2E-06				
Glucuronate NA								1.9E-05				
Guanine								1.6E-06				
Guanosine					3.5E-05							
Hypoxanthine			3.0E-05	1.5E-05				2.2E-06				
5-Me-deoxycytidine						4.1E-07						
D-Ribose								3.3E-06				
Thymidine			3.0E-06	1.5E-06	4.1E-05	4.1E-05						
Thymine								2.4E-06				
TPN								1.3E-06				
Uracil								2.7E-06				
Uridine					4.1E-05							
UTP								1.8E-06				
Xanthine								2.0E-06				
Energy metabolism												
Cocarboxylase								2.2E-06				
Coenzyme A								3.3E-06				
D-galactose										5.0E-02		
D-glucose	5.6E-03	2.5E-02	1.0E-02	1.8E-02	5.6E-03	5.6E-03	1.1E-02	5.6E-03		1.7E-02	5.6E-03	2.8E-02
Sodium acetate						6.1E-04		4.5E-04				
Sodium pyruvate		1.0E-03	1.0E-03	1.0E-03	1.0E-03					5.0E-03		
Lipids and precursors												
Cholesterol						5.2E-07		5.2E-07				
Ethanol (solvent)						3.5E-04						
Linoleic acid			3.0E-07	1.5E-07								8.9E-05
Lipoic acid			1.0E-06	5.1E-07	9.7E-07							

(续表)

Component	MEM	DMEM	F12	DMEM/F12	aMEM	CMRL 1066	RPMI 1640	M199	L15	McCoy's 5A	Fischer	MB 752/1
Tween 80				1.8E-05	1.8E-05							
Other components												
Peptone, mg/mL										0.6		
Phenol red	2.7E-05	4.0E-05	3.2E-05	3.6E-05	2.9E-05	5.3E-05	1.3E-05	4.5E-05	2.7E-05	2.9E-05	1.3E-05	2.7E-05
Putrescine				1.0E-06	5.0E-07							
Gas Phase												
CO_2	5%	10%	2%	7%	5%	5%	5%	5%	Air	5%	2%	5%

All concentrations are molar, and computer-style notation is used (e.g., 3.0E-2 = 3.0×10^{-2} = 30 mM). Molecular weights are given for root compounds(根化合物); Although some recipes use salts or hydrated forms, molarities will, of course, remain the same. Synonyms and abbreviations: AMP, adenosine monophosphate; ATP, adenosine triphosphate; biotin = vitamin H; calciferol = vitamin D2; FAD, flavine adenine dinucleotide; lipoic acid = thioctic acid; menadione = vitamin K; myo-inositol = I-inositol; nicotinamide = niacinamide; nicotinic acid = niacin; pyridoxine HCl = vitamin B_6; thiamin = vitamin B; α-tocopherol = vitamin E; retinol = vitamin A; TPN, triphosphopyridine nucleotide; UTP, uridine triphosphate; vitamin B_{12} = cobalamin.

Antibiotics

Antibiotics were originally introduced into culture media to reduce the frequency of contamination. However, the use of laminar-flow hoods, coupled with strict aseptic technique, makes antibiotics unnecessary. Indeed, antibiotics have a number of significant disadvantages:

(1) They encourage the development of antibiotic-resistant organisms.

(2) They hide the presence of low-level, cryptic contaminants that can become fully operative if the antibiotics are removed, the culture conditions change, or resistant strains develop.

(3) They may hide mycoplasma(支原体) infections.

(4) They have antimetabolic effects that can cross-react with mammalian cells.

(5) They encourage poor aseptic technique.

For all these reasons, it is strongly recommended that routine culture be performed in the absence of antibiotics and their use be restricted to primary-culture or large-scale labor-intensive experiments with a high cost of consumables(消耗品). If conditions demand the use of antibiotics, then they should be removed as soon as possible, or, if they are used in the long term, parallel cultures should be maintained free of antibiotics. A number of antibiotics used in tissue culture are moderately effective in controlling bacterial infections (Table 20.3). However, a significant number of bacterial strains are resistant to antibiotics, either naturally or by selection, so the control that they provide is never absolute. Fungal and yeast contaminations are particularly hard to control with antibiotics; they may be held in check, but are seldom eliminated.

Table 20.3 Antibiotics used in tissue culture (By R. Ian Freshney)

Antibiotic	Concentration, μg/mL (unless otherwise stated)		Activity against
	Working	Cytotoxic	
Amphotericin B(Fungizone)(两性霉素 B)	2.5	30	Fungi, yeasts
Ampicillin	2.5		Bacteria, gram positive and gram negative
Ciprofloxacin(环丙沙星)	100		Mycoplasma(支原体)
Erythromycin(红霉素)	50	300	Mycoplasma
Gentamycin(庆大霉素)	50	>300	Bacteria, gram positive and gram negative; mycoplasma
Kanamycin(卡那霉素)	100	10 mg/mL	Bacteria, gram positive and gram negative; mycoplasma
MRA (ICN)	0.5		Mycoplasma
Neomycin(新霉素)	50	3,000	Bacteria, gram positive and gram negative
Nystatin(制霉菌素)	50		Fungi, yeasts
Penicillin-G(青霉素 G)	100 U/mL	10,000 U/mL	Bacteria, gram positive
Polymixin B(多粘菌素 B)	50	1 mg/mL	Bacteria, gram negative
Streptomycin SO$_4$(硫酸链霉素)	100	20 mg/mL	Bacteria, gram positive and gram negative
Tetracyclin(四环素)	10	35	Bacteria, gram positive and gram negative
Tylosin(泰乐菌素)	10	300	Mycoplasma

Serum

Serum contains growth factors, which promote cell proliferation, and adhesion factors and antitrypsin activity (抗胰蛋白酶活性), which promote cell attachment. Serum is also a source of minerals, lipids, and hormones, many of which may be bound to protein (Table 20.4).

1. Protein. Although proteins are a major component of serum, the functions of many proteins in vitro remain obscure; it may be that relatively few proteins are required other than as carriers for minerals, fatty acids, and hormones. Those proteins for which requirements have been found are albumin(血清白蛋白), which may be important

as a carrier of lipids or minerals and globulins; fibronectin (cold-insoluble globulin), which promotes cell attachment; and α2-macroglobulin(巨球蛋白), which inhibits trypsin. Fetuin(胎球蛋白) in fetal serum enhances cell attachment, and transferrin(转铁蛋白)binds iron, making it less toxic but bioavailable. Other proteins, as yet uncharacterized, may be essential for cell attachment and growth. Protein also increases the viscosity of the medium, reducing shear stress during pipetting and stirring, and may add to the medium's buffering capacity.

2. Growth Factors. Natural clot serum stimulates cell proliferation more than serum from which the cells have been removed physically (e.g., by centrifugation). This increased stimulation appears to be due to the release of platelet-derived growth factor (PDGF) from the platelets during clotting. PDGF is one of a family of polypeptides with mitogenic activity and is probably the major growth factor in serum. PDGF stimulates growth in fibroblasts and glia, but other platelet-derived factors, such as TGF-β, may inhibit growth or promote differentiation in epithelial cells. Other growth factors, such as fibroblast growth factors (FGFs), epidermal growth factor (EGF), endothelial cell growth factors such as vascular endothelial growth factor (VEGF) and angiogenin (血管生长素), and insulin-like growth factors IGF-1 and IGF-2, which have been isolated from whole tissue or released into the medium by cells in culture, have varying degrees of specificity and are probably present in serum in small amounts. Many of these growth factors are available commercially as recombinant proteins, some of which also are available in long-form analogues with increased mitogenic activity and stability.

Table 20.4　Constituents of serum (By R. Ian Freshney)

Constituent	Range of concentration a
Proteins and Polypeptides	40-80 mg/mL
Albumin	20-50 mg/mL
Fetuin b	10-20 mg/mL
Fibronectin	1-10 μg/mL
Globulins	1-15 mg/m
Protease inhibitors: α1-antitrypsin, α2-macroglobulin	0.5-2.5 mg/mL
Transferrin	2-4 mg/mL
Growth factors: EGF, PDGF, IGF-1 and 2, FGF, IL-1, IL-6	1-100 ng/mL
Amino acids	0.01-1.0 μM
Lipids	2-10 mg/mL
Cholesterol	10 μM
Fatty acids	0.1-1.0 μM
Linoleic acid	0.01-0.1 μM

(续表)

Constituent	Range of concentration a
Phospholipids	0.7-3.0 mg/mL
Carbohydrates	1.0-2.0 mg/mL
Glucose	0.6-1.2 mg/mL
Hexosamine c	6-1.2 mg/mL
Lactic acids d	0.5-2.0 mg/mL
Pyruvic acid	2-10 μg/mL
Polyamines:	
Putrescine, spermidine	0.1-1.0 μM
Urea	170-300 μg/mL
Inorganics	0.14-0.16 M
Calcium	4-7 mM
Chlorides	100 μM
Iron	10-50 μM
Potassium	5-15 mM
Phosphate	2-5 mM
Selenium	0.01 AM
Sodium	135-155 mM
Zinc	0.1-1.0 AM
Hormones	0.1-200 nM
Hydrocortisone	10-200 nM
Insulin	1-100 ng/mL
Triiodothyronine	20 nM
Thyroxine	100 nM
Vitamins	10 ng-10 μg/mL
Vitamin A	10-100 ng/mL
Folate	5-20 ng/mL

[a] The range of concentrations is very approximate and is intended to convey only the order of magnitude.

[b] In fetal serum only.

[c] Highest in human serum.

[d] Highest in fetal serum.

3. Hormones. Insulin promotes the uptake of glucose and amino acids and may owe its mitogenic effect to this property or to activity via the IGF-1 receptor. IGF-1 and IGF-

2 bind to the insulin receptor, but also have their own specific receptors, to which insulin may bind with lower affinity. IGF-2 also stimulates glucose uptake. Growth hormone may be present in serum particularly fetal serum and, in conjunction with the somatomedins (生长激素介质,IGFs), may have a mitogenic effect. Hydrocortisone(氢化可的松) is also present in serum (particularly fetal bovine serum) in varying amounts and it can promote cell attachment and cell proliferation, but under certain conditions (e. g., high cell density) may be cytostatic(抑制细胞)and can induce cell differentiation.

4. Nutrients and Metabolites. Serum may also contain amino acids, glucose, ketoacids(酮酸), nucleosides(核苷), and a number of other nutrients and intermediary metabolites. These may be important in simple media but less so in complex media, particularly those with higher amino acid concentrations and other defined supplements.

5. Lipids. Linoleic acid(亚油酸), oleic acid(油酸), ethanolamine(乙醇胺), and phosphoethanolamine(磷酸乙醇胺) are present in serum in small amounts, usually bound to proteins such as albumin.

6. Minerals. Serum replacement experiments have also suggested that trace elements and iron, copper, and zinc may be bound to serum protein.

7. Inhibitors. Serum may contain substances that inhibit cell proliferation. Some of these may be artifacts (假象) of preparation (e. g., bacterial toxins from contamination prior to filtration, or antibodies, contained in the γ-globulin fraction, that cross-react with surface epitopes on the cultured cells), but others may be physiological negative growth regulators, such as TGF-β. Heat inactivation removes complement(补体)from the serum and reduces the cytotoxic action of immunoglobulins without damaging polypeptide growth factors, but it may also remove some more labile constituents and is not always as satisfactory as untreated serum.

The sera used most in tissue culture are calf (bovine), fetal bovine, horse, and human serum. Calf (CS) and fetal bovine serum (FBS) are most widely used, the latter particularly for more demanding cell lines and for cloning. Human serum is sometimes used in conjunction with some human cell lines, but it needs to be screened for viruses, such as HIV and hepatitis B. Horse serum is preferred to calf serum by some workers, as it can be obtained from a closed herd(群) and is often more consistent from batch to batch. Horse serum may also be less likely to metabolize polyamines(聚胺), due to lower levels of polyamine oxidase; polyamines are mitogenic for some cells.

Considerable variation may be anticipated between batches of serum. Such variation results from differing methods of preparation and sterilization, different ages and storage conditions, and variations in animal stocks from which the serum was derived, including different strains and disparities (不一致) in pasture(牧场), climate, and other environmental conditions. It is important to select a batch, use it for as long as possible, and replace it, eventually, with one as similar to it as possible. Serum standardization is difficult, as batches vary considerably, and one batch will last only about six months to a year, stored at $-20°C$. Select the type of serum that is most appropriate for your purposes, and request batches to test from a number of suppliers.

Originally, heating was designed to inactivate complement for immunoassays, but it may achieve other effects not yet documented. Often, heat-inactivated serum is used because of the adoption of a previous protocol, without any concrete evidence that it is beneficial. Claims that heat inactivation removes mycoplasma are probably unfounded, although heat treatment may reduce the titer(滴度) for some mycoplasma. Serum is heat inactivated by incubating it for 30 min at 56°C. It may then be dispensed into aliquots and stored at $-20°C$.

The quality of a given serum is assured by the supplier, but the firm's quality control is usually performed with one of a number of continuous cell lines. If your requirements are more discriminating, then you will need to do your own testing. There are four main parameters for testing serum.

(1) Plating Efficiency. During cloning, the cells are at a low density and hence are at their most sensitive, making this a very stringent test. Plate the cells out at 10 to 100 cells/mL, and look for colonies after 10 days to two weeks. Stain and count the colonies, and look for differences in plating efficiency (survival) and colony size (cell proliferation).

(2) Growth curve. A growth curve should be plotted for cell growth in each serum, so that the lag period, doubling time, and saturation density (density at "plateau") can be determined. A long lag implies that the culture have to adapt to the serum, short doubling times are preferable if you want a lot of cells quickly, and a high saturation density will provide more cells for a given amount of serum and will be more economical.

(3) Preservation of Cell Culture Characteristics. Clearly, the cells must do what you require of them in the new serum, whether they are acting as host to a given virus, producing a certain cell product, differentiating, or expressing a characteristic sensitivity to a given drug.

(4) Sterility. Serum from a reputable supplier will have been tested and shown to be free of microorganisms. However, in the unlikely event that a sample of serum is contaminated, but has escaped quality control, the fact that it is contaminated should show up in mycoplasma screening.

Other supplements(其他添加成分)

In addition to serum, tissue extracts and digests have traditionally been used as supplements to tissue culture media. Many such supplements are derived from microbiological culture techniques and autoclavable broths. Bactopeptone(细菌蛋白胨), tryptose (胰蛋白), and lactalbumin hydrolysate(乳白蛋白水解产物)are proteolytic digests(蛋白

裂解消化产物) of beef heart or lactalbumin(乳白蛋白) and contain mainly amino acids and contain nucleosides and other heat-stable tissue constituents, such as fatty acids and carbohydrates.

1. Embryo Extract(胚胎抽提液). Embryo extract is a crude homogenate of 10-day-old chick embryo that is clarified by centrifugation. The crude extract was fractionated to give fractions of either high or low molecular weight. The low-molecular-weight fraction promoted cell proliferation, while the high-molecular-weight fraction promoted pigment and cartilage(软骨) cell differentiation. Embryo extract was originally used as a component of plasma clots(血浆凝块) to promote cell migration from the explant. It has been retained in some organ culture techniques and can still be used in nerve and muscle culture, although in the latter it is gradually being replaced with defined growth factors and matrix components.

2. Conditioned Medium (条件培养基). The survival of low-density cultures could be improved by growing the cells in the presence of feeder layers. This effect is probably due to a combination of effects including conditioning of the substrate and conditioning of the medium by the release into it of small molecular metabolites and growth factors. Using feeder layers and conditioning the medium with embryonic fibroblasts or other cell lines remains a valuable method of culturing difficult cells. Conditioned medium contains both substrate-modifying matrix constituents, like collagen, fibronectin, and proteoglycans, and growth factors, such as those of the heparin-binding group (FGF, etc.), insulin-like growth factors (IGF-1 and-2), PDGF, and several others, in addition to the intermediary metabolites previously proposed. However, conditioned medium adds undefined components to medium and should be eliminated after the active constituents are determined. Still, they often constitute an easy alternative of many months of tedious attempts to optimize the medium and may be the only economical route to successful culture, given that you wish to invest your time dealing with the specific problem of interest, rather than developing culture conditions.

SELECTION OF MEDIUM

All 12 media described in Table 20.2 were developed to support particular cell lines or conditions. Many were developed with L929 or HeLa cells, and Ham's F12 was designed for Chinese hamster ovary (CHO) cells; all now have more general applications and have become classical formulations. Among them, data from suppliers would indicate that RPMI 1640, DMEM, and MEM are the most popular, making up about 75% of sales. Other formulations seldom account for more than 5% of the total; most constitute 2%-3%, although blended DMEM/F12 comes closer, with over 4%.

Eagle's Minimal Essential Medium (MEM) was developed from Eagle's Basal Medium (BME) by increasing the range and concentration of the constituents. For many years, Eagle's MEM had the most general use of all media. Dulbecco's modification of BME (DMEM) was developed for mouse fibroblasts for transformation and virus propagation studies. It has twice the amino acid concentrations of MEM, has four times the vitamin concentrations, and uses twice the HCO_3^- and CO_2 concentrations to achieve better buffering. α-MEM has additional amino acids and vitamins, as well as nucleosides and lipoic acid(硫辛酸); it has been used for a wide range of cell types, including hematopoietic cells. Ham's F12 was developed to clone CHO cells in low serum; it is also used widely, particularly for clonogenic assays and primary culture. Ham's F12 has also been combined with DMEM, 50 : 50 (v/v), to produce a compromise between high concentrations and a wide range of ingredients. The combination has been used for many primary cultures, for more fastidious cell lines, and as a basis for serum-free media.

CMRL 1066, M199, and Waymouth's were all developed to grow L929 cells in a serum-free medium, but have been used alone or in combination with other media, such as DMEM or F12, for a variety of more demanding conditions. RPMI 1640 and Fischer's were developed for lymphoid cells and Fischer's is specifically for L5178Y lymphoma, which has a high folate(叶酸) requirement. RPMI 1640 in particular has quite widespread use, often for attached cells, in spite of being designed for suspension culture and lacking calcium. L15 medium was developed specifically to provide buffering in the absence of HCO_3^- and CO_2. It is often used as a transport and primary culture medium for this reason, but its value was diminished by the introduction of HEPES and the demonstration that HCO_3^- and CO_2 are often essential for optimal cell growth, regardless of the requirement for buffering.

Information regarding the selection of the appropriate medium for a given type of cell is usually available in the literature in articles on the origin of the cell line or the culture of similar cells. Information may also be obtained from the source of the cells. Cell banks, such as ATCC and ECACC, provide information on media used for currently available cell lines, and data sheets can be accessed from their Web sites. Failing this, the choice is made either empirically or by comparative testing of several media as for selection of serum.

Many continuous cell lines, primary cultures of human, rodent, and avian fibroblasts, and cell lines derived from them can be maintained on a relatively simple medium such as Eagle's MEM, supplemented with calf serum. More complex media may be required when a specialized function is being expressed or when cells are subcultured at low seeding density ($<1\times10^3/mL$), as in cloning. Frequently, the more demanding culture conditions that require complex media also require fetal bovine serum rather than calf or horse serum, unless the formulation specifically allows for the omission of serum.

Finally, you may have to compromise in your choice of medium or serum because of cost. If fetal bovine serum seems essential, try mixing it with calf serum. This may allow you to reduce the concentration of the more expensive fetal serum. If you can, leave out serum altogether, or reduce the concentration, and use a serum-free formulation.

QUESTIONS FOR DISCUSSION

1. What physicochemical properties are required for animal cell culture medium?
2. What components is a complete medium for animal cell culture composed of?
3. What is conditioned medium?
4. How to select an appropriate medium for a given type of cell?
5. What is balanced salt solution used for?

知识要点

早期的动物细胞培养采用天然培养基，包括组织或胚胎抽提液、血清和淋巴液。现在的动物细胞培养则采用在成分明确的人工培养基中添加5%～20%的血清来实现。动物细胞培养基中添加一定量的碳酸氢钠使其pH维持在7.0～7.4，平衡盐溶液是最简单的培养基。动物培养基的主要成分有氨基酸、维生素、无机盐、葡萄糖、激素和生长因子、血清和其他有机添加物等。血清中含有促进细胞生长的激素和生长因子以及促进细胞贴附的因子。胚胎抽提液和条件培养基常用于克隆筛选和不易培养的细胞。目前已有多种商品化的基础培养基可供选用，如MEM、DMEM、F12和RPMI-1640等。

CHAPTER 21 PREPARATION AND STERILIZATION OF MEDIA(培养基的配制和灭菌)

All stocks of chemicals and glassware used in tissue culture should be reserved for that purpose alone. Traces of heavy metals or other toxic substances can be difficult to remove and are detectable only by a gradual deterioration in the culture. The requirements of tissue culture washing are higher than those for general glassware. Whenever glass is used, either for storage or for culture, the problem of chemical contaminants leaching out into media or reagents remains, and absolute cleanliness is therefore essential. If the glass surface is to be used for cell propagation, it must not only be clean, but also carry the correct charge. Caustic alkaline detergents render the surface of the glass unsuitable for cell attachment and require subsequent neutralization with dilute HCl or H_2SO_4, but many modern detergents do not alter the glass surface and can be removed completely.

All new apparatus and materials (silicone tubing, filter holders, instruments, etc.) should be soaked in detergent overnight, thoroughly rinsed, and dried. Anything that will corrode in the detergent (mild steel(软钢), aluminum, copper, brass(黄铜), etc.) should be washed directly by hand without soaking and then rinsed and dried. Used items should be rinsed in tap water and immersed in detergent immediately after use. Allow them to soak overnight, and then rinse and dry them. Again, do not expose materials that might corrode to detergent for longer than 30 min.

Ideally, all apparatus used for sterilization should be wrapped in a covering that will allow steam to penetrate, but that will be impermeable to dust, microorganisms, and mites(螨虫). Proprietary bags, bearing sterile-indicating marks that show up after sterilization, are available from clinical sterile-supply services. Tubes and orifices(口) should be covered with tape and paper before packaging, and needles or other sharp points should be shrouded with a glass test tube or other appropriate guard.

All apparatus and liquids that come in contact with cultures or other reagents must be sterilized. Sterilization procedures are designed not just to kill replicating microorganisms but also to eliminate the more resistant spores. Moist heat is more effective than dry heat; however, it does carry a risk of leaving a residue. Dry heat is preferable, but a minimum temperature of 160℃ maintained for 1 h is required. Moist heat (for fluids and perishable(易腐蚀) items) should be maintained at 121℃ for 15-20 min. For moist heat to be effective, steam penetration must be assured, which means that the sterilization chamber must be evacuated prior to steam injection or the air must be completely replaced with steam by downward displacement.

The type of sterilization used will depend on the material. Metallic items are best sterilized by dry heat. Silicone rubber(硅橡胶), polycarbonate(聚碳酸酯), cellulose acetate(醋酸纤维素), and cellulose nitrate(硝酸纤维素) filters, etc., should be autoclaved for 20 min at 121℃ and 100 kPa (1 bar, 15 lb/in^2) with preevacuation and postevacuation steps in the cycle, except when filters are sterilized in a filter assembly. In small bench-top autoclaves and pressure cookers, make sure that the autoclave boils vigorously for 10-15 min before pressurizing to displace all the air. (Take care that enough water is put in at the start to allow for evaporation.) After sterilization, the steam is released and the items are removed to dry off in an oven or rack. To avoid burns, take care releasing steam and handling hot items. Wear elbow-length insulated gloves, and keep your face well clear of escaping steam when you open doors, lids, etc. Use safety locks on autoclaves.

Plastic culture flasks are, on the whole, meant for single use, as detergent renders them unsuitable for cell propagation (in monolayer) and resterilization is difficult. Cells may be reseeded back into the same flask after subculturing them, but this tends to increase the risk of contamination. However, with care, cells that grow in suspension can be subcultured in the same flask for many generations.

BALANCED SALT SOLUTIONS

The formulation of BSS has been discussed previously. The formula for Hanks's BSS contains magnesium chloride in place of some of the sulfate originally recommended; it should be autoclaved below pH 6.5 to prevent calcium and magnesium phosphates from precipitating and should be neutralized before use. Similarly, Dulbecco's PBS is made up without calcium and magnesium (PBSA), which are made up separately (PBSB) and added just before use if required. PBS is often used without the addition of the Ca^{2+} and Mg^{2+} component, and in that form it should be referred to as PBSA or PBS without Ca^{2+} and Mg^{2+}. Most balanced salt solutions contain glucose, which, because it may caramelize(焦糖化) on autoclaving, is best omitted at first and added later. If glucose is prepared as a 100 × concentrate (200 g/L), caramelization during autoclaving is reduced, and it can be used at 5-25 mL/L BSS to give 1-5 g/L.

MEDIA

During the preparation of complex solutions, care must be taken to ensure that all of the constituents dissolve and do not get filtered out during sterilization and that they remain in solution after autoclaving or storage. Concentrated media are often prepared at a low pH (between 3.5 and 5.0) to keep all the constituents in solution, but even then, some precipitation may occur.

Commercial media are supplied as (1) working strength solutions, with or without sodium bicarbonate and glutamine; (2) 10 × concentrates, usually without $NaHCO_3$ and glutamine, which are available as separate concentrates; or (3) powdered media, with or without $NaHCO_3$ and glutamine. Powdered media are the cheapest and not a great deal more expensive than making up medium from your own chemical constituents. Powdered media are quality controlled by the manufacturer for their growth-promoting properties, but not, of course, for sterility. They are mixed very efficiently by ball milling, so, in theory, a pack may be subdivided for use at different times. However, in practice, it is better to match the size of the pack to the volume that you intend to prepare, because once the pack is opened, the contents may deteriorate and some of the constituents may settle.

Preparation of Medium from Powdered Media

Outline

Dissolve the entire contents of the pack in the correct volume of high-purity water, using a magnetic stirrer and adding the powder gradually with constant mixing. When all the constituents have dissolved completely, the medium should be filtered immediately and not allowed to stand, in case any of the constituents precipitate or microbial contamination progresses. The pH is adjusted better when the final constituents (e.g., glutamine, $NaHCO_3$ or serum) have been added to the medium.

Protocol

1. Add appropriate volume of ultrapure water to container.
2. Place container on magnetic stirrer and set to around 200 rpm.
3. Open packet of powder and add contents slowly to container while mixing.
4. Stir until powder is completely dissolved.
5. Check pH and conductivity of a sample and enter in record:
   (a) pH should be within 0.1 unit of expected level for particular medium
   (b) Conductivity should be within 2% of expected value for particular medium
   (c) Discard sample; do not add back to main stock.
6. Connect container with medium to filter via tubing and peristaltic pump.
7. Turn on pump and dispense medium into graduated bottles.
8. Collect samples from beginning, middle, and end of run to test for sterility.
9. Cap and seal bottles.
10. Store at 4℃.
11. Add serum, and correct pH to 7.4 as required, just before use.

For people using smaller amounts (<1.0 l/week) or several different types of medium, smaller volumes may be prepared, complete with glutamine, and filtered directly into storage bottles using a bottle-top filter sterilizer or filter flasks. With this and other negative-pressure filtration systems, some dissolved CO_2 may be lost during filtration, and the pH may rise. Provided that the correct amount of $NaHCO_3$ is in the medium to suit the gas phase, the medium will reequilibrate in the incubator, but this should be confirmed the first time the medium is used.

For large-scale requirements ($>$ 10 1/week), medium can be prepared in a pressure vessel, checked at intervals with a large pipette to determine whether solution is complete, and sterilized by positive pressure through an in-line disposable or reusable filter into a receiver vessel (Figs. 19. 10, 19. 11).

SERUM

Preparing serum is one of the more difficult procedures in tissue culture, because of variations in the quality and consistency of the raw materials and because of the difficulties encountered in sterile filtration. Moreover, serum is also one of the most costly constituents of tissue culture, accounting for 20-30% of the total budget if it is bought from a commercial supplier. Buying sterile serum is certainly the best approach from the point of view of consistency and quality control, but the next protocol is suggested if the serum has to be prepared in the laboratory

Collection and Sterilization of Serum

Outline

Collect blood, allow it to clot, and separate the serum. Filter serum, through filters of gradually reducing porosity. Bottle and freeze filtered serum.

Protocol

1. Collection. Arrangements may be made to collect whole blood from a slaughter (屠宰) house. The blood should be collected directly from the bleeding carcass (尸体) and not allowed to lie around after collection. Alternatively, blood may be withdrawn from live animals under proper veterinary supervision. If the procedure is done carefully, blood may be collected aseptically.

2. Clotting. Allow the blood to clot by having it stand overnight in a covered container at 4°C. This so-called natural clot serum is superior to serum that is physically separated from the blood cells by centrifugation and defibrination (脱纤维蛋白), as platelets release growth factor into the serum during clotting. Separate the serum from the clot, and centrifuge the serum at 2,000 g for 1 h to remove sediment.

3. Sterilization. Serum is usually sterilized by filtration through a sterilizing filter of 0.1 μm porosity, but because of its viscosity and high particulate content, the serum should be passed through a graded series of fiber glass or other prefilter before passing

through the final sterilizing filter. Only the last filter, a 142-350 mm in-line disc filter or equivalent disposable filter, need be sterile.

QUESTIONS FOR DISCUSSION

1. How to prepare and sterilize a medium?
2. How to prepare and sterilize serum?

知识要点

用于配制动物细胞培养基的器具器皿必须清洗干净,并采用适当的方法进行灭菌消毒,以避免化学污染和微生物污染。平衡盐溶液一般采用湿热灭菌的方法,而培养基和血清则需要采用微孔滤膜过滤除菌。干粉培养基的配制很简单,首先将所有粉末溶解于水中,然后抽滤除菌。自制血清的主要过程包括抽血、凝固、收集血清、离心和多次抽滤除菌等步骤。

CHAPTER 22　SERUM-FREE MEDIA(无血清培养基)

Although most cell lines still require that the medium in which they are growing be supplemented with serum, in many instances cultures may now be propagated in serum-free media. While a degree of selection may have been involved in the adaptation of continuous cell lines to serum-free conditions, the MCDB series of media, Sato's DMEM/F12-based media, and others based on RPMI 1640 demonstrated that serum could be reduced or omitted without apparent cellular adaptation if appropriate nutritional and hormonal modifications were made to the media. These also provided selective conditions for primary culture of particular cell types. Specific formulations (e.g., MCDB 110) were derived to culture human fibroblasts, many normal and neoplastic murine and human cells, lymphoblasts, and several different primary cultures in the absence of serum, with, in several cases, some protein added. This list now covers a wide range of cell types, and many of the media are available commercially.

DISADVANTAGES OF SERUM

Using serum in a medium has a number of disadvantages:

1. Physiological Variability. The major constituents of serum such as albumin(血清白蛋白) and transferrin(转铁蛋白), are known, but serum also contains a wide range of minor components that may have a considerable effect on cell growth. These components include nutrients (amino acids, nucleosides, sugars, etc.), peptide growth factors, hormones, minerals, and lipids, the concentrations and actions of which have not been fully determined.

2. Shelf Life and Consistency. Serum varies from batch to batch, and at best a batch will last one year, perhaps deteriorating during that time. It must then be replaced with another batch that may be selected as similar, but will never be identical, to the first batch.

3. Quality Control. Changing serum batches requires extensive testing to ensure that the replacement is as close as possible to the previous batch. This can involve several tests (for growth, plating efficiency, and special functions) and a number of different cell lines.

4. Specificity. If more than one cell type is used, each type may require a different batch of serum, so that several batches must be held on reserve simultaneously. Cocul-

turing different cell types will present an even greater problem.

5. Availability. Periodically, the supply of serum is restricted due to drought in the cattle-rearing areas, the spread of disease among the cattle, or economic or political reasons. Today, demand is increasing, and it will probably exceed supply unless the majority of commercial users are able to adopt serum-free media.

6. Downstream Processing. To anyone interested in recovering cell products, the presence of serum creates a major obstacle to purification and may even limit the pharmaceutical acceptance of the product.

7. Contamination. Serum is frequently contaminated with viruses, many of which may be harmless to cell culture, but represent an additional unknown factor outside the operator's control. Fortunately, improvements in serum sterilization techniques have virtually eliminated the risk of mycoplasma infection from sera from most reputable suppliers, but this cannot be guaranteed for viral infection, in spite of claims that some filters may remove viruses. Because of the risk of spreading bovine spongiform encephalitis（疯牛病）among cattle, cell cultures and serum shipped to the United States or Australia require information on the country of origin and the batch number of the serum.

8. Cost. Cost is often cited as a disadvantage of serum supplementation. Certainly, serum constitutes the major part of the cost of a bottle of medium, but if it is replaced by defined constituents, the cost of these may be as high as that of the serum.

9. Growth Inhibitors. As well as its growth-promoting activity, serum contains growthinhibiting activity, and although stimulation usually predominates, the net effect of the serum is an unpredictable combination of both inhibition and stimulation of growth. While substances such as PDGF may be mitogenic to fibroblasts, other constituents of serum can be cytostatic. Hydrocortisone, present at around 10^{-8} M in fetal serum, is cytostatic（抑制细胞）to many cell types, such as glia and lung epithelium, at high cell densities (though it may be mitogenic at low cell densities), and TGF-β, released from platelets, is cytostatic to many epithelial cells.

10. Standardization. Standardization of experimental and production protocols is difficult, both at different times and among different laboratories, due to batch-to-batch variations in serum.

ADVANTAGES OF SERUM-FREE MEDIA

As well as eliminating the foregoing（前述）disadvantages of serum, serum-free media have two major positive benefits.

Selective Media

One of the major advantages of the control over growth-promoting activity afforded by serum-free media is the ability to make a medium selective for a particular cell type.

The long-standing (长期) problem of overgrowth by stromal fibroblasts can now be tackled (解决) effectively in breast and skin cultures by using MCDB 170 and 153, melanocytes can be cultivated in the absence of fibroblasts and keratinocytes, and separate lineages and even stages of development may be selected in hematopoietic cells by choosing the correct growth factor or group of growth factors.

Regulation of Proliferation and Differentiation

Add to the ability to select for a specific cell type the possibility of switching from a growth factor, after necessary amplification of the culture, to a differentiation factor or set of factors, and the amplified culture may then be made to perform one or more specialized functions.

DISADVANTAGES OF SERUM-FREE MEDIA

Serum-free media are not without disadvantages:

1. Multiplicity of Media. Each cell type appears to require a different recipe, and cultures from malignant tumors may vary in requirements from tumor to tumor, even within one class of tumors. While this degree of specificity may be an advantage to those isolating specific cell types, it presents a problem for laboratories maintaining cell lines of several different origins.

2. Selectivity. Unfortunately, the transition to serum-free conditions, however desirable, is not as straightforward as it seems. Some media may select a sub-lineage that is not typical of the whole population, and even in continuous cell lines, some degree of selection may still be required. Cells at different stages of development (e. g., stem cells vs. committed precursor cells) may require different formulations, particularly in the growth factor and cytokine components.

3. Reagent Purity. The removal of serum also requires that the degree of purity of reagents and water and the degree of cleanliness of all apparatus be extremely high, as the removal of serum also removes the protective, detoxifying action that some serum proteins may have.

4. Cell Proliferation. Growth is often slower in serum-free media, and fewer generations are achieved with finite cell lines.

5. Availability. Although improving steadily, the availability of properly controlled serum-free media is quite limited, and the products are often more expensive than conventional media.

REPLACEMENT OF SERUM

The essential factors in serum have been described and include (1) adhesion factors such as fibronectin; (2) peptides, such as insulin, PDGF, and TGF-β, that regulate growth and differentiation; (3) essential nutrients, such as minerals, vitamins, fatty acids, and intermediary metabolites; and (4) hormones, such as insulin, hydrocortisone, estrogen(雌激素), and triiodothyronine(三碘甲状腺氨酸), that regulate membrane transport, phenotypic status, and the constitution of the cell surface. While some of these are catered for(供养) in the formulation of serum-free media others are not and may require an alteration in procedures.

Hormones

Hormones that have been used to replace serum include growth hormone(生长激素, somatotropin) at 50 ng/mL, insulin at 1-10 U/mL, which enhances plating efficiency in a number of different cell types, and hydrocortisone, which improves the cloning efficiency of glia and fibroblasts and has been found necessary for the maintenance of epidermal keratinocytes and some other epithelial cells. Various combinations of estrogen (雌激素), androgen(雄激素), or progesterone(黄体素) with hydrocortisone(氢化可的松) and prolactin(催乳素) at around 10 nM can be shown to be necessary for the maintenance of mammary epithelium.

Growth Factors

The family of polypeptides that has been found to be mitogenic(促有丝分裂) *in vitro* is now quite extensive and includes the heparin-binding growth factors(肝素结合生长因子), EGF, PDGF, IGF-1 and-2, and the interleukins that are active in the 1-10 ng/mL range. Growth factors and cytokines tend to have a wide-ranging specificity, except for some that are active in the hematopoietic system. Keratinocyte growth factor (KGF), besides showing activity with epidermal keratinocytes, will also induce proliferation and differentiation in prostatic(前列腺) epithelium. Hepatocyte growth factor (HGF) is mitogenic for hepatocytes, but is also morphogenic for kidney tubules. Growth factors and cytokines acquire their specificity by virtue of the fact that their production is localized and that they have a limited range. Most act as paracrine(旁分泌) factors (they are active on adjacent cells) and not by systemic distribution in the blood. Growth factors may act synergistically(协同) or additively with each other or with other hormones and paracrine(旁分泌)factors, such as prostaglandin(前列腺素) $F_{2\alpha}$ and hydrocortisone.

Nutrients in Serum

Iron, copper, and a number of minerals have been included in serum-free recipes, although evidence that some of the rarer minerals are required is still lacking. Selenium (Na_2SeO_3), at around 20 nM, is found in most formulas, and there appears to be some requirement for lipids or lipid precursors such as choline(胆碱), linoleic acid(亚油酸), ethanolamine(乙醇胺), or phosphoethanolamine(磷酸乙醇胺).

Proteins and Polyamines(聚胺)

The inclusion in medium of proteins such as bovine serum albumin (BSA) or tissue extracts often increases cell growth and survival, but adds undefined constituents to the medium. BSA, fatty acid free, is used at 1-10 mg/mL. Transferrin, at around 10 ng/mL, is required as a carrier for iron and may also have a mitogenic role. Putrescine (腐胺) has been used at 100 nM.

Matrix

One of the properties of serum is to provide a number of proteins, such as fibronectin, that coat the plastic and make it more adhesive. In the absence of serum, the plastic substrate may need to be coated with fibronectin or polylysine(多聚赖氨酸).

SUBCULTURE OF CELLS IN SERUM-FREE MEDIUM

Adhesion Factors

When serum is removed, it may be necessary to treat the plastic growth surface with fibronectin (25-50 μg/mL) or laminin (1-5 μg/mL), added directly to the medium. Pretreating the plastic with poly-L-lysine, 1 mg/mL, and washing off was shown to enhance the survival of human diploid fibroblasts.

Protease Inhibitors

Following trypsin-mediated subculture, the addition of serum inhibits any residual proteolytic activity. Consequently, protease inhibitors such as soya bean trypsin inhibitor(大豆胰蛋白酶抑制剂)or 0.1 mg/mL aprotinin(抑肽酶) must be added to serum-free media after subculture. Furthermore, because crude trypsin is a complex mixture of proteases, some of which may require different inhibitors, it is preferable to use pure trypsin followed by a trypsin inhibitor. Alternatively, one may wash cells by centrifugation to remove trypsin. It is possible that Pronase(链霉蛋白酶)can be inactivated by dilution without subsequent neutralization in serum-free conditions.

Trypsin Temperature

Special care may be required when trypsinizing cells from serum-free media, as the cells are more fragile and may need to be chilled to reduce damage.

SELECTION OF SERUM-FREE MEDIUM

If the reason for using a serum-free medium is to promote the selective growth of a particular type of cell, then that reason will determine the choice of medium. If the reason is simply to avoid using serum with continuous cell lines, such as CHO cells or hybridomas(杂交瘤), in order to reduce the likelihood of contamination or serum proteins in the cell product, then the choice will be wider, and there will be several commercial sources to choose. When a cell line is obtained from the originator or a reputable cell bank, the supplier will recommend the appropriate medium, and the only reason to change will be if the medium is unavailable or is incompatible with other stocks. If possible, it is best to stay with the originator's recommendation, as this may be the only way to ensure that the line exhibits its specific properties.

DEVELOPMENT OF SERUM-FREE MEDIUM

There are two general approaches to the development of a serum-free medium for a particular cell line or primary culture. The first is to take a known recipe for a related cell type, with or without 10%-20% dialyzed serum(透析血清), and alter the constituents individually or in groups, while reducing the serum, until the medium is optimized for your own particular requirement. This was the approach adopted by Ham and co-workers and generally will provide optimal conditions. If a group of compounds is found to be effective in reducing serum supplementation, the active constituents may be identified by the systematic omission of single components and then the concentrations of the essential components optimized. However, this is a very time-consuming and laborious process, involving growth curves and clonal growth assays at each stage, and it is not unreasonable to expect to spend at least three years developing a new medium for a new type of cell. There is no doubt, however, that the need for consistent and defined conditions for the investigation of regulatory processes governing growth and differentiation, the pressure from biotechnology to make the purification of products easier, and the need to eliminate all sources of potential infection will eventually force the adoption of serum-free media on a more general scale.

QUESTIONS FOR DISCUSSION

1. List some disadvantages of serum when used in cell culture.
2. What is a serum-free medium? List its advantages and disadvantages.
3. How to develop a serum-free medium?

知识要点

血清是动物细胞培养基中促进细胞生长和贴壁所必不可少的添加成分，但是血清中也存在一些抑制细胞生长的细胞因子和补体成分，是支原体和病毒污染的主要来源，而且血清的成分复杂，品质和质量也不稳定，限制了血清的应用。因而人们开始研制无血清培养基，使培养基的成分能够完全明确，实现对细胞生长和分化的人工调控。但是，无血清培养基的成本往往更高，要添加种类繁多的生长因子和营养成分，对试剂纯度的要求更严格，对细胞的选择性更强，甚至使传代更复杂，需要额外添加胰蛋白酶抑制剂，这些都限制了无血清培养基的普及和应用。

CHAPTER 23　PRIMARY CELL CULTURE(原代细胞培养)

TYPES OF PRIMARY CELL CULTURE

　　A primary culture is that stage of the culture following isolation of the cells, but before the first subculture. There are three stages to consider: (a) isolation of the tissue, (b) dissection and/or disaggregation, and (c) culture following seeding into the culture vessel. Following isolation, a primary cell culture may be obtained either by allowing cells to migrate out from fragments of tissue adhering to a suitable substrate or by disaggregating the tissue mechanically or enzymatically to produce a suspension of cells, some of which will ultimately attach to the substrate. It appears to be essential for most normal untransformed cells-with the exception of hematopoietic cells(造血细胞)- to attach to a flat surface in order to survive and proliferate with maximum efficiency. Transformed cells, on the other hand, particularly cells from transplantable animal tumors, are often able to proliferate in suspension.

　　The enzymes used most frequently are crude preparations of trypsin(胰蛋白酶), collagenase(胶原酶), elastase(弹性蛋白酶), hyaluronidase(透明质酸酶), DNase(脱氧核糖核酸酶), pronase(链霉蛋白酶), dispase(分散酶), alone or in various combinations. Crude preparations are often more successful than purified enzyme preparations, as the former contain other proteases as contaminants, although the latter are generally less toxic and more specific in their action. Trypsin and pronase give the most complete disaggregation, but may damage the cells. Collagenase and dispase, on the other hand, give incomplete disaggregation, but are less harmful. Hyaluronidase can be used in conjunction with collagenase to digest the intracellular matrix, and DNase is employed to disperse DNA released from lysed cells, as it tends to impair proteolysis and promote reaggregation.

　　Although each tissue may require a different set of conditions, certain requirements are shared by most primary cultures:

　　(1) Fat and necrotic tissue is best removed during dissection.

　　(2) The tissue should be chopped finely with sharp instruments to cause minimum damage.

　　(3) Enzymes used for disaggregation should be removed subsequently by gentle

centrifugation.

(4) The concentration of cells in the primary culture should be much higher than that normally used for subculture, since the proportion of cells from the tissue that survives in primary culture may be quite low.

(5) A rich medium, such as Ham's F12, is preferable to a simple medium, such as Eagle's MEM, and, if serum is required, fetal bovine often gives better survival than does calf or horse. Isolation of specific cell types will probably require selective media.

(6) Embryonic tissue is preferable, as it disaggregates more readily, yields more viable cells, and proliferates more rapidly in primary culture than adult tissue does.

ISOLATION OF THE TISSUE

Before attempting to work with human or animal tissue, make sure that your work fits within medical ethical rules or current legislation on experimentation with animals. For example, in the United Kingdom, the use of embryos or fetuses beyond 50% gestation(怀孕) or incubation is regulated under the Animal Experiments Act of 1986. Work with human biopsies or fetal material usually requires the consent of the local ethical committee and the patient and/or his or her relatives.

An attempt should be made to sterilize the site of the dissection with 70% alcohol if the site is likely to be contaminated (e.g., skin). Remove the tissue aseptically and transfer it to the tissue culture laboratory in BSS or medium as soon as possible. Do not dissect animals in the tissue culture laboratory, as the animals may carry microbial contamination. If a delay in transferring the tissue is unavoidable, it can be held at 4℃ for up to 72 h, although a better yield will result from a quicker transfer.

Isolation of Mouse Embryo

Mouse embryos are a convenient source of cells for undifferentiated fibroblastic cultures. They are often used as feeder layers.

Outline

Remove uterus(子宫) aseptically from timed pregnant mouse and dissect out embryos.

Protocol

1. Induction of estrus (发情). If males and females are housed separately, then when they are put together for mating, estrus will be induced in the female 3 d later, when the maximum number of successful matings will occur. This process enables the planned production of embryos at the appropriate time. The timing of successful matings may be determined by examining the females' vaginas(阴道) each morning for a hard mucous plug(黏液栓).

2. Dating the embryos. The day of detection of a vaginal plug (阴道栓), or the

"plug date," is noted as day zero, and the development of the embryos is timed from this date. Full term is about 19-21 d. The optimal age for preparing cultures from a whole disaggregated embryo is around 13 d, when the embryo is relatively large but still contains a high proportion of undifferentiated mesenchyme, which is the main source of the culture.

3. Kill the mouse by cervical dislocation(颈椎脱位法), and swab the ventral surface liberally with 70% alcohol. (Fig. 23.1A)

4. Tear the ventral skin transversely at the median line just over the diaphragm(横膈膜)(Fig. 23.1B), and, grasping the skin on both sides of the tear, pull in opposite directions to expose the untouched ventral surface of the abdominal wall (Fig. 23.1C).

5. Cut longitudinally along the median line of the exposed abdomen with sterile scissors, revealing the viscera(内脏). At this stage, the uteri filled with embryos are obvious in the posterior abdominal cavity.

6. Dissect out the uteri into a 25-ml or 50-ml screw-capped tube containing 10 or 20 mL BSS. (Fig. 23.1D-F) Antibiotics may be added to the BSS when there is a high risk of infection.

7. Take the intact uteri to the tissue culture laboratory, and transfer them to a fresh dish of sterile DBSS.

8. Dissect out the embryos (Fig. 23.1):

(1) Tear the uterus with two pairs of sterile forceps, keeping the points of the forceps close together to avoid distorting the uterus and bringing too much pressure to bear on the embryos.

(2) Free the embryos from the membranes and placenta and place them to one side of the dish to bleed.

9. Transfer the embryos to a fresh dish. If a large number of embryos is required, it may be helpful to place the dish on ice.

Isolation of Chick Embryo

Chick embryos are easier to dissect, as they are larger than the equivalent stage of mouse embryo. Like mouse embryos, chick embryos are used to provide predominantly mesenchymal cell primary cultures for cell proliferation analysis, to provide feeder layers, and as a substrate for viral propagation. Because of their larger size, it is also easier to dissect out individual organs to generate specific cell types, such as hepatocytes, cardiac muscle, and lung epithelium.

Outline

Remove embryo aseptically from the egg and transfer to dish.

Protocol

1. Incubate the eggs at 38.5℃ in a humid atmosphere, and turn the eggs through

180° daily. Although hen's eggs hatch at around 20 to 21 d, the lengths of their developmental stages are different from those of mouse embryos. For a culture of dispersed cells from the whole embryo, the egg should be taken at about 8 d, and for isolated-organ rudiments, at about 10-13 d.

2. Swab the egg with 70% alcohol, and place it with its blunt end facing up in a small beaker. (Fig. 23.2a)

3. Crack the top of the shell and peel the shell off to the edge of the air sac using sterile forceps. (Fig. 23.2b)

4. Resterilize the forceps (i.e., dip them in alcohol, burn off the alcohol, and cool the forceps in sterile BSS), and then use the forceps to peel off the white shell membrane to reveal the chorioallantoic membrane(尿囊绒毛膜,CAM) below, with its blood vessels. (Fig. 23.2c, d)

5. Pierce the CAM with sterile curved forceps, and lift out the embryo by grasping it gently under the head. Do not close the forceps completely, or else the neck will be severed(切断). (Fig. 23.2e-g)

6. Transfer the embryo to a 9-cm Petri dish containing 20 mL of DBSS.

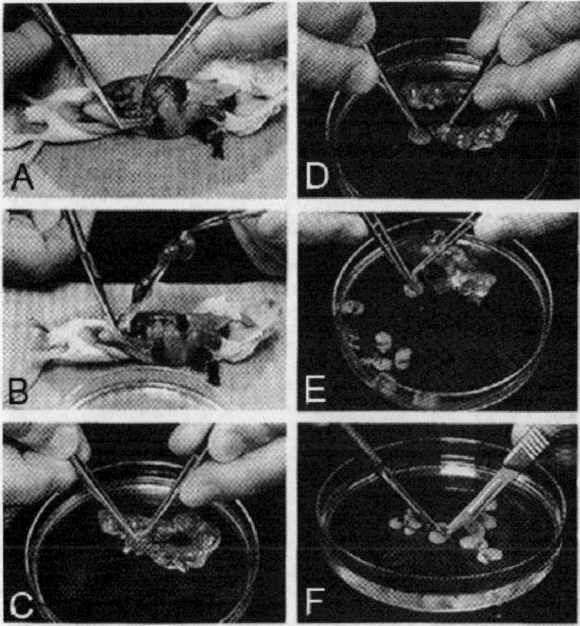

Fig. 23.1 Mouse dissection. Stages in the aseptic removal of mouse embryos for primary culture. A, swabbing the abdomen, tearing the skin to expose the abdominal wall, opening the abdomen and finding the uterus *in situ*. B, removing the uterus. C and D, dissecting the embryos from the uterus. E, removing the membranes. F, chopping the embryos. (By R. Ian Freshney)

CYTOTECHNOLOGY（细胞工程技术）

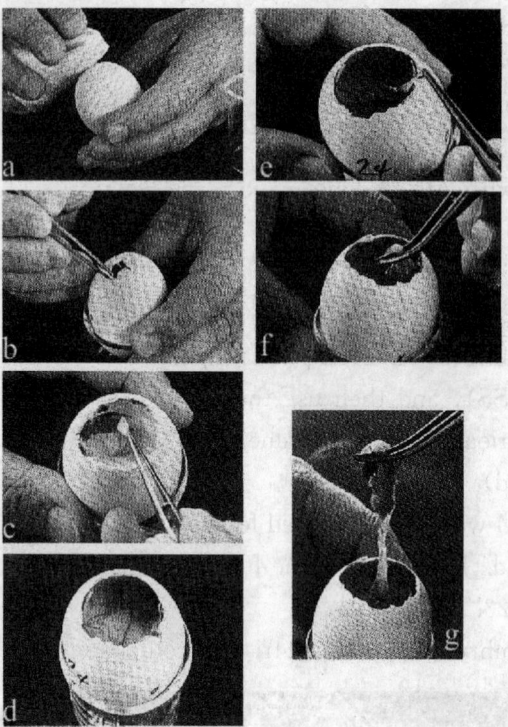

Fig. 23.2 Removing a chick embryo from an egg. (By R. Ian Freshney)

PRIMARY CULTURE

Several techniques have been devised for primary culture of isolated tissue. These techniques can be divided into purely mechanical techniques, involving dissection with or without some form of maceration(浸软), and techniques utilizing enzymic disaggregation. (Fig. 23.5)

Primary Explant Culture

The primary-explant technique was the original method developed by Harrison [1907], Carrel [1912], and others for initiating a tissue culture. As originally performed, a fragment of tissue was embedded in blood plasma or lymph, mixed with heterologous serum and embryo extract, and placed on a coverslip which was inverted over a concavity(中央凹陷) slide. The clotted plasma held the tissue in place, and the explant could be examined with a conventional microscope. The embryo extract and serum, together with the plasma, supplied nutrients and stimulated migration out of the explant across the solid substrate. The heterologous serum was used to promote clotting of the plasma. This technique is still used, but has been largely replaced by the simplified method described in Protocol 23.3.

Outline

The tissue is chopped finely and rinsed, and the pieces are seeded onto the surface of a culture flask or Petri dish in a small volume of medium with a high concentration of serum, such that surface tension holds the pieces in place until they adhere spontaneously to the surface. (Fig. 23.3) Once this is achieved, outgrowth of cells usually follows. (Fig. 23.4)

Protocol

1. Transfer tissue to fresh, sterile BSS, and rinse.

2. Transfer the tissue to a second dish; dissect off unwanted tissue, such as fat or necrotic material; and chop finely with crossed scalpels (see Fig. 23.3) to about 1-mm cubes.

3. Transfer by pipette (10-20 mL, with wide tip) to a 15-or 50-ml sterile centrifuge tube or universal container. (Wet the inside of the pipette first with BSS, or else the pieces will stick.) Allow the pieces to settle.

4. Wash by resuspending the pieces in BSS, allowing the pieces to settle, and removing the supernatant fluid. Repeat this step two more times.

5. Transfer the pieces (remember to wet the pipette) to a culture flask, with about 20-30 pieces per 25-cm^2 flask.

6. Remove most of the fluid, and add about 1 mL of growth medium per 25-cm^2 growth surface. Tilt the flask gently to spread the pieces evenly over the growth surface.

7. Cap the flask, and place it in an incubator or hot room at 37℃ for 18-24 h.

8. If the pieces have adhered, then the medium volume may be made up gradually over the next 3-5 d to 5 mL per 25 cm^2 and then changed weekly until a substantial outgrowth of cells is observed. (See Fig. 23.4)

9. The explants may then be picked off from the center of the outgrowth with a scalpel and transferred by prewetted pipette to a fresh culture vessel. (Then return to step 7)

10. Replace the medium in the first flask until the outgrowth has spread to cover at least 50% of the growth surface, at which point the cells may be subcultured.

This technique is particularly useful for small amounts of tissue, such as skin biopsies, for which there is a risk of losing cells during mechanical or enzymatic disaggregation. Its disadvantages lie in the poor adhesiveness of some tissues and the selection of cells in the outgrowth. In practice, however, most cells-fibroblasts, myoblasts, glia, and epithelia particularly those from the embryo, migrate out successfully.

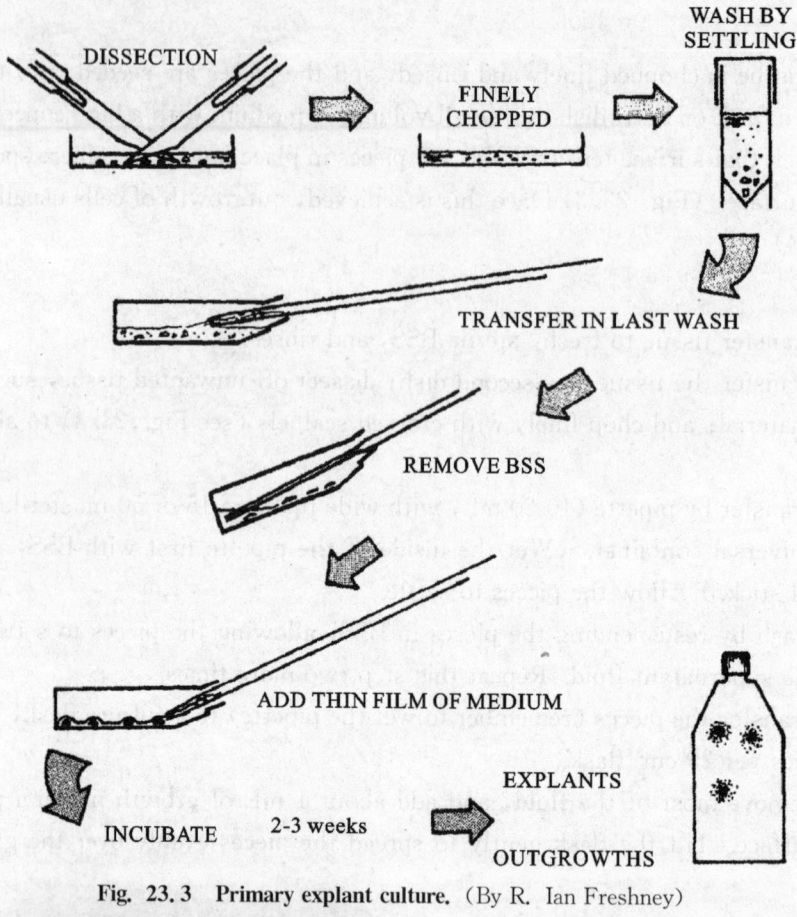

Fig. 23.3 Primary explant culture. (By R. Ian Freshney)

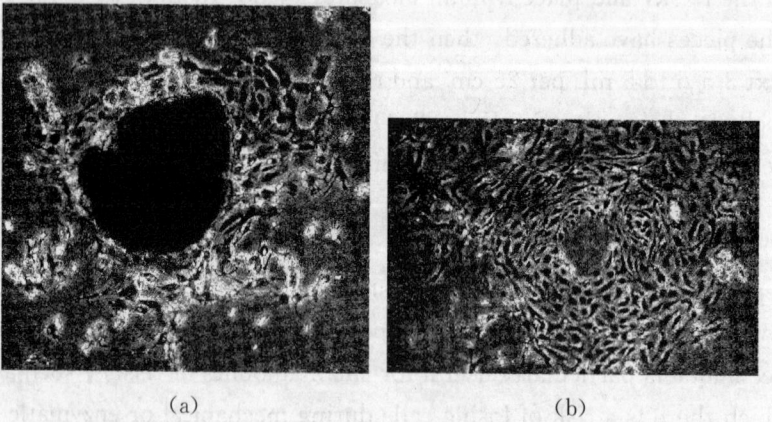

Fig. 23.4 **Primary explant.** Primary explant culture from mouse squamous skin carcinoma. (a) Explant and early stage outgrowth about 3 d after explantation. (b) Outgrowth after removal of explant, about 7 d after explantation. (By R. Ian Freshney)

Enzymatic Disaggregation Culture

Cell-cell adhesion in tissues is mediated by a variety of homotypic interacting glycopeptides (cell adhesion molecules, or CAMs), some of which are calcium dependent (cadherins(钙粘素) and hence are sensitive to chelating agents such as EDTA or EGTA. Integrins, which bind to the RGD motif in extracellular matrix, also have Ca^{2+}-binding domains and are affected by Ca^{2+} depletion. Intercellular matrix and basement membranes also contain other glycoproteins, such as fibronectin and laminin, which are protease sensitive, and proteoglycans(蛋白聚糖), which are less so, and can sometimes be degraded by glycanases(糖苷酶), such as hyaluronidase(透明质酸酶)or heparinase (肝素酶). The easiest approach is to proceed from a simple disaggregation solution to a more complex solution with trypsin alone or trypsin/EDTA as a starting point, adding other proteases to improve disaggregation, and deleting trypsin if necessary to increase viability. In general, increasing the purity of an enzyme will give better control and less toxicity with increased specificity, but may result in less disaggregation activity.

Mechanical and enzymatic disaggregation of the tissue avoids problems of selection by migration and yields a higher number of cells that are more representative of the whole tissue in a shorter time. However, just as the primary-explant technique selects on the basis of cell migration, dissociation techniques select cells resistant to the method of disaggregation and still capable of attachment.

Embryonic tissue disperses more readily and gives a higher yield of proliferating cells than newborn or adult tissue does. The increasing difficulty in obtaining viable proliferating cells with increasing age is due to several factors, including the onset of differentiation, an increase in fibrous connective tissue and extracellular matrix, and a reduction of the undifferentiated proliferating cell pool. When procedures of greater severity are required to disaggregate the tissue (e. g. longer trypsinization or increased agitation), the more fragile components of the tissue may be destroyed. In fibrous tumors, for example, it is very difficult to obtain complete dissociation with trypsin while still retaining viable carcinoma cells.

Trypsin

The choice of trypsin grade to use has always been difficult, as there are two opposing trends: (1) The purer the trypsin, the less toxic it becomes, and the more predictable its action; (2) the cruder the trypsin, the more effective it may be, due to other proteases. In practice, a preliminary test experiment may be necessary to determine the optimum grade for viable cell yield, as the balance between sensitivity to toxic effects and disaggregation ability may be difficult to predict. Crude trypsin is by far the most common *enzyme* used in tissue disaggregation, as it is tolerated quite well by many cells, it is effective for many tissues, and any residual activity left after washing is neutralized by the serum of the culture medium, or by a trypsin inhibitor (e. g. , soya bean trypsin inhibitor) when serum-free medium is used.

It is important to minimize the exposure of cells to active trypsin in order to preserve maximum viability. Hence, when whole tissue is being trypsinized at 37℃, dissociated cells should be collected every half hour, and the trypsin should be removed by centrifugation and neutralized with serum in medium. Soaking the tissue for 6-18 h in trypsin at 4℃ allows penetration with minimal tryptic activity, and digestion may then proceed for a much shorter time (i.e., 20-30 min) at 37℃. Although the cold-trypsin method gives a higher yield of viable cells and requires less effort, the warm trypsin method is still used extensively (Fig. 23.5).

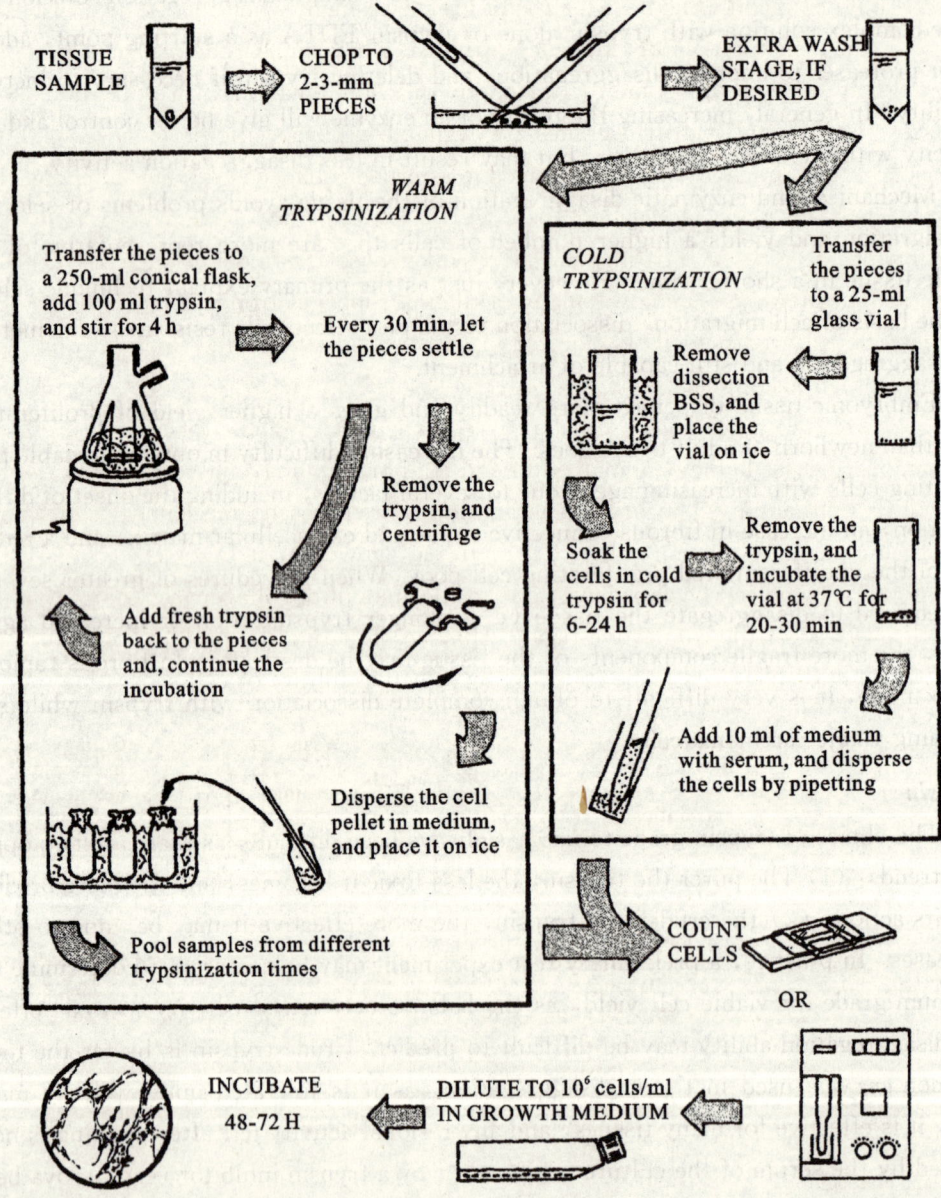

Fig. 23.5 Trypsin disaggregation. Preparation of primary culture by disaggregation in trypsin. The warm-trypsin method is shown on the left, and the cold-trypsin method is shown on the right. (By R. Ian Freshney)

This technique is useful for the disaggregation of large amounts of tissue in a relatively short time, particularly for whole mouse embryos or chick embryos. It does not work as well with adult tissue, in which there is a lot of fibrous connective tissue, and mechanical agitation can be damaging to some of the more sensitive cell types, such as epithelium.

Collagenase

This technique is very simple and effective for many tissues: embryonic, adult, normal, and malignant (Fig. 23.6). It is of greatest benefit when the tissue is either too fibrous or too sensitive to allow the successful use of trypsin. Crude collagenase is often used and may depend, for some of its action, on contamination with other nonspecific proteases. More highly purified grades are available if nonspecific proteolytic activity is undesirable, but they may not be as effective as crude collagenase. Some cells, particularly macrophages, may adhere to the first flask during the collagenase incubation. Transferring the cells to a fresh flask after collagenase treatment (and subsequent removal of the collegenase) removes many of the macrophages from the culture. The first flask may be cultured as well, if required. Light trypsinization will remove any adherent cells other than macrophages.

Disaggregation in collagenase has proved particularly suitable for the culture of human tumors, mouse kidney, human adult and fetal brain, lung, and many other tissues, particularly epithelium. The process is gentle and requires no mechanical agitation or special equipment. With more than 1 g of tissue, however, it becomes tedious at the dissection stage and can be expensive, due to the amount of collagenase required. It will also release most of the connective tissue cells, accentuating the problem of fibroblastic outgrowth, so it may need to be followed by selective culture or cell separation.

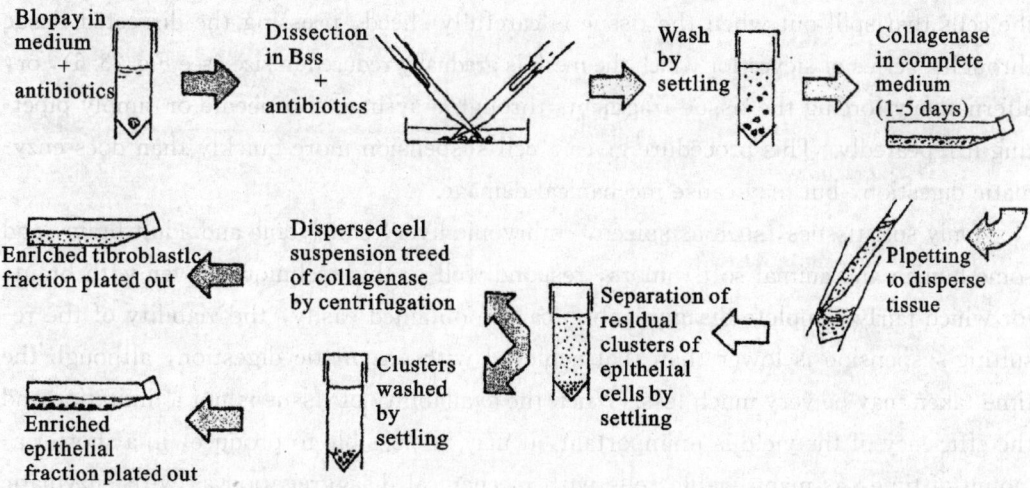

Fig. 23.6 **Disaggregation of tissue for primary culture by collagenase.** (By R. Ian Freshney)

Other Enzymatic Procedures

Disaggregation in trypsin can be damaging (e. g. , to some epithelial cells) or ineffective (e. g. , for very fibrous tissue, such as fibrous connective tissue), so attempts have been made to utilize other enzymes. Since the extracellular matrix often contains collagen, particularly in connective tissue and muscle, collagenase has been the obvious choice; Other bacterial proteases, such as pronase(链霉蛋白酶) and dispase(分散酶) have also been used with varying degrees of success. The participation of carbohydrate in intracellular adhesion has led to the use of hyaluronidase(透明质酸酶)and neuraminidase(神经氨酸苷酶) in conjunction with collagenase(胶原酶). With the selection now available, screening available samples is the only option if trypsin, collagenase, dispase, pronase, hyaluronidase, and DNase(脱氧核糖核酸酶), alone and in combinations, do not prove to be successful.

The addition of hyaluronidase aids disaggregation by attacking terminal carbohydrate residues on the surface of the cells. This combination has been found to be particularly effective for dissociating rat or rabbit liver, by perfusing the whole organ *in situ* and completing the disaggregation by stirring the partially digested tissue in the same enzyme solution for a further 10-15 min, if necessary. This technique gives a good yield of viable hepatocytes and is a good starting point for further culture.

Mechanical Disaggregation Culture

The outgrowth of cells from primary explants is a relatively slow process and can be highly selective. Enzymatic digestion, discussed earlier in this chapter, is rather more labor intensive, though, potentially, it gives a culture that is more representative of the tissue. As there is a risk of damaging cells during enzymatic digestion, many people have chosen to use the alternative of mechanical disaggregation-for example, collecting the cells that spill out when the tissue is carefully sliced, pressing the dissected tissue through a series of sieves for which the mesh is gradually reduced in size (see Fig. 23. 6), or, alternatively forcing the tissue fragments through a syringe and needle or simply pipetting it repeatedly. This procedure gives a cell suspension more quickly than does enzymatic digestion, but may cause mechanical damage.

Only soft tissues, such as spleen, embryonic liver, embryonic and adult brain, and some human and animal soft tumors, respond well to this technique. Even with brain, for which fairly complete disaggregation can be obtained easily, the viability of the resulting suspension is lower than that achieved with enzymatic digestion, although the time taken may be very much less. When the availability of tissue is not a limitation and the efficiency of the yield is unimportant, it may be possible to produce, in a shorter amount of time, as many viable cells with mechanical disaggregation as with enzymatic digestion, but at the expense of very much more tissue.

PART Ⅱ　CYTOTECHOLOGY IN ANIMALS（动物细胞工程）

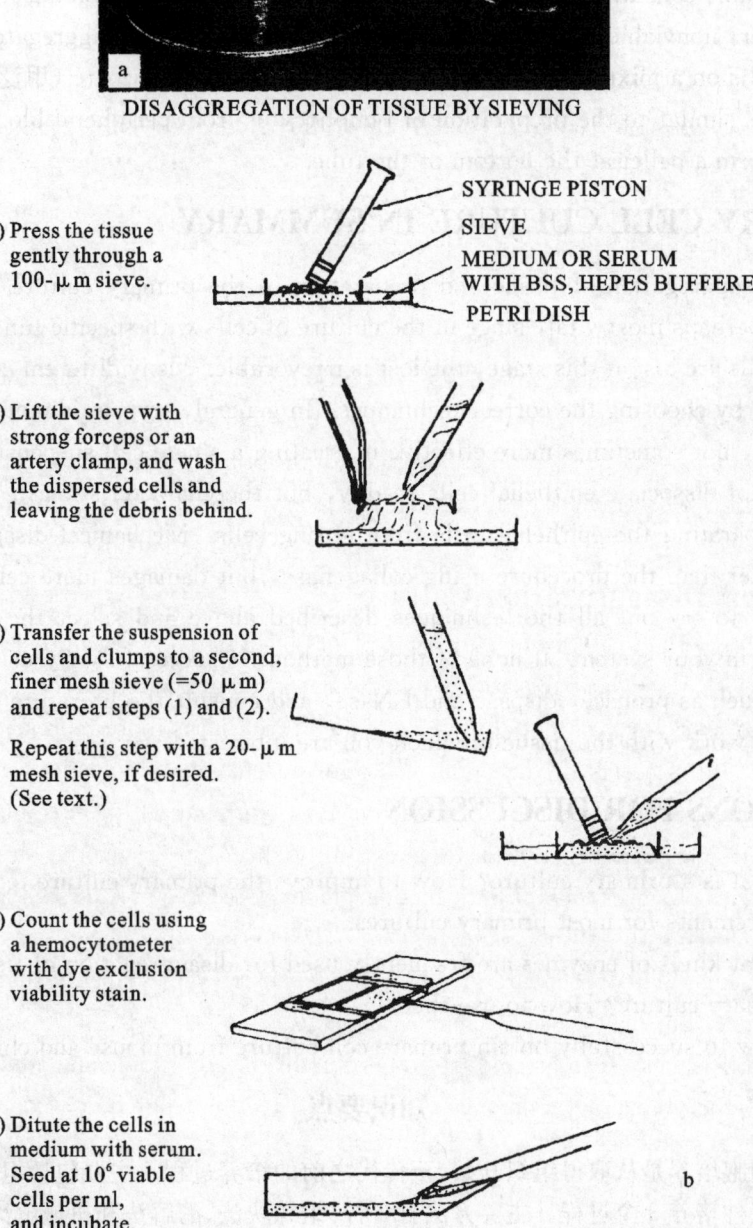

Fig. 23.6　**Disaggregation by sieving.** (a) Stainless-steel sieves suitable for disaggregating tissue. (b) Disaggregation of tissue by sieving. (c) Falcon disposable sieve; it nests in the top of a 50-ml centrifuge tube. Can be used for mechanical disaggregation, as in (b), or for filtering aggregates out of a disaggregated suspension. (By R. Ian Freshney)

· 169 ·

SEPARATION OF VIABLE AND NONVIABLE CELLS

When an adherent primary culture is prepared from dissociated cells, nonviable cells are removed at the first change of medium. With primary cultures maintained in suspension, nonviable cells are gradually diluted out when cell proliferation starts. If necessary, however, nonviable cells may be removed from the primary disaggregate by centrifuging the cells on a mixture of Ficoll(聚蔗糖) and sodium metrizoate(甲泛影钠). This technique is similar to the preparation of lymphocytes from peripheral blood. The dead cells will form a pellet at the bottom of the tube.

PRIMARY CELL CULTURE IN SUMMARY

The disaggregation of tissue and preparation of the primary culture make up the first, and perhaps most vital, stage in the culture of cells with specific functions. If the required cells are lost at this stage, the lost is irrevocable. Many different cell types may be cultured by choosing the correct techniques. In general, trypsin is more severe than collagenase, but sometimes more effective in creating a single-cell suspension. Collagenase does not dissociate epithelial cells readily, but the characteristic can be an advantage for separating the epithelial cells from strong cells. Mechanical disaggregation is much quicker than the procedure using collagenase, but damages more cells. The best approach is to try out all the techniques described above and select the method that works best in your system. If none of those methods is successful, try using additional enzymes, such as pronase, dispase and DNase, and consult the literature for examples of previous work with the tissue in which you are interested.

QUESTIONS FOR DISCUSSION

1. What is a primary culture? How to improve the primary culture technique? List some requirements for most primary cultures.

2. What kinds of enzymes are frequently used for disaggregation of tissue explants during primary culture? How to use them?

3. How to successfully obtain primary cell culture from mouse and chick embryos?

知识要点

原代细胞培养是从取得组织开始,至传代之前的培养过程。主要包括组织取材、剪碎或酶解和接种培养3个过程。主要方法有3种:组织块法、酶解法和机械解离法。常用于酶解组织的酶有胰蛋白酶和胶原酶。胰蛋白酶多用于酶解胚胎组织和软的组织,活性强,但对细胞的伤害大,酶解过程中要每隔半个小时回收细胞1次;胶原酶多用于硬的组织,活性不如胰蛋白酶,但对细胞伤害小。当以上两种酶的消化效果不好时,可考虑选用其他种类的酶,如弹性蛋白酶、透明质酸酶、链霉蛋白酶和分散酶等。

CHAPTER 24   CELL LINES(细胞系)

SUBCULTURE AND PROPAGATION

The first subculture (传代) represents an important transition for a culture. The need to subculture implies that the primary culture has increased to occupy all the available substrate. Hence, cell proliferation has become an important feature. While the primary culture may have a variable growth fraction, depending on the type of cells present in the culture, after the first subculture, the growth fraction is usually high (80% or more). From a very heterogeneous primary culture, containing many of the cell types present in the original tissue, a more homogeneous cell line emerges. Once a primary culture is subcultured (or passaged, or transferred), it becomes known as a cell line. This term implies the presence of several cell lineages of either similar or distinct phenotypes. If one cell lineage is selected, by cloning, by physical cell separation, or by any other selection technique, to have certain specific properties that have been identified in the bulk of the cells in the culture, this cell line becomes known as a cell strain.

The first subculture gives rise to a secondary culture, the secondary to a tertiary(三级), and so on. The passage number is the number of times that the culture has been subcultured, while the generation number is the number of doublings that the cell population has undergone, given that the number of doublings in the primary culture is very approximate. Cell lines with limited culture life spans are known as finite cell lines and behave in a fairly reproducible fashion. They grow through a limited number of cell generations, usually between 20 and 80 population doublings, before extinction. It is important that reference to a cell line should express the approximate generation number or number of doublings since explanation. If a cell line transforms in vitro, it gives rise to a continuous cell line. The relative advantages and disadvantages of finite cell lines and continuous cell lines are listed in Table 24.1.

Table 24.1 Properties of finite and continuous cell lines (By R. Ian. Freshney)

Properties	Finite	Continuous (transformed)
Ploidy	Euploid, diploid	Aneuploid, heteroploid
Transformation	Normal	Immortal, growth-control altered, and tumorigenic

(续表)

Properties	Finite	Continuous (transformed)
Anchorage dependence	Yes	No
Contact inhibition	Yes	No
Density limitation of cell proliferation	Yes	Reduced or lost
Mode of growth	Monolayer	Monolayer or suspension
Maintenance	Cyclic	Steady state possible
Serum requirement	High	Low
Cloning efficiency	Low	High
Markers	Tissue specific	Chromosomal, enzymic, antigenic
Special functions (e. g., virus susceptibility, differentiation)	May be retained	Often lost
Growth rate	Slow	Rapid
Yield	Low	High
Control parameters	Generation No.; Tissue-specific markers	Stain characteristics markers

SELECTION OF CELL LINE

Apart from specific functional requirements, there are a number of general parameters to consider in selecting a cell line:

1. Finite vs. continuous. A continuous cell line generally is easier to maintain, grows faster, clones more easily, and produces a higher cell yield per flask.

2. Normal or Transformed. Some cell lines are malignantly transformed (恶性转化). Some are immortalized but not tumorigenic.

3. Species. Non-human cell lines have fewer biohazard restrictions and have the advantage that the original tissue may be more accessible.

4. Growth characteristics. In terms of growth rate, yield, plating efficiency, and ease of harvesting, you will need to consider the following parameters: population-doubling time, saturation density (yield per flask), cloning efficiency, and ability to grow in suspension.

5. Availability. If you have to use a finite cell line, are there sufficient stocks available, or will you have to generate your own line(s)? If you choose a continuous cell line, are authenticated stocks available?

6. Validation. How well characterized is the line, if it exists already, or, if not, can you do the necessary characterization?

7. Phenotypic expression. Can the line be made to express the right characteristics?

8. Control cell line. If you are using a mutant, transfected, transformed, or abnormal cell line, is there a normal equivalent available?

9. Stability. How stable is the cell line? Has it been cloned? If not, can you clone it, and how long would this cloning process take to generate sufficient frozen and usable stocks?

ROUTINE MAINTENANCE

Once a culture is initiated, whether it is a primary culture or a subculture of a cell line, it will need a periodic medium change, or "feeding," followed eventually by subculture if the cells are proliferating. In non-proliferating cultures, the medium will still need to be changed periodically, as the cells will still metabolize and some constituents of the medium will become exhausted or will degrade spontaneously. Intervals between medium changes and between subcultures vary from one cell line to another, depending on the rate of growth and metabolism; rapidly growing transformed cell lines, such as HeLa, are usually subcultured once per week, and the medium should be changed four days later. More slowly growing, particularly non-transformed, cell lines may need to be subcultured only every two, three, or even four weeks, and the medium should be changed weekly between subcultures.

Examination of Cell Morphology

Whatever procedure is undertaken, it is vital that the culture be examined regularly to confirm the absence of contamination, the healthy status of the cells, and the lack of any signs of deterioration, such as granularity around the nucleus, cytoplasmic vacuolation, and rounding up of the cells with detachment from the substrate. (Fig. 24.1) Such signs may imply that the culture requires a medium change, or may indicate a more serious problem, e.g., inadequate or toxic medium or serum, microbial contamination, or senescence of the cell line. During routine maintenance, the medium change or subculture frequency should prevent such deterioration, as it is often difficult to reverse.

Replacement of Medium

Four factors indicate the need for the replacement of culture medium:

A drop in pH

The rate of fall and absolute level should be considered. Most cells stop growing as the pH falls from pH 7.0 to pH 6.5 and start to lose viability between pH 6.5 and pH 6.0, so if the medium goes from red through orange to yellow, the medium should be changed.

Cell concentration

Cultures at a high cell concentration exhaust the medium faster than those at a low concentration. This factor is usually evident in the rate of change of pH, but not always.

Cell type

Normal cells usually stop dividing at a high cell density, due to cell crowding, growth factor depletion, and other reasons. The cells block in the G1 phase of the cell cycle and deteriorate very little, even if left for two to three weeks or longer. Transformed cells, continuous cell lines, and some embryonic cells, however, deteriorate rapidly at high cell densities unless the medium is changed daily or they are subcultured.

Morphological deterioration

This factor must be anticipated by regular examination and familiarity with the cell line. If deterioration is allowed to progress too far, it will be irreversible.

Note

When a culture is at a low density and growing slowly, it may be preferable to half-feed it-i. e. , to remove only half of the medium and replace it with the same volume as was removed.

Volume, Depth, and Surface Area

The usual ratio of medium volume to surface area is 0.2-0.5 mL/cm^2. The upper limit is set by gaseous diffusion through the liquid layer, and the optimum ratio depends on the oxygen requirement of the cells. Cells with a high O_2 requirement do better in shallow medium (e. g. , 2 mm), and those with a low requirement may do better in deep medium (e. g. , 5 mm). If the depth of the medium is greater than 5 mm, then gaseous diffusion may become limiting. With monolayer cultures, this problem can be overcome by rolling the bottle or perfusing the culture with medium and arranging for gas exchange in an intermediate reservoir.

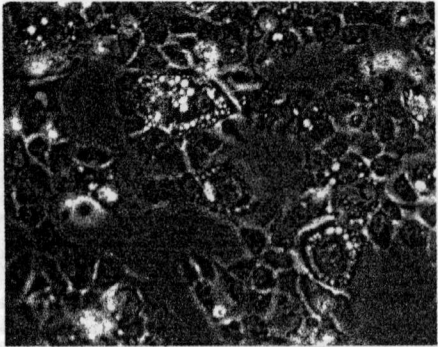

Fig. 24.1 Unhealthy cells. Vacuolation and granulation in bronchial epithelial cells (BEAS-2B). The cytoplasm of the cells becomes granular, particularly around the nucleus, and vacuolation occurs. The cells may become more refractile at the edge if cell spreading is impaired. (By R. Ian Freshney)

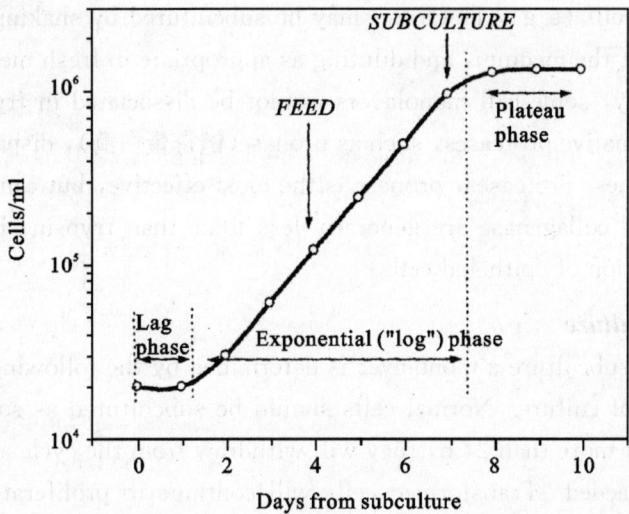

Fig. 24.2 Growth curve and maintenance. Semilog plot of cell concentration versus time from subculture, showing the lag phase, the exponential phase, and a plateau, and indicating times at which subculture and feeding should be performed. (By R. Ian Freshney)

Holding Medium（维持培养基）

A holding medium may be used when stimulation of mitosis, which usually accompanies a medium change, even at high cell densities, is undesirable. Holding media are usually regular media with the serum concentration reduced to 0.5% or 2% or eliminated completely. For serum-free media, growth factors and other mitogens are omitted. This omission inhibits mitosis in most untransformed cells. Transformed cell lines are unsuitable for this procedure, as either they may continue to divide successfully or the culture may deteriorate, because transformed cells do not block in a regulated fashion in G_1 of the cell cycle. Holding media are used to maintain cell lines with a finite life span without using up the limited number of cell generations available to them. Reduction of serum and cessation of cell proliferation also promote expression of the differentiated phenotype in some cells.

Subculture

The growth of cells in culture usually follows a standard pattern. (Fig. 24.2) A lag following seeding is followed by a period of exponential growth, called the log phase. When the cell density (cells/cm² substrate) reaches a level such that all of the available substrate is occupied, or when the cell concentration (cells/mL medium) exceeds the capacity of the medium, growth ceases or is greatly reduced. Then either the medium must be changed more frequently or the culture must be divided. For an adherent cell line, dividing a culture, or subculture as it is called, usually involves removal of the medium and dissociation of the cells in the monolayer with trypsin, although some loosely adherent cells (e. g. , HeLa-S_3) may be subcultured by shaking the bottle, collecting the cells in the medium, and diluting as appropriate in fresh medium in new bottles. Exceptionally, some cell monolayers cannot be dissociated in trypsin and require the action of alternative proteases, such as pronase(链霉蛋白酶), dispase(分散酶), and collagenase. Of these proteases, pronase is the most effective, but can be toxic to some cells. Dispase and collagenase are generally less toxic than trypsin, but may not give complete dissociation of epithelial cells.

Criteria for Subculture

The need to subculture a monolayer is determined by the following criteria:

(1) Density of culture. Normal cells should be subcultured as soon as they reach confluence. If left more than 24 h, they will withdraw from the cycle and take longer to recover when reseeded. Transformed cells will continue to proliferate beyond confluence, but will start to deteriorate after about two doublings, and reseeding efficiency will decline. These cells also should be subcultured on reaching confluence.

(2) Exhaustion of medium. Exhaustion of the medium usually indicates that the medium requires replacement. Usually, a drop in pH is accompanied by an increase in cell density, which is the prime indicator of the need to subculture.

(3) Time since last subculture. Routine subculture is best performed according to a strict schedule, so that reproducible behavior is achieved and monitored.

(4) Requirements for other procedures. When cells are required for purposes other than routine propagation, they also have to be subcultured, in order to increase the stock or to change the type of culture vessel or medium.

Outline

Remove the medium. Expose the cells briefly to trypsin. Incubate the cells. Disperse the cells in medium. Count the cells. Dilute and reseed the subculture.

Protocol

1. Prepare the hood, and bring the reagents and materials to the hood to begin the procedure.

2. Examine the culture carefully for signs of contamination or deterioration. (See Fig. 24.1). Check the culture and decide whether or not to subculture. If subculture is required, proceed as follows.

3. Take the culture to a sterile work area, remove and discard the medium.

4. Add PBSA prewash (0.2 mL/cm^2) to the side of the flask opposite the cells so as to avoid dislodging cells, rinse the cells, and discard the rinse. This step is designed to remove traces of serum that would inhibit the action of the trypsin.

5. Add trypsin (0.1 mL/cm^2) to the side of the flask opposite the cells. Turn the flask over and lay it down. Ensure that the monolayer is completely covered. Leave the

flask stationary for 15-30 s, and then withdraw all but a few drops of the trypsin, making sure that the monolayer has not detached. Using trypsin at 4℃ helps to prevent premature detachment.

6. Incubate, with the flask lying flat, until the cells round up; when the bottle is tilted, the monolayer should slide down the surface. (This usually occurs after 5-15 min.) Do not leave the flask longer than necessary, but on the other hand, do not force the cells to detach before they are ready to do so, or else clumping may result.

7. Add medium (0.1-0.2 mL/cm^2), and disperse the cells by repeated pipetting over the surface bearing the monolayer. Finally, pipette the cell suspension up and down a few times, with the tip of the pipette resting on the bottom corner of bottle, taking care not to create foams. A single-cell suspension is desirable at subculture to ensure an accurate cell count and uniform growth on reseeding.

8. Count the cells using a hemocytometer or an electronic particle counter. And dilute the suspension to the appropriate seeding concentration.

(1) by adding the appropriate volume of cells to a premeasured volume of medium in a culture flask or

(2) by diluting the cells to the total volume required and distributing that volume among several flasks.

9. Cap the flask(s), and return them to the incubator. Check after about 1 h for a change in the pH.

As a general rule, most continuous cell lines subculture satisfactorily at a seeding concentration of between 1×10^4 and 5×10^4 cells/mL, finite fibroblast cell lines subculture at about the same concentration, and more fragile cultures, such as endothelium and some early-passage epithelia, subculture at around 1×10^5/mL. For a new culture, start at a high seeding concentration and gradually reduce until a convenient growth cycle is achieved without any deterioration in the culture.

Propagation in Suspension

Cells that grow continuously in suspension, either because they are nonadhesive (e.g., many leukemias and murine ascites tumors) or because they have been kept in suspension mechanically, or selected, may be subcultured like bacteria or yeast. Replacement of the medium is not usually carried out with suspension cultures, and instead, the culture is either diluted and expanded, or some cell suspension is withdrawn and the residue is diluted back to an appropriate seeding concentration. In either case, a growth cycle will result, similar to that for monolayer cells, but usually with a shorter lag period.

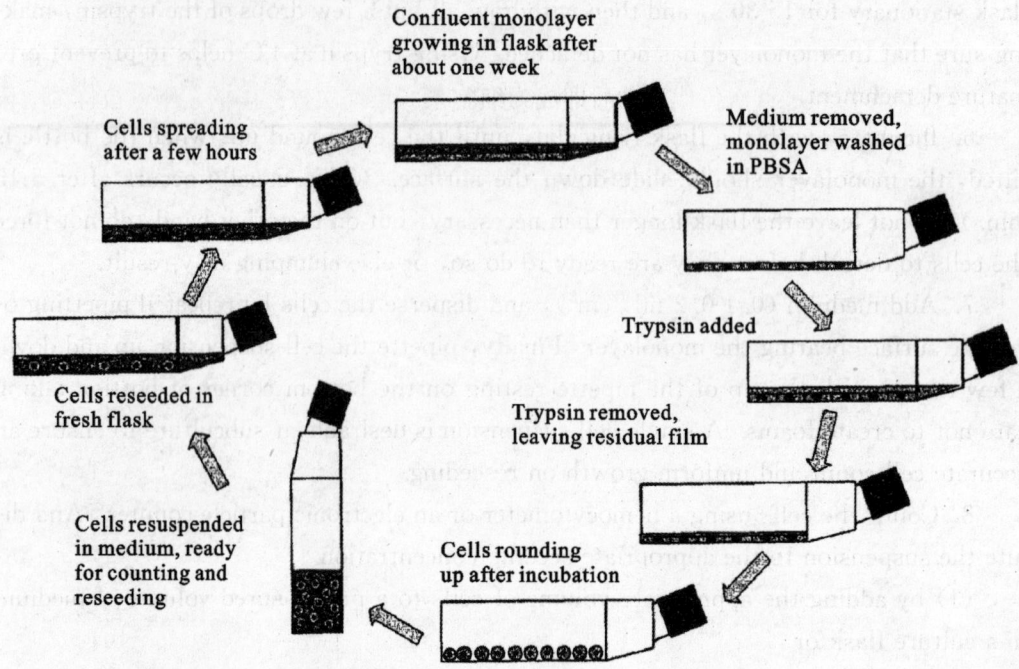

Fig. 24.3 Subculture of the monolayer. Stages in the subculture and growth cycle of monolayer cells following trypsinization. (By R. Ian Freshney)

Use of Antibiotics

The continuous use of antibiotics encourages cryptic (隐蔽) contaminations, particularly mycoplasma (支原体), and the development of antibiotic-resistant organisms. It may also interfere with cellular processes under investigation. If they are used, then it is important to maintain some antibiotic-free stocks in order to reveal any cryptic contaminations; these stocks can be maintained in parallel, and stock may be alternated in and out of antibiotics until antibiotic-free culture is possible.

Maintenance Records

Details of routine maintenance, including feeding and subculture, should be kept, and deviations or changes should be added to the database record for that cell line.

QUESTIONS FOR DISCUSSION

1. Explain the following concepts: cell line, finite cell line, continuous cell line, passage number, generation number, subculture, holding medium.
2. How to know the need for replacement of culture medium?
3. How to know the need for subculture?
4. How to subculture a monolayer?

PART Ⅱ　CYTOTECHOLOGY IN ANIMALS（动物细胞工程）

知识要点

原代培养物经过首次传代成功后即成为细胞系。传代次数有限的细胞系被称为有限细胞系，少数细胞系发生了转化，获得了永生性，成为连续性细胞系或永生性细胞系。细胞系可通过不断传代来进行长期保种。当细胞长满瓶底或培养基pH值偏酸变黄时，一定要及时传代，以利于保持细胞旺盛的生长活力。

· 179 ·

CHAPTER 25   CELL CLONING AND SELECTION
(细胞克隆和筛选)

Cell cloning and selection can be very useful in the isolation of a specific cell type and maintaining its specialized properties. While environmental conditions undoubtedly play a significant role in maintaining the differentiated properties of specialized cells in a culture, the selective overgrowth of unspecialized cells and cells of the wrong lineage is still a major problem.

CELL CLONING

The traditional microbiological approach to the problem of culture heterogeneity (异质性) is to isolate pure cell strains by cloning, but, while this technique is relatively easy for continuous cell lines, its success in most primary cultures is limited by poor cloning efficiencies. A further problem of cultures derived from normal tissue is that they may survive only for a limited number of generations, and by the time that a clone has produced a usable number of cells, it may already be near to senescence. Although cloning of continuous cell lines is more successful than cloning finite cell lines considerable heterogeneity may still arise within the clone as it is grown up for use. Nevertheless, cloning may help to reduce the heterogeneity of a culture. Cloning is also used as a survival assay for optimizing growth conditions (selection of medium and serum) and for determining chemosensitivity and radiosensitivity.

Cloning may be carried out in monolayer culture, using Petri dishes, multiwell plates, or flasks. Since the cells remain attached, it is relatively easy to discern individual colonies. Micromanipulation is the only conclusive method for determining genuine clonality(克隆性质) of a colony (i. e. , that the colony was derived from one cell). Cloning can also be carried out in suspension by seeding cells into a gel, such as agar, or a viscous solution, such as Methocel(甲基纤维素). The stability of the gel, or viscosity of the Methocel ensures that daughter cells do not break away from the colony as it forms. Hematopoietic cells are usually cloned in suspension; depending on the cells and growth factors used, the colony generates undifferentiated cells with high repopulation efficiency, *in vivo* or *in vitro*, or may mature into colonies of differentiated hematopoietic cells with very little repopulation efficiency. Cloning then becomes an assay for reproductive potential and stem cell identity.

Continuous cell lines generally have a high plating efficiency in monolayer and in suspension due to their transformed status, while normal cells, which may have a moderately high cloning efficiency in monolayer, have a very low cloning efficiency in suspension. This distinction has allowed suspension cloning to be used as an assay for transformation. Dilution cloning is the technique that is used most widely, based on the observation that cells diluted below a certain density form discrete colonies.

Dilution Cloning

Outline

Seed the cells at low density. Incubate them until colonies form. Stain the cells (if they used as a survival assay) or isolate them, and propagate the cells into a cell strain if they are being used for selection (Fig. 25.1).

Protocol

Trypsinize the cells to produce a single-cell suspension. Under-trypsinizing will produce clumps and over-trypsinizing will reduce the viability of the cells, but it is fundamental to the concept of cloning that the cells be singly suspended.

While the cells are trypsinizing, number the dishes, and measure out medium for the dilution steps. Up to four dilution steps may be necessary to reduce a regular monolayer to a concentration suitable for cloning.

When the cells round up and start to detach, disperse the monolayer in medium containing serum or trypsin inhibitor.

Count the cells, and dilute the cell suspension to the desired seeding concentration. If cloning the cells for the first time, choose a range of 10, 50, 100, and 200 cells/mL.

Seed the Petri dishes with the requisite amount of medium containing cells.

Place the dishes in a transparent plastic box.

Put the box in a humid CO_2 incubator or gassed sealed container (2%-10% CO_2).

Leave the cultures untouched for 1 week. If colonies have formed:

(a) For plating efficiency assay, stain and count the colonies.

(b) For clonal selection, isolate individual colonies. If no colonies are visible, replace medium and continue to culture for another week. Feed the dishes again and culture them for a third week if necessary. If no colonies appear by 3 weeks, then it is unlikely that they will appear at all.

1. Feeding. As the density of cells during cloning is very low, the need to feed the dishes after one week is debatable. Feeding mainly counteracts the loss of nutrients (such as glutamine), which are unstable, and replaces growth factors that have degraded. However, it also increases the risk of contamination, so it is reasonable to leave dishes for two weeks without feeding. If it is necessary to leave the dishes for a third week, then the medium should be replaced, or at least half of it. The preferential formation of colonies at the center of the plate can be due to incorrect seeding, either from

seeding the cells into the center of a plate that already contains medium or from swirling the plate such that the cells tend to focus in the center, but it can also be due to resonance (共振) in the incubator.

2. Microtitration Plates (微量滴定板). If the prime purpose of cloning is to isolate colonies, then seeding into microtitration plates can be an advantage. When the clones grow up, isolation is easy, but the plates have to be monitored at the early stages in order to mark which wells genuinely have single clones. The statistical probability of a well having a single clone can be increased by reducing the seeding density to a level such that only 1 in 5 or 10 wells would be expected to have a colony.

STIMULATION OF PLATING EFFICIENCY AT LOW CELL DENSITY

When cells are seeded at low densities, the survival rate of cells falls in all but a few cell lines. This does not usually present a severe problem with continuous cell lines, for which the plating efficiency seldom drops below 10%, but with primary cultures and finite cell lines, the plating efficiency maybe quite low (0.5%-5%), or even zero. Numerous attempts have been made to improve plating efficiencies, based on the assumption either that cells require a greater range of nutrients at low densities, because of loss by leakage, or that cell-derived diffusible signals or conditioning factors are present in high-density cultures and are absent or too dilute at low densities. The intracellular metabolic pool of a leaky cell in a dense population will soon reach equilibrium with the surrounding medium, while that of an isolated cell never will. This principle was the basis of the capillary technique of Sanford et al. [1948], by which the L929 clone of L-cells was first produced. The confines of the capillary tube allowed the cell to create a locally enriched environment that mimicked the higher cell-density state. In microdrop techniques developed later, the cells were seeded as a microdrop under liquid paraffin(石蜡). By keeping one colony separate from another, as in the capillary techniques, colonies could be isolated subsequently. As media improved, however, plating efficiencies increased, and Puck and Marcus [1955] were able to show that cloning cells by simple dilution in association with a feeder layer of irradiated mouse embryo fibroblasts gave acceptable cloning efficiencies, although subsequent isolation required trypsinization from within a collar (圈) placed over each colony.

To improve the clonal growth, several considerations should be included as followed:

1. Medium. Choose a rich medium, such as Ham's F12, or a medium that has been optimized for the cell type in use (e.g., MCDB 110 for human fibroblasts, Ham's F12 or MCDB 302 for CHO).

PART II　CYTOTECHOLOGY IN ANIMALS（动物细胞工程）

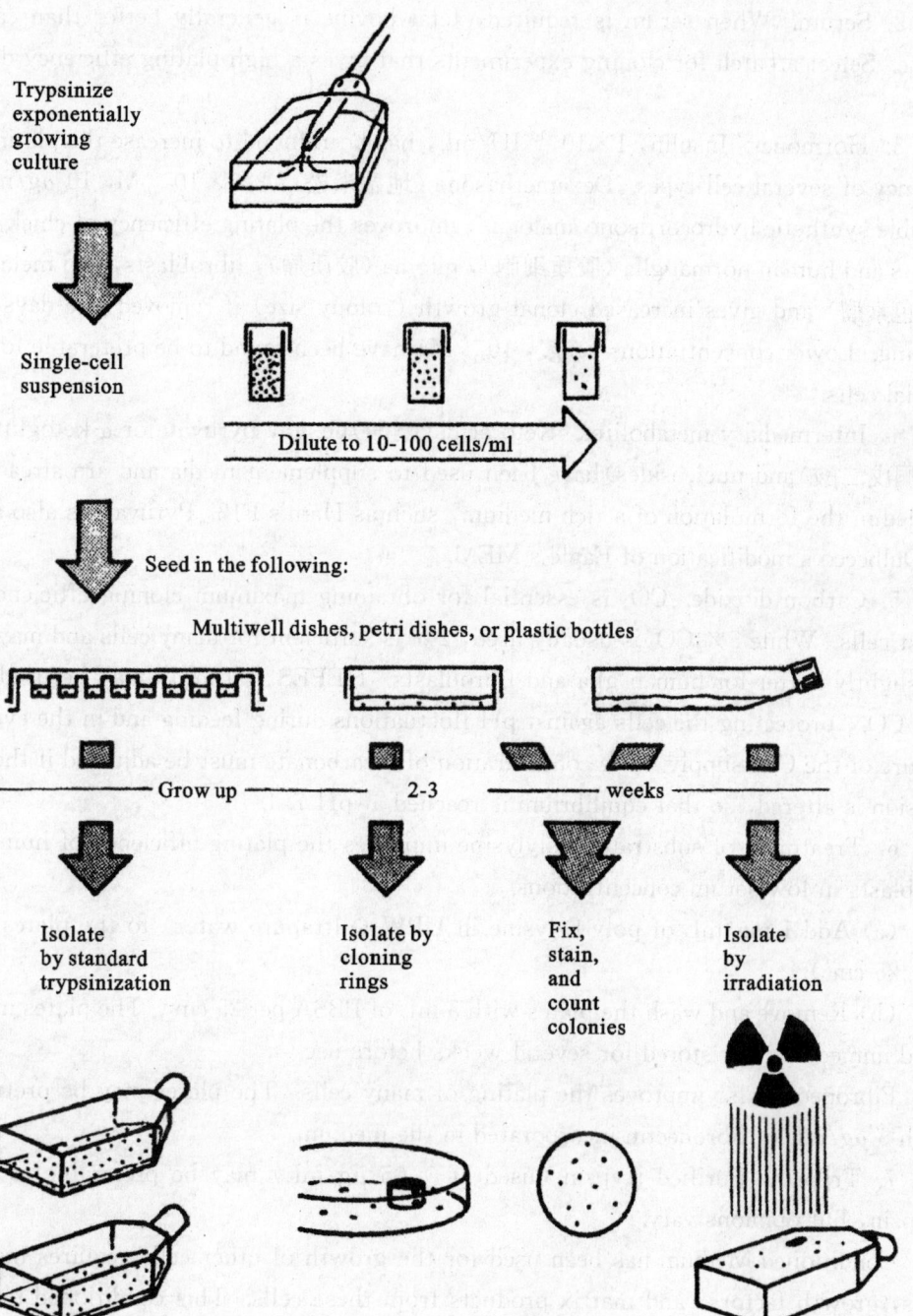

Fig. 25.1 Dilution cloning. Cloning cells from a monolayer culture. When clones form, they may be isolated directly from multiwell dishes (center left and lower left of the figure), by the cloning ring technique (center and lower center), or by irradiating the flask while shielding one colony (center right and lower right). If isolation is not required and the cloning is being performed for quantitative assay, then the colonies are fixed, stained, and counted. (By R. Ian Freshney)

2. Serum. When serum is required, fetal bovine is generally better than calf or horse. Select a batch for cloning experiments that gives a high plating efficiency during tests.

3. Hormones. Insulin, 1×10^{-10} IU/mL, has been found to increase the plating efficiency of several cell types. Dexamethasone（地塞米松）, 2.5×10^{-5} M, 10 μg/mL, a soluble synthetic hydrocortisone analogue, improves the plating efficiency of chick myoblasts and human normal glia（胶质细胞）, glioma（胶质瘤）, fibroblasts, and melanoma （黑色素瘤）and gives increased clonal growth（colony size）if removed five days after plating. Lower concentrations (e.g., 10^{-7} M) have been found to be preferable for epithelial cells.

4. Intermediary metabolites. Keto acids（酮酸）(e.g., pyruvate or α-ketoglutarate （α-酮戊二酸）and nucleosides)have been used to supplement media and are already included in the formulation of a rich medium, such as Ham's F12. Pyruvate is also added to Dulbecco's modification of Eagle's MEM.

5. Carbon dioxide. CO_2 is essential for obtaining maximum cloning efficiency for most cells. While 5% CO_2 is usually used, 2% is sufficient for many cells and may even be slightly better for human glia and fibroblasts. HEPES (20 mM) may be used with 2% CO_2, protecting the cells against pH fluctuations during feeding and in the event of failure of the CO_2 supply. The concentration of bicarbonate must be adjusted if the CO_2 tension is altered, so that equilibrium is reached at pH 7.4.

6. Treatment of substrate. Polylysine improves the plating efficiency of human fibroblasts in low serum concentrations.

(a) Add 1 mg/mL of poly-D-lysine in UPW (ultrapure water) to the plates (~5 mL/25 cm^2).

(b) Remove and wash the plates with 5 mL of PBSA per 25 cm^2. The plates may be used immediately or stored for several weeks before use.

Fibronectin also improves the plating of many cells. The plates may be pretreated with 5 μg/mL of fibronectin incorporated in the medium.

7. Trypsin. Purified trypsin (used at 0.05 μg/mL) may be preferable to crude trypsin, but opinions vary.

Conditioned Medium has been used for the growth of other cells acquires metabolites, growth factors, and matrix products from these cells. This conditioned medium can improve the plating efficiency of some cells if it is diluted into the regular growth medium. Outline for the preparation of conditioned medium is as followed: Harvest medium from homologous cells, or a different cell line, from the late log phase. Filter the medium, and dilute it with fresh medium as required.

The reason that some cells do not clone well is related to their inability to survive at low densities. One way to maintain cells at clonogenic densities, but, at the same time, to mimic high cell densities, is to clone the cells onto a growth-arrested feeder layer (Fig. 25. 2). The feeder cells may provide nutrients, growth factors, and matrix constituents that enable the cloned cells to survive more readily. There are two ways to prepare the feeder layers: growth-arrest by irradiation or mitomycin C (丝裂霉素 C).

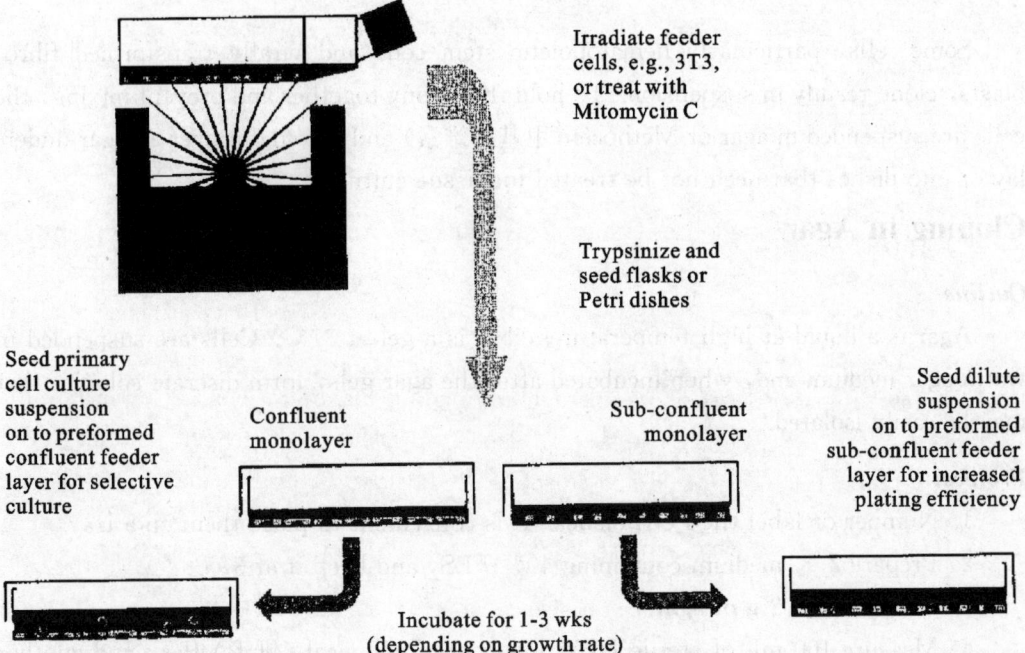

Fig. 25. 2 **Feeder layers.** Cells are irradiated and trypsinized (or may be trypsinized first and then irradiated in suspension) and plated at a low density to enhance cloning efficiency or at high density to provide a confluent monolayer for selective growth. (By R. Ian Freshney)

Preparation of Feeder Layers

Outline

Plant homologous (同源) or heterologous (异源) cells rendered nonproliferative by irradiation or drug treatment at medium density before the cloning of test cells. X-ray or ^{60}Co source capable of delivering 30 Gy in 30 min or less (instead of mitomycin C)

Protocol

1. Trypsinize embryo fibroblasts, from the primary culture, and reseed the cells at 10^5 cells/mL.

2. At 50% confluence, add mitomycin C, 2 μg/10^6 cells, 0.25 μg/mL, overnight, or irradiate the culture with 30 Gy.

3. Change the medium after treatment, and after a further 24 h, trypsinize the cells and reseed them in fresh medium at 5×10^4 cells/mL (10^4 cells/cm^2).

4. Incubate the culture for a further 24-48 h, and then seed the cells for cloning.

The feeder cells will remain viable for up to 3 weeks, but will eventually die out and are not carried over if the colonies are isolated. Cell lines that have been used for feeder cells include 3T3, MRC-5, and STO cells.

SUSPENSION CLONING

Some cells, particularly hematopoietic stem cells and virally transformed fibroblasts, clone readily in suspension. To hold the colony together and prevent mixing, the cells are suspended in agar or Methocel (甲基纤维素) and plated out over an agar underlay or into dishes that need not be treated for tissue culture.

Cloning in Agar

Outline

Agar is a liquid at high temperatures, but is a gel at 37℃. Cells are suspended in warm agar medium and, when incubated after the agar gels, form discrete colonies that may be easily isolated.

Protocol

1. Number or label the Petri dishes. It is convenient to place them on a tray.
2. Prepare 2 × medium containing 40% FBS, and keep it at 37℃.
3. Weigh out 1.2 g of agar.
4. Measure 100 mL of sterile UPW into a sterile conical (锥形) flask and another 100 mL into a sterile bottle. Add 1.2 g of agar to the flask. Cover the flask, and boil the solution for 2 min. Alternatively, the agar may be sterilized in the autoclave in advance, but, if subsequently stored, it will still need to be boiled, in order to melt it for use.
5. Transfer the boiled agar and the bottle of sterile UPW to a water bath at 55℃.
6. Prepare a 0.6% agar underlay by combining an equal volume of 2 × medium and 1.2% agar. Keep the underlay at 37℃. If any growth factors, hormones or other supplements are being used, they should be added to the underlay medium at this point.
7. Add 1 mL of 0.6% agar medium to the dishes, mix, and ensure that the medium covers the base of the dish. Leave the dishes at room temperature to set.
8. Prepare the cell suspension, and count the cells.
9. Prepare 0.3% agar medium, and keep it at 37℃. This medium may be prepared by diluting 2 × medium at 37℃ with 1.2% agar at 55℃ and UPW at 55℃ in the respective proportions of 2:1:1.
10. Prepare the following cell dilutions, making the top concentration of cells 1×10^5/mL:

    (a) 1×10^5/mL.

(b) Dilute 1×10^5/mL by 1/3 to give 3.3×10^4/mL.

(c) Dilute 3.3×10^4/mL by 1/3 to give 1.1×10^4/mL.

(d) Dilute 1.1×10^4/mL by 1/3 to give 3.7×10^3/mL.

11. Label four bottles or tubes one for each dilution and pipette 40 μL of each cell dilution, including the 1×10^5/mL concentration, into the respective container. Add 4 mL of 0.3% agar medium at 37℃ to each container, mix, and pipette 1 mL from each container onto each of three Petri dishes. This will give final concentrations as follows: (a) 1×10^3/mL/dish; (b) 330/mL/dish; (c) 110/mL/dish; (d) 37/mL/dish.

12. Allow the solution in the Petri dishes to gel at room temperature.

13. Put the Petri dishes into a clean plastic box with a lid, and incubate them at 37℃ in a humid incubator for 10 d.

ISOLATION OF CLONES

When cloning is used for the selection of specific cell strains, the colonies that form need to be isolated for further propagation. If monolayer cells are cloned directly into multiwell plates, then colonies may be isolated by trypsinizing individual wells. It is necessary to confirm the clonal origin of the colony during its formation by regular microscopic observation. If, however, cloning is performed in Petri dishes, then there is no physical separation between colonies. This separation must be created by removing the medium and placing a stainless steel or ceramic(陶瓷) ring around the colony to be isolated. Then the colony is trypsinized from within the stainless steel or porcelain(瓷器) ring and transferred to one of the wells of a 24-or 12-well plate, or directly to a 25-cm² flask. The clony can also be isolated by irradiation. Invert the flask under an X-ray machine or ^{60}Co source after the desired colony is covered by a piece of lead of appropriate size. The isolation of colonies growing in suspension is simple, but requires a dissection microscope. Draw the colony into a pipettor or Pasteur pipette, and transfer the colony to a flask or the well of a multiwell plate and pick the colonies using a dissecting microscope.

Selective Inhibitors

Manipulating the conditions of a culture by using a selective medium is a standard method for selecting microorganisms. Its application to animal cells in culture is limited, however, by the basic metabolic similarities of most cells isolated from one animal, in terms of their nutritional requirements. The problem is accentuated(突出)by the effect of serum, which tends to mask the selective properties of different media. Most selective media that have been shown to be generally successful have been serum-free formulations. A number of metabolic inhibitors, however, have had recurrent success. One of the more successful approaches was the development of a monoclonal antibody to the stromal cells of a human breast carcinoma. Used with complement(补体), this antibody proved to be cytotoxic to fibroblasts from several tumors and helped to purify a number of malignant cell lines.

Selective media are also commonly used to isolate hybrid clones from somatic hybridization experiments. HAT medium, a combination of hypoxanthine(次黄嘌呤), aminopterin(氨基喋呤), and thymidine(胸腺嘧啶), selects hybrids with both hypoxanthine guanine phosphoribosyl transferase(次黄嘌呤鸟嘌呤磷酸核糖转移酶) and thymidine kinase(胸腺嘧啶激酶) from parental cells deficient in one or the other enzyme.

Transfected cells are also selected by resistance to a number of drugs, such as neomycin(新霉素), its analogue Geneticin (G418), hygromycin(潮霉素), and methotrexate(氨甲喋呤), by including a resistance-conferring gene in the construct used for transfection (e.g., *neo* (aminoglycoside phosphotransferase), *hph* (hygromycin B phosphotransferase), or *dhfr* (dihydrofolate reductase). Culture in the correct concentration of the selective marker, determined by titration against the transfected and nontransfected controls, selects for stable transfectants(转染子). Selection with methotrexate(氨甲喋呤) has the additional advantage that increasing the methotrexate concentration leads to amplification of the *dhfr* gene and can coamplify other genes in the construct. Negative selection is also possible by using the Herpes simplex virus (HSV) *TK* gene, which activates Ganciclovir(更昔洛韦) into a cytotoxic product. Transfected cells will be sensitized to the drug.

Selective Adhesion

Different cell types have different affinities for the culture substrate and attach at different rates. If a primary cell suspension is seeded into one flask and transferred to a second flask after 30 min, a third flask after 1 h, and so on for up to 24 h, then the most adhesive cells will be found in the first flask and the least adhesive in the last. Macrophages will tend to remain in the first flask, fibroblasts in the next few flasks, epithelial cells in the next few flasks, and, finally, hematopoietic cells in the last flask. If collagenase in complete medium is used for primary disaggregation of the tissue, then most of the cells that are released will not attach within 48 h unless the collagenase is removed. However, macrophages migrate out of the fragments of tissue and attach during this period and can be removed from other cells by transferring the disaggregate to a fresh flask after 48-72 h of treatment with collagenase. This technique works well during disaggregation of biopsy specimens from human tumors.

Selective Detachment

Treatment of a heterogeneous monolayer with trypsin or collagenase will remove some cells more rapidly than others. Periodic brief exposure to trypsin removed fibroblasts from cultures of fetal human intestine and skin. Dispase II selectively dislodges sheets of epithelium from human cervical (子宫颈) cultures grown on feeder layers of 3T3 cells without dislodging the 3T3 cells. This technique may be effective in subculturing epithelial cells from other sources, excluding stromal fibroblasts.

Nature of Substrate

The hydrophilic(亲水) nature of most culture substrates appears to be necessary for cell attachment, but little is known about variations in charge distribution on the cell surface and how different receptor arrays might interact with a complex charge array on the substrate. The selective effect of substrates on growth may depend on both differential rates of attachment and growth, although in practice the two are indistinguishable. Macrophages also attach to Teflon(一种涂料), but do not proliferate. Collagen has been used in gel form to favor epithelial cell growth and in its denatured form to support endothelial outgrowth from aorta (大动脉) into a fibrin clot.

Selective Feeder Layers

As well as conditioning the substrate, feeder layers can also be used for the selective growth of epidermal cells and for repressing stromal overgrowth in cultures of breast and colon carcinoma. The role of the feeder layer is probably quite complex; it provides not only extracellular matrix for adhesion of the epithelium, but also positively acting growth factors and negative regulators that inactivate TGF-β.

Semisolid Media

The transformation of many fibroblast cultures reduces the anchorage dependence of cell proliferation. By culturing the cells in agar after viral transformation, it is possible to isolate colonies of transformed cells and exclude most of the normal cells. Normal cells will not form colonies in suspension with the high efficiency of virally transformed cells.

Generally, cell cloning and the use of selective conditions have a significant advantage over physical cell separation techniques, in that contaminated cells are either eliminated entirely by clonal selection or repressed by constant or repeated application of selective conditions. Even the best physical cell separation techniques still allow some overlap between cell populations, such that overgrowth recurs. A steady state cannot be achieved, and the constitution of the culture changes continuously.

QUESTIONS FOR DISCUSSION

1. What is a feeder layer? How to prepare a feeder layer?
2. How to isolate cell lones by dilution cloning?
3. How to isolate cell clones by selective inhibitors or substrates?

知识要点

　　细胞克隆和筛选是分离和培养特定性状细胞的重要手段。稀释克隆法是最常用的方法，尤其适用于增殖能力强的转化细胞，但对原代培养物的克隆效果差。对于生长能力差的细胞，需要改进培养基和培养基质等培养条件，或者添加条件培养基和利用饲养层来促进克隆细胞的生长。软琼脂克隆法常被用来克隆失去贴壁依赖性的细胞。对已形成的细胞克隆可以通过毛细管、金属圈等的帮助进行分离培养。在培养基中添加选择性抑制剂可以有效筛选出阳性细胞克隆。通过改变培养基质，选择性贴附或选择性地从基质上解离也可以有效分离和克隆某些类型的细胞。

CHAPTER 26 TRANSFORMATION OF CULTURED CELLS(培养细胞的转化)

WHAT IS TRANSFORMATION?

In microbiology, where the term was first employed in this context, *transformation* implies a change in phenotype that is dependent on the uptake of new genetic material. Although this process is now possible in mammalian cells, it is called *transfection* or *DNA transfer* in this case to distinguish it from transformation. Transformation of cultured cells implies a spontaneous or induced permanent phenotypic change resulting from a heritable change in DNA and gene expression. Although transformation can arise from infection with a transforming virus, such as polyoma(多形瘤), or from gene transfection, it can also arise spontaneously or following exposure to ionizing radiation or chemical carcinogens.

Transformation is associated with *genetic instability* and three major classes of phenotypic change, one or all of which may be expressed in one cell strain: (1) immortalization(永生化), the acquisition of an infinite life span, (2) aberrant growth control(生长控制异常), the loss of contact inhibition(接触抑制) and anchorage dependence(贴壁依赖性), and (3) malignancy(恶性化), as evidenced by the tumorigenic potential of the cells. The term *transformation is* used here to imply all three of these processes. The acquisition of an infinite life span alone is referred to as *immortalization*, since it can be achieved without grossly(显著) aberrant growth control and malignancy, which are usually correlated. The criteria that determine whether cells are transformed are listed in Table 26.1.

Transformation is seen as a particular event or series of events that depends on and promotes genetic instability. It alters many of the cell line's properties, including growth rate, mode of growth, specialized product formation, longevity, and tumorigenicity(致瘤性) (see Table 26.1). It is therefore important that these characteristics are included when a cell line is characterized. The transformation status is a vital characteristic that is required when culturing cells from tumors, in order to confirm that the cells are derived from the neoplastic(肿瘤) component of the tumor, rather than from normal infiltrating (浸润性)fibroblasts, blood vessel cells, and inflammatory cells(炎症细胞).

More than one criterion is necessary to confirm neoplastic status, as most of the aforementioned characteristics are expressed in normal cells at particular stages of development. The exceptions are gross aneuploidy(非整倍体), heteroploidy(异倍体), and tumorigenicity, which are regarded as conclusive positive indicators of malignant transformation(恶性转化). However, some tumor cells can be near euploid and not tumorigenic, and other criteria are therefore required.

GENETIC INSTABILITY

The characteristics of a cell line do not always remain stable. In addition to the selective and adaptive processes already described cell lines are also prone to genetic instability. Normal, human finite cell lines are usually genetically stable, but cell lines from other species, particularly the mouse, are genetically unstable and transform quite readily. Continuous cell lines, particularly from tumors of all species, are very unstable, not surprisingly, as this instability was a major reason for their undergoing the necessary mutations to become continuous.

Evidence of genetic rearrangement can be seen in chromosome counts and karyotype analysis. While the mouse karyotype is made up exclusively of small telocentric(端着丝粒) chromosomes, several metacentrics(中着丝粒) are apparent in many continuous murine cell lines. Furthermore, while virtually every cell in the animal has the normal diploid set, this condition is more variable in culture. In extreme cases-e. g., continuous cell strains, such as HeLa, less than half of the cells will have exactly the same karyotype; i. e., they are heteroploid.

Most continuous cell strains, even after cloning, contain a range of genotypes that are constantly changing. Because transformation often involves chromosomal rearrangement, it is probable that it can occur only in cells with the capacity for chromosomal alterations. Additionally, transformation may cause genetic instability to arise in a previously stable genotype. Hence, transformed continuous lines retain a capacity for genetic variation that is not apparent *in vivo* or in many finite cell lines.

There are two main causes of genetic variation: (a) The spontaneous mutation rate appears to be higher *in vitro*, associated, perhaps, with the high rate of cell proliferation, and (b) mutant cells are not eliminated unless their growth capacity is impaired. It is not surprising that phenotypic variation will arise as a result of this genetic variation.

Table 26.1 Properties of transformed cells (By R. Ian Freshney)

Property	Assay
Growth	
Immortal	Grow beyond 100 population doublings
Anchorage independent	Clone in agar; may grow in stirred suspension
Loss of contact inhibition	Microscopic observation; time lapse(定时)

(续表)

Property	Assay
Growth on confluent monolayers of homologous cells	Focus formation
Reduced density limitation of growth	High saturation density; high growth fraction at saturation density
Low serum requirement	Clone in limiting serum
Growth factor independent	Clone in limiting serum
Production of autocrine growth factors	Immunostaining; clone in limiting serum with conditioned medium; receptor-blocking antibody or peptide inhibitor
Transforming growth factor production	Suspension cloning
High plating efficiency	Clone in limiting serum
Shorter population-doubling time	Growth curve
Genetic	
High spontaneous mutation rate	Sister chromosome exchange
Aneuploid	Chromosome content
Heteroploid	Chromosome content
Overexpressed or mutated oncogenes	Southern blot; FISH; Immunostaining
Deleted or mutated suppressor genes	Southern blot; FISH; Immunostaining
Gene and chromosome translocation	FISH, chromosome paints
Structural	
Modified actin cytoskeleton	Immunostaining
Loss of cell-surface-associated fibronectin	Immunostaining
Modified extracellular matrix	Immunostaining; DEAE chromatography
Altered expression of cell adhesion molecules	Immunostaining
Disruption in cell polarity	Immunostaining; polarized transport in filter wells
Neoplastic	
Tumorigenic	Xenograft in nude or scid mice
Angiogenic	CAM assay; filter wells
Enhanced protease secretion	Plasminogen activator assay
Invasive	Organoid confrontation; filter-well invasion assay

Chromosomal Aberrations

Both variations in ploidy and increases in the frequency of individual chromosomal aberrations can be found, and variations in chromosome number are found in most tumor cultures. The incidence of genetic instability and frequency of chromosomal rearrangement can be determined by the sister chromatid exchange assay. Some specific aberrations are associated with particular types of malignancy. These aberrations constitute tumor-specific markers that can be extremely valuable in cell line characterization and confirmation of neoplasia.

DNA Content

Flow cytometry(流式细胞仪) shows that the DNA content of tumor cells mimics chromosomal aberrations-i. e. , it may vary from the normal somatic cell DNA content and show marked heterogeneity within a population. DNA analysis does not substitute for chromosome analysis. However, as cells with an apparently normal DNA content can still have an aneuploid karyotype. Deletions and polysomy may cancel out, or translocations may occur without net loss of DNA.

IMMORTALIZATION

Most normal cells have a finite life span of 20-100 generations, but some cells, notably those from rodents and from most tumors, can produce continuous cell lines with an infinite life span. The rodent cells are karyotypically normal at isolation and appear to go through a crisis after about 12 generations; most of the cells die out in this crisis, but a few survive with an enhanced growth rate and give rise to a continuous cell line. If continuous cell lines from mouse embryos (e. g. , the various 3T3 cell lines) are maintained at a low cell density and are not allowed to remain at confluence for any length of time, they remain sensitive to contact inhibition and density limitation of growth. If, however, they are allowed to remain at confluence for extended periods, foci of cells appear with reduced contact inhibition, begin to pile up, and will ultimately overgrow. (Fig. 26.1) The fact that these cells are not apparent at low densities or when confluence is first reached suggests that they arise *de novo*, by a further transformation event. They appear to have a growth advantage, and subsequent subcultures will rapidly be overgrown by the randomly growing cell. This cell type is often found to be *tumorigenic*.

Control of Senescence

The finite life span of cells in culture is regulated by a group of 10 or more dominantly acting senescence genes, the products of which negatively regulate cell cycle progression. Somatic hybridization experiments between finite and immortal cell lines usually generate hybrids with a finite life span, suggesting that the senescence genes are dominant. It is likely that one or more of these genes negatively regulate the expression of telomerase (端粒酶), required for the terminal synthesis of telomeric DNA, which otherwise becomes progressively shorter during a finite life span, until the chromosomal DNA can no longer replicate. Telomerase is expressed in germ cells and has moderate activity in stem cells, but is absent from somatic cells. Deletions and/or mutations within senescence genes, or overexpression or mutation of one or more oncogenes that override the action of the senescence genes, can allow cells to escape from the negative control of the cell cycle and reexpress telomerase.

It has been assumed that immortalization is a multistep process involving the inactivation of a number of cell cycle regulatory genes, such as Rb and p53. The SV40 LT gene is often used to induce immortalization. The product of this gene, T antigen, is known to bind Rb and p53. By doing so, it not only allows an extended proliferative life span, but also restricts the DNA surveillance(监督) activity of genes like p53, thereby allowing an increase in genomic instability and an increased chance of generating further mutations favorable to immortalization (e. g., the up regulation of telomerase or the down regulation of one of the telomerase inhibitors). Recent studies have shown that transfection of the telomerase gene with a regulatable promoter is sufficient to immortalize cells.

Immortalization per se(本身) does not imply the development of aberrant growth control and malignancy, since a number of immortal cell lines, such as 3T3 cells and BHK21-C13, retain contact inhibition of cell motility, density limitation of cell proliferation, and anchorage dependence, and are not tumorigenic. It must be assumed, however, that some aspects of growth control are abnormal and that there is a likely increase in genomic instability. Furthermore, immortalized cell lines often lose the ability to differentiate.

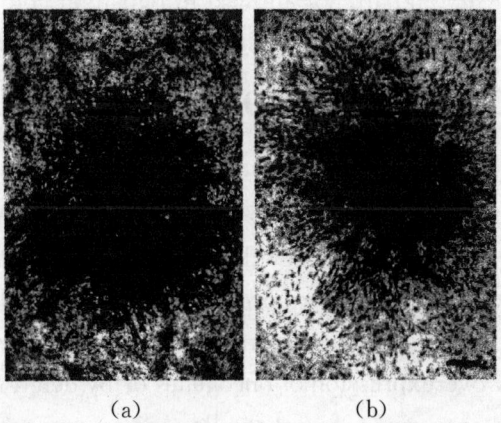

(a)　　　　　　(b)

Fig. 26.1　Transformation foci. A monolayer of normal, contact-inhibited NIH3T3 mouse fibroblasts. (a) NIH3T3 mouse fibroblasts transformed by transfection with bovine papilloma virus DNA cloned in bacterial plasmid pAT-153, coprecipitated with calcium phosphate. (b) Spontaneous transformant arising when cells are maintained at a high density. (By D. Spandidos)

Immortalization with Viral Genes

A number of viral genes have been used to immortalize cells. It has been recognized for some time that SV40 can be used to immortalize cells, and the gene responsible for this appears to be the large T (LT) gene. Other viral genes that have been used to immortalize cells are adenovirus Ela, human papilloma virus (人乳头状瘤病毒, HPV) E6 and E7, and Epstein Barr virus (爱波斯坦巴尔病毒, EBV). Most of these genes probably act by blocking the inhibition of cell cycle progression by inhibiting the activity of genes such as CIP-1/WAF-1/p21, Rb, p53, and p16, thus giving an increased life span and enhanced opportunity for further mutations. Those genes that have been used most extensively are EBV for lymphoblastoid cells(类淋巴母细胞) and SV40LT for adherent cells such as fibroblasts, keratinocytes, and endothelial cells. Endothelial cells have also been immortalized by fusion with A549 cells and by irradiation. Typically, cells are transfected or retrovirally infected with the immortalizing gene before they enter senescence. This extends their proliferative life span for another 20-30 population doublings, whereupon the cells cease proliferation and enter *crisis*. After a variable period in crisis (up to several months), a subset of immortal cells overgrows. The proportion of cells that eventually immortalize can be 1×10^{-5} to 1×10^{-9}.

Telomerase-Induced Immortalization

The primary cause of senescence appears to be telomeric shortening, followed by telomeric fusion and the formation of dicentric(双着丝粒) chromosomes and subsequent apoptosis. Transfecting cells with the telomerase gene *htrt* extends the life span of the cell line, and a proportion of these cells become immortal, but not malignantly transformed. As a high proportion of the $htrt^+$ clones become immortal, this appears to be a promising technique for immortalization, although the functionality of these lines has yet to be demonstrated.

ABERRANT GROWTH CONTROL

Cells cultured from tumors, as well as cultures that have transformed *in vitro*, show aberrations in growth control, such as growth to higher saturation densities, clonogenicity(克隆形成) in agar, and growth on confluent monolayers of homologous cells. These cell lines exhibit lower serum or growth factor dependence, usually form clones with a higher efficiency, and are assumed to have acquired some degree of autonomous growth control by overexpression of oncogenes or by deletion of suppressor genes. Growth control is often autocrine-i. e., the cells secrete mitogens(促有丝分裂原) for which they possess receptors, or the cells express receptors or stages in signal transduction that are permanently active and unregulated. Although immortalization does not necessarily imply a loss of growth control, many cells progress readily from immortalization to aberrant growth, perhaps due to genetic instability that is intrinsic to the immortalized genotype.

Anchorage Independence

Many of the properties associated with neoplastic transformation *in vitro* are the result of cell surface modifications-e. g. , changes in the binding of plant lectins(凝集素) and in cell surface glycoproteins-which may be correlated with the development of invasion and metastasis *in vivo*. Fibronectin (large extracellular transformation-sensitive protein) is lost from the surface of transformed fibroblasts due to alterations in integrins. This loss may contribute to a decrease in cell-cell and cell-substrate adhesion and to a decreased requirement for attachment and spreading for the cells to proliferate. Transformed cells may lack specific CAMs (e. g. , L-CAM), which, when transfected back into the cell, regenerate the normal, noninvasive phenotype, and, as such, they may be recognized as tumor suppressor genes. Other CAMs may be overexpressed, such as N-CAM in small-cell lung cancer. The expression of and degree of phosphorylation of integrins may also change, potentially altering cytoskeletal interactions, the regulation of gene transcription, the substrate adhesion of the cells, and the relationship between cell spreading and cell proliferation. In addition, the loss of cell-cell recognition, a product of reduced adhesion, leads to a disorganized growth pattern and the loss of contact inhibition of cell motility and density limitation of cell proliferation. Cells can grow detached from the substrate, either in stirred suspension culture or suspended in semisolid media, such as agar or Methocel.

Suspension Cloning

Polyoma-transformed BHK21 cells could be grown preferentially in soft agar, while untransformed cells cloned very poorly. Subsequently, it has been shown that colony formation in suspension is frequently enhanced following viral transformation. There is a close correlation between tumorigenicity and suspension cloning. Although many human tumors contain a small percentage of cells ($< 1.0\%$) that are clonogenic in agar, a number of normal cells will also clone in suspension with equivalent efficiency. However, it remains a valuable technique for assaying neoplastic transformation *in vitro* by tumor viruses and was used extensively to assay for carcinogenesis.

Contact Inhibition

The loss of contact inhibition may be detected morphologically by the formation of a disoriented monolayer of cells or rounded cells in foci within the regular pattern of normal surrounding cells (see Fig. 26.1).

Serum Dependence

Transformed cells have lower serum dependence than that of their normal counterparts, due, in part, to the secretion of growth factors by tumors. These factors have been collectively described as autocrine(自分泌) growth factors. Implicit in this definition is that (1) the cell produces the factor; (2) the cell has receptors for the factor; and (3) the cell responds to the factor by entering mitosis. Some of these factors may have an apparent transforming activity on normal cells (e.g., TGFα) binding to the EGF receptor and inducing mitosis although, unlike true transformation, this type is probably reversible. These factors also cause nontransformed cells to adopt a transformed phenotype and grow in suspension. Autonomous growth control is also achieved in transformed cells by the expression of modified receptors, such as the *erb-B*2 oncogene product, and the modified G protein, such as mutant *ras*, or by the overexpression of genes regulating stages in signal transduction (e.g., *src* kinase) or transcriptional control (e.g., *myc*, *fos*, and *jun*). In many cases, the gene product is permanently active and is unable to be regulated.

TUMORIGENICITY

Transformation is a multistep process that often culminates in the production of neoplastic cells. However, cell lines derived from malignant tumors, presumably already transformed, can undergo further transformation with an increased growth rate, reduced anchorage dependence, more pronounced aneuploidy, and immortalization. This suggests that a series of steps, not necessarily coordinated or interdependent and not necessarily individually tumorigenic, is required for malignant transformation. Furthermore, all cell lines need not have the same transformed properties, and the same set of properties need not be expressed in every cell lines. Progression may imply the expression of new properties or the deletion of old ones that may induce metastasis(转移)or even spontaneous remission(缓解). There are, therefore, several steps in transformation, the sequence of which may be determined by environmental selective pressure. *In vitro*, where little restriction on growth is imposed, the events need not necessarily follow in the same sequence as *in vivo*.

Malignancy

Malignancy implies that the cells have developed the capacity to generate invasive tumors if implanted *in vivo* into an isologous(同源) host or if transplanted as a xenograft(异种移植) into an immune-deprived animal. While the development of malignancy can be recognized as a discrete phenotypic event, it often accompanies the development of aberrant growth control, suggesting that some of the lesions responsible for aberrant growth control also cause malignancy. An obvious candidate for such lesions is a deficit in cell-cell interaction that deprives the cell of control of proliferation and of motility control. The first, and most important, property of malignancy is tumorigenesis(致瘤性), which is universally accepted as a reliable criterion for malignant transformation, but is subject to a significant number of false negatives, due to problems with the implantation technique and rejection by the immune system, even in immunocompromised animals.

Tumorigenesis

The only generally accepted sign of malignancy is the demonstration of the formation of invasive or metastasizing tumors *in vivo*. Transplantable tumor cells ($\sim 1 \times 10^6$) injected into isogeneic hosts will produce invasive tumors in a high proportion of cases, while 10^6 normal cells of similar origin will not. Models have been developed, using immune-suppressed or immune-deficient host animals, to study the tumorigenicity of human tumors. The genetically athymic(无胸腺) "nude" mouse and thymectomized(切除胸腺) irradiated mice have both been used extensively as hosts for xenografts. In spite of the frequency of false negatives, tumorigenesis remains a good indicator of malignancy.

Invasiveness

Tumorigenesis assays should always be accompanied by histology of the tumor to confirm its histpathological similarity to the original tumor and to demonstrate that it is invasive. However, if the cells are not tumorigenic, or if transplantation facilities are not available or not considered desirable, then it is possible to utilize a number of *in vitro* assays, such as Chick chorioallantoic membrane(尿囊绒膜) (see Fig. 26.2) and filter well. The subsequent histology may reveal whether the tumor cells have penetrated the underlying basement membrane or filter.

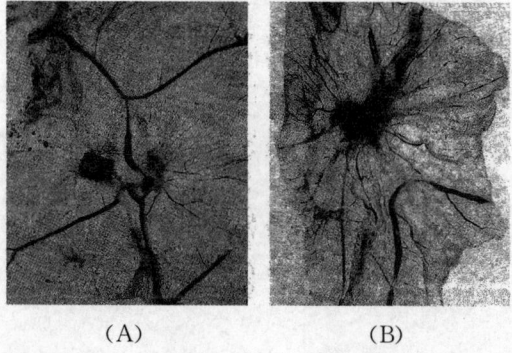

(A)   (B)

Fig. 26.2 Induction of angiogenesis. Neovascularization(新生血管) was induced in chick chorioallantoic membrane by crude extracts of normal glial cells (A) and Walker 256 carcinoma cells (B). The crude extract was placed on the chorioallantoic membrane of chick embryos at 10d incubation, and the membrane was removed 2 weeks later. (By Margaret Frame)

Angiogenesis(血管生成)

Tumor cells release factors, including VEGF, FGF-2, and angiogenin(血管生成素), that are capable of inducing neovascularization. Fragments of tumor, pellets of cultured cells, or cell extracts, implanted on the surface of the chorioallantoic membrane (尿囊膜) of a hen's egg, promote an increase in vascularization(血管生成) that is apparent to the naked eye 6-8 d later. (Fig. 26.2)

QUESTIONS FOR DISCUSSION

1. Explain the following concepts: transformation of cell line, malignant transformation, tumorigenicity, contact inhibition.
2. What is transformation? Describe the properties of transformed cells.
3. How to immortalize a finite cell line?

知识要点

与原代培养细胞相比,细胞系在体外培养过程中发生的在形态、生长特性和遗传特性等方面的可遗传的变化称为细胞系的转化。培养细胞的转化可以是自然发生的,也可以通过多形瘤病毒感染和肿瘤基因转染来诱导发生,或者通过电离辐射和致癌化合物诱变产生。转化细胞的生长特性表现为获得永生性、失去贴壁依赖性和接触抑制、高饱和密度、可集中生长和堆积生长、低血清依赖性、低生长因子依赖性、高铺板率和倍增时间缩短等方面。转化细胞的遗传物质不稳定,表现为非整倍性和异倍体性、高突变率、癌基因过表达、抑癌基因失活等方面。恶性转化细胞的细胞极性和贴附能力改变,具有了动物致瘤性以及血管生成性和侵染性。

CHAPTER 27 CHARACTERIZATION OF CELL LINES(细胞系的鉴定)

THE NEED FOR CHARACTERIZATION

When a new cell line is derived, either from a primary culture or from an existing cell line, it is difficult to assess its future value. Often, it is only after a period of use and dissemination that the true importance of the cell line becomes apparent, and, at that point, details of its origin are required. However, by that time, it is too late to collect information retrospectively(回顾). It is, therefore, vital that adequate records are kept, from the time of isolation of the tissue, or of the receipt of a new cell line, detailing the origin and handling of the cell line, and the more detailed the provenance(出处), the more valuable the cell line. This aspect of cell culture has become particularly important with the widespread dissemination(传播)of cell lines through cell banks and personal contacts to research laboratories and commercial companies far removed from their origin. In particular, if a cell line becomes incorporated into a procedure that requires validation of its components, then the authentication(鉴定)of the cell line becomes crucial. Authentication requires that the cell line be characterized on receipt, and periodically during use, and that these data are compatible with, and added to, the existing provenance.

Special attention must be paid to the possibility that the cell line has become cross-contaminated. The demonstration that the majority of continuous cell lines in use in the United States in the late 1960s had become cross-contaminated with HeLa cells first brought this serious problem to light, but the continued use of the lines 30 years later indicates that many people are still unaware, or are unwilling to accept, that many lines in common use are not authentic. Some of the methods in general use for cell line characterization are listed in Table 27.1.

There are six main requirements for cell line characterization: (1) Confirmation of the species of origin; (2) Correlation with the tissue of origin, which comprises the following charcteristics: (a) identification of the lineage to which the cell belongs and (b) position of the cells within that lineage (i.e., the stem, precursor, or differentiated status; (3) Whether the cell line is transformed or not: (a) is the cell line finite or continuous? (b) Does it express properties associated with malignancy? (4) Confirmation of the absence of cross-contamination (5) Indication of whether the cell line is prone to genetic instability, transformation, and phenotypic variation; (6) Identification of specific cell lines within a group from the same origin, selected cell strains, or hybrid cell lines, all of which require demonstration of features unique to that cell line or cell strain.

Table 27.1 Characterization of cell Lines and cell strains (By R. Ian Freshney)

Criterion	Method
Karyotype	Chromosome spread with banding
Isoenzyme analysis	Agar gel electrophoresis
Cell surface antigens	Immunohistochemistry
Cytoskeleton	Immunocytochemistry with antibodies to specific cytokeratins
DNA fingerprint	Restriction enzyme digest; PAGE; satellite DNA probes

SPECIES IDENTIFICATION

Chromosomal analysis is one of the best methods for distinguishing between species. Isoenzyme electrophoresis is also a good diagnostic test and is quicker than chromosomal analysis, but requires the appropriate apparatus and reagents. In practice, a combination of the two methods is often used and gives unambiguous results. Recently, techniques have been introduced for "chromosome painting"-i. e. , using combinations of specific molecular probes to hybridize to individual chromosomes. These probes identify individual chromosome pairs and are species specific. Thus chromosome painting is a good method for distinguishing between human and mouse chromosomes in potential cross-contaminations and interspecific hybrids. Certain chromosome banding patterns can also be used to distinguish human and mouse chromosomes.

LINEAGE OR TISSUE MARKERS

Cell Surface Antigens

These markers are particularly useful in sorting hematopoietic cells and have also been effective in discriminating epithelium from stroma with antibodies.

Intermediate Filament Proteins

These are among the most widely used lineage or tissue markers. Glial fibrillary acidic protein(神经胶质酸性蛋白) for astrocytes(星形胶质细胞) and desmin(肌间线蛋白) for muscle are the most specific, while cytokeratin(细胞角蛋白) marks epithelial cells and mesothelium(间皮). Vimentin(波形纤维蛋白), though usually restricted to mesodermally(中胚层) derived cells *in vivo*, can appear in other cell types *in vitro*.

Differentiated Products and Functions

Hemoglobin(血红蛋白)for erythroid cells(血红细胞), myosin(肌球蛋白)or tropomyosin(原肌球蛋白)for muscle, melanin(黑色素)for melanocytes(黑色素细胞), and serum albumin(血清白蛋白)for hepatocytes(肝细胞)are among the best examples of specific cell type markers, but, like all differentiation markers, they depend on the complete expression of the differentiated phenotype.

Transport of inorganic ions, and the resultant transfer of water, is characteristic of some epithelia; grown as monolayers, they will produce *domes*(瘤状突起), which are hemicysts(半瘤状物) in the monolayer caused by accumulation of water on the underside of the monolayer. (Fig. 27.1)

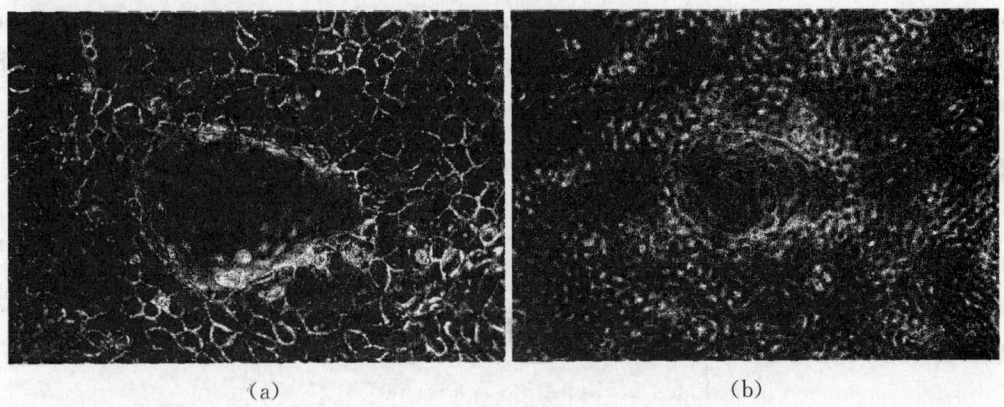

(a) (b)

Fig. 27.1 Domes. (a) Dome, or hemicyst, formed in an epithelial monolayer by downward transport of ions and water, lower focus (on monolayer). (b) Upper focus (top of dome). (By R. Ian Freshney)

Enzymes

Three parameters are available in enzymic characterization: (1) the constitutive level; (2) the response to inducers and repressors; and (3) isoenzyme polymorphisms. Creatine kinase(肌酸激酶)BB isoenzyme is characteristic of neuronal and neuroendocrine(神经内分泌)cells, as is neuron-specific enolase(烯醇化酶); lactic dehydrogenase(乳酸脱氢酶) is present in most tissues, but as different isoenzymes, and a high level of tyrosine aminotransferase(酪氨酸氨基转移酶), inducible by dexamethasone(地塞米松), is generally regarded as specific to hepatocytes.

Regulation

Although differentiation is usually regarded as an irreversible process, the level of expression of many differentiated products is under the regulatory control of environmental influences, such as hormones, the matrix, and adjacent cells. Hence, the measurement of specific lineage markers may require preincubation of the cells in, for example, a hormone such as hydrocortisone, specific growth factors, or growth of the cells on extracellular matrix of the correct type.

Lineage Fidelity

Although many of the markers described above have been claimed as lineage markers, they are more properly regarded as tissue or cell type markers, as they are often more characteristic of the function of the cell than its embryologic origin. Cytokeratins occur in mesothelium and kidney epithelium, although both of these tissues derive from the mesoderm.

Unique Markers

Unique markers include specific chromosomal aberrations (e. g., deletions, translocations, polysomy(多染色体性)); major histocompatibility (组织相容性) group antigens (e. g., HLA in human), which are highly polymorphic(多态); and DNA fingerprinting. Enzymic deficiencies (e. g., thymidine kinase deficiency (TK^-)) and drug resistance (e. g., vinblastine (长春花碱) resistance) are not truly unique, but may be used to distinguish among cell lines from the same tissues but different donors.

TRANSFORMATION

Transformation forms a major element in cell line characterization and is dealt with separately.

MORPHOLOGY

Observation of morphology is the simplest and most direct technique used to identify cells. It has, however, certain shortcomings that should be recognized. Most of these are related to the plasticity(可塑性)of cellular morphology in response to different culture conditions; For example, epithelial cells growing in the center of a confluent sheet are usually regular, polygonal, and with a clearly defined edge, while the same cells growing at the edge of a patch may be more irregular and distended(肿胀) and, if transformed, may break away from the patch and become fibroblast-like in shape. Subconfluent fibroblasts from hamster kidney or human lung or skin assume multipolar or bipolar shapes and are well spread on the culture surface, but at confluence they are bipolar and less well spread. They also form characteristic parallel arrays and whorls (漩涡) that are visible to the naked eye. Mouse 3T3 cells and human glial cells grow like multipolar fibroblasts at low cell density, but become epithelial-like at confluence. (Fig. 27.2) Alterations in the substrate, and the constitution of the medium, can also affect cellular morphology. Hence, comparative observations of cells should always be made at the same stage of growth and cell density in the same medium, and growing on the same substrate. The terms "fibroblastic"(成纤维细胞的) and "epithelial" (上皮的)are used rather loosely in tissue culture and often describe the appearance rather than the origin of the cells. Thus, a bipolar or multipolar migratory cell, the length of which is usually more than twice its width, would be called "fibroblastic," while a monolayer cell that is polygonal, with more regular dimensions, and that grows in a discrete patch along with other cells is usually regarded as "epithelial." However, when the identity of the cells has not been confirmed, the terms "fibroblast-like"(成纤维细胞样) and "epithelial-like"(上皮样)should be used.

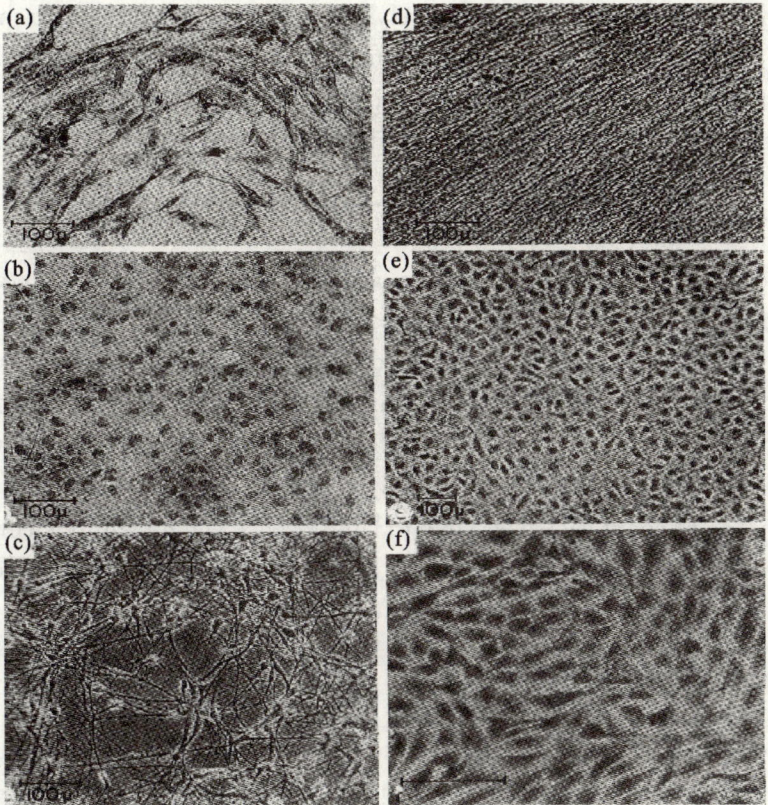

Fig. 27.2 **Cell morphology.** (a) BHK-21 (baby hamster kidney fibroblasts), clone 13, in log growth. The culture is not confluent, and the cells are well spread and randomly oriented (although some orientation is beginning to appear). (b) Cells of an epithelial-like morphology from fetal human intestine (FHI). (c) Astrocytes from human astrocytoma. This pattern is quite characteristic, but is lost as the cells are passaged, and a morphology not unlike that shown in (b), (e), or (f) develops. (d) Plateau-phase BHK-21 C13 cells. The cells are smaller, more highly condensed, and have assumed a parallel orientation with each other. (e) Bovine aortic endothelium. This morphology has a regular appearance similar to that shown in (b), though here the cells are more closely packed. (f) Again, a pavementlike appearance similar to that found in (b) and (c), but now produced by 3T3 cells, mouse fibroblasts. With experience, these cell types may be distinguished, but their similarity underlines the need for criteria for identification other than morphology. (a, b, d) Giemsa stained. (c, e-h) Phase contrast. (By R. Ian Freshney)

CHROMOSOME ANALYSIS

Chromosome content is one of the most characteristic and well-defined criteria for identifying cell lines and relating them to the species and sex from which they were derived. Chromosome analysis can also distinguish between normal and malignant cells, since the chromosome number is more stable in normal cells.

DNA CONTENT

The amount of DNA per cell is relatively stable and is characteristic to the species in normal cell lines, but varies in cell lines from the mouse and from many neoplasms(肿瘤). Analysis of DNA content is particularly useful in the characterization of transformed cells that are often aneuploid(非整倍体) and heteroploid(异倍体).

DNA HYBRIDIZATION

It is possible to hybridize specific molecular probes to unique DNA sequences (Southern blotting) and detect the hybrids by radioisotopic(放射性同位素), fluorescent (荧光), or luminescent(发光) labels. Extracted DNA, whole or cut with restriction endonucleases, electrophoresed(电泳), blotted onto nitrocellulose(硝酸纤维素膜), and hybridized to a particular probe or set of probes, provides information about species-specific regions, amplified regions of the DNA, or altered base sequences that are characteristic to that cell line. It is also possible to label a cell strain for future identification by transfecting(转染) in a reporter gene, such as β-galactosidase, and then to detect it by Southern blotting or by chromogenic assay for the gene product.

DNA FINGERPRINTING

DNA contains regions known as satellite DNA that are apparently not transcribed. These regions are not highly conserved, and give rise to regions of hyper-variability. When analyzed by polyacrylamide electrophoresis(聚丙烯酰胺凝胶电泳), each individual's DNA gives a specific hybridization pattern as revealed by autoradiography with radioactive probes. These patterns have come to be known as DNA fingerprints and are cell line specific. DNA fingerprints appear to be quite stable in culture, and cell lines from the same origin, but maintained separately in different laboratories for many years, still retain the same or very similar DNA fingerprints. Furthermore, if a cross-contamination is suspected, this can be confirmed or denied by fingerprinting the cells and all potential contaminants.

RNA AND PROTEIN

Cells of a particular characteristic phenotype can be recognized by analysis of gene expression by Northern blotting using radioactive, fluorescent, or luminescent probes. This procedure can be carried out at the cellular level, as in FISH, or *by in situ* hybridization with radioactive probes. Qualitative analysis of total cell protein reveals differences between cells when whole cells, or cell membrane extracts, are run on two-dimensional gels. This technique produces a characteristic "fingerprint" similar to polypeptide maps of protein hydrolysates. This protein fingerprint contains enough information to distinguish among cell lines.

ENZYME ACTIVITY

Specialized functions *in vivo* are often expressed in the activity of specific enzymes. Unfortunately, many enzyme activities are lost *in vitro* and are no longer available as markers of tissue specificity-e. g., liver parenchyma loses arginase activity within a few days of culture. However, some cell lines do express specific enzymes, such as tyrosine aminotransferase in the rat hepatoma HTC cell lines. When looking for specific marker enzymes, the constitutive (uninduced) level and the induced level should be measured and compared with a number of control cell lines.

ISOENZYMES

Enzyme activities can also be compared qualitatively between cell strains, due to enzyme protein polymorphisms among species and, sometimes, among races, individuals, and tissues within a species. These socalled *isoenzymes*, or *isozymes*, may be separated chromatographically(色谱法) or electrophoretically(电泳), and the distribution patterns (酶谱, zymograms) may be found to be characteristic of species or tissue. (Fig. 27.3) The species of origin of a cell line can often be determined by seven isoenzymes: nucleoside phosphorylase (核糖核苷酸磷酸化酶, NP; E. C. 2. 4. 2. 1); glucose-6-phosphate dehydrogenase (6-磷酸葡萄糖脱氢酶, G6PD; E. C. 1. 1. 1. 49); malate dehydrogenase (苹果酸脱氢酶, MD, E. C. 1. 1. 1. 37); lactate dehydrogenase (乳酸脱氢酶, LD; E. C. 1. 1. 1. 27); aspartate aminotransferase (天冬氨酸转氨酶, AST, E. C. 2. 6. 1. 1); mannose-6-phosphate isomerase (6-磷酸甘露糖异构酶, MPI, E. C. 5. 3. 1. 8); and peptidase B (Pep B, E. C. 3. 4. 11. 4). In most cases, the species of origin can be determined by using only four of the seven isoenzymes listed: nucleoside phosphorylase, glucose-6-phosphate dehydrogenase, malate dehydrogenase, and lactate dehydrogenase. Similarly, interspecies cell line cross-contamination can be detected in most cases using these four isoenzymes.

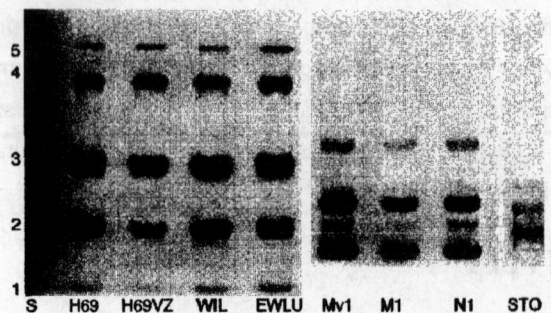

Fig. 27.3 Isoenzyme Patterns. Agarose electrophoresis of lactate dehydrogenase isoenzymes in four human cell lines (NCI-H69; H69VZ, an adherent derivative of NCI-H69; WIL, a human lung adenocarcinoma; and EWLU, human fetal lung fibroblasts), two from mink lung-MV1Lu (MV1) and M1, *a myc* transfected derivative of MV1Lu-and STO mouse fibroblasts. (By R. Ian Freshney)

ANTIGENIC MARKERS

As a result of the abundance of antibodies and kits from commercial suppliers, immunostaining and Elisa assays are among the most useful techniques available for cell line characterization. Regardless of the source of the antibody, however, it is essential to be certain of its specificity by using appropriate control material. This is true for monoclonal antibodies and polyclonal antisera alike; a monoclonal antibody is highly specific for a particular epitope, but the generality or specificity of the expression of the epitope must still be demonstrated.

DIFFERENTIATION

Many of the characteristics described under antigenic markers or enzyme activities may also be regarded as markers of differentiation, and as such, they can help to correlate cell lines with their tissue of origin.

AUTHENTICATION

Characterization of a cell line is vital, not only in determining its functionality, but also in proving its authenticity. Cell culture has been plagued since the 1960s with repeated examples of cross contamination, mostly, but by no means invariably, with HeLa cells. Characterization studies, particularly with continuous cell lines, have thus become a process of authentication that is vital to the validation of the data derived from these cells.

QUESTIONS FOR DISCUSSION

1. Why should a cell line be characterized?

2. How to characterize a newly established cell line?

知识要点

细胞系建立之后,必须对细胞系进行鉴定和特性分析,以确定细胞系的种属来源和细胞谱系来源、细胞系是否发生交叉污染、是否发生转化,以及遗传物质是否稳定等,这对于细胞系的进一步应用具有重要意义。鉴定方法主要有形态学分析、生长特性分析、染色体分析、同工酶分析、转化特征分析、细胞谱系或组织的遗传标记分析等。

CHAPTER 28 CONTAMINATION OF CULTURED CELLS(培养细胞的污染)

SOURCES OF CONTAMINATION

Maintaining asepsis is still one of the most difficult challenges to the newcomer to tissue culture. Part of the difficulty, of course, is due to awkwardness during early training, and experience will eventually cure the problem. However, in certain situations, even the most experienced worker will suffer from contamination. There are several potential routes to contamination (Table 28.1) including failure in the sterilization procedures for glassware and pipettes, turbulence and particulates (dust and spores) in the air in the room, poorly maintained incubators and refrigerators, faulty laminar-flow hoods, the importation of contaminated cell lines or biopsies, and poor technique. The last of these is probably the most significant.

Operator Technique

If reagents are sterile and equipment is in proper working order, contamination depends on the interaction of the operator's technique with environmental conditions. If the skill and level of care of the operator is high and the atmosphere is clean, free of dust, and still, contamination as a result of manipulation will be rare.

Table 28.1 Routes to contamination

Route or cause	Prevention
Technique	
Manipulations, pipetting, dispensing, etc	
Nonsterile surfaces and equipment	Clear work area of items not in immediate use
Spillage on necks and outside of bottles and on work surface	Swab regularly with 70% alcohol. Do not pour liquids. Dispense or transfer by pipette, autodispenser, or transfer device. If pouring is unavoidable: (1) do so in one smooth movement, (2) discard the bottle that you pour from, and (3) wipe up any spillage

(续表)

Route or cause	Prevention
Touching or holding pipettes too low down, touching necks of bottles, inside screw caps	Hold pipettes above graduations. Do not work over open vessels
Splash-back from waste beaker	Discard waste into a beaker with a funnel or, preferably, by drawing off the waste into a reservoir by means of a vacuum pump
Sedimentary dust or particles of skin settling on the culture or bottle. Hands or apparatus held over an open dish or bottle	Do not work over (vertical laminar flow and open bench) or behind and over (horizontal laminar flow) an open bottle or dish
Work surface	
Dust and spillage	Swab the surface with 70% alcohol before, after, and during work. Mop up spillage immediately
Operator hair, hands, breath, clothing	
Dust from skin, hair, or clothing dropped or blown into the culture	Wash hands thoroughly or wear gloves. Wear a lint-free lab coat with tight cuffs and gloves overlapping them
Aerosols from talking, coughing, sneezing, etc	Keep talking to a minimum and face away from work when you talk. Avoid working with a cold or throat infection, or wear a mask. Tie back long hair or wear a cap. Wear a lab coat different from the one you wear in the general lab area or animal house
Materials and reagents	
Solutions	
Nonsterile reagents and media	Filter or autoclave solutions before using them
Dirty storage conditions	Clean up storage areas and disinfect regularly
Inadequate sterilization procedures	Monitor the performance of the autoclave with a recording thermometer or sterility indicator. Check the integrity of filters with a bubble-point or microbial assay after using them. Test all solutions after sterilization
Poor commercial supplier	Test solutions; change suppliers
Glassware and screw caps	
Dust and spores from storage	Shroud caps with foil. Wipe bottles with EtOH before taking them into the hood. Replace stocks from the back of the shelf. Do not store anything unsealed for more than 24 h

CYTOTECHNOLOGY（细胞工程技术）

（续表）

Route or cause	Prevention
Ineffective sterilization (e. g., an overfilled oven or sealed bottles, preventing the ingress of steam)	Check the temperature of the load throughout the cycle. In the autoclave, keep caps slack on empty bottles. Stack oven and autoclave correctly
Instruments, pipettes	
Ineffective sterilization	Sterilize items by dry heat before using them. Monitor the performance of the oven
Contact with a nonsterile surface or some other material	Resterilize instruments using 70% alcohol. Do not grasp any part of an instrument or pipette that will pass into a culture vessel
Invasion by insects, mites, or dust	Do not store instruments for more than 24 h, unless sealed with tape
Culture flasks and media bottles in use	
Dust and spores from incubator or refrigerator	Use screw caps instead of stoppers. Swab bottles before placing in hood. Box plates and dishes
Dirty storage or incubation conditions	Cover caps and necks of bottles with aluminum foil during storage or incubation. Wipe flasks and bottles with 70% alcohol before using them. Clean out stores and incubators regularly
Media under the cap and spreading to the outside of the bottle	Discard all bottles that show spillage on the outside of the neck. Do not pour
Equipment and Facilities	
Room air	
Drafts, eddies(漩涡), turbulence, dust, aerosols	Clean filtered air. Reduce traffic and extranegus activity. Wipe the floor and work surfaces regularly
Laminar-Flow Hoods	
Perforated(穿孔) filter	Check filters regularly for holes and leaks
Change of filter needed	Check the pressure drop across the filter
Spillages, particularly in crevices or below a work surface	Clear around and below the work surface regularly. Let alcohol run into crevices
Dry incubators	
Growth of molds and bacteria on spillages	Wipe up any spillage with 70% alcohol on a swab. Clean out incubators regularly

(续表)

Route or cause	Prevention
Humidified CO_2 incubators	
Growth of molds and bacteria on walls and shelves in a humid atmosphere	Clean out with detergent followed by 70% alcohol
Spores, etc., carried on forced-air circulation	Enclose open dishes in plastic boxes with close-fitting lids. Swab incubators with 70% alcohol before opening them. Put a fungicide or bacteriocide in humidifying water
Mites, insects, and other infestations ( ) in wooden furniture, or benches, in incubators, and on mice, etc., taken from the animal house	
Entry of mites, etc., into sterile packages	Seal all sterile packs. Avoid wooden furniture if possible; use plastic laminate, one-piece, or stainless-steel bench tops. Keep animals out of the tissue culture lab
Importation of Biological Materials	
Tissue samples	
Infected at source or during dissection	Do not bring animals into the tissue culture lab. Incorporate antibiotics into the dissection fluid. Dip all potentially infected large-tissue samples in 70% alcohol for 30 s
Incoming cell lines	
Contaminated at the source or during transit	Handle these cell lines alone, preferably in quarantine, after all other sterile work is finished. Swab down the bench or hood after use with 2% phenolic disinfectant in 70% alcohol, and do not use it until the next morning. Check for contamination by growing a culture for two weeks without antibiotics. Check for contamination visually, by phase-contrast microscopy and Hoechst stain for mycoplasma. Using indicator cells allows screening before first subculture

Environment

It is fairly obvious that the environment in which tissue culture is carried out must be as clean as possible and free from disturbance and through traffic. Equipments brought in and air currents from doors, refrigerators, centrifuges, and the movement of staff increase the risk of contamination. Accordingly, one should keep the area clean and free of through traffic and wipe down anything that is brought in.

Use and Maintenance of Laminar-flow Hood(层流式超净工作台)

The most common form of poor technique is improper use of the laminar-flow hood. If it becomes overcrowded with bottles and equipment, the laminar airflow is disrupted, and the protective boundary layer between operator and room is lost. This in turn leads to the entry of nonsterile air into the hood and the release of potentially biohazardous materials into the room. One should bring into the hood only those items that are directly involved in the current operation. Laminar-flow hoods also must be maintained regularly, and the integrity of their filters and the security of their cabinets should be checked at least twice a year by a competent engineer.

Humid Incubators

A major source of contamination stems from the use of humid incubators. High humidity is not required, unless open vessels are being used; sealed flasks are better kept in a dry incubator or the hot room. Using permeable(透气) caps minimizes the risk of contamination, but increases the unit cost and still exposes the flask to a higher risk atmosphere than in a dry incubator. Copper-lined incubators have reduced fungal growth, but are usually about 20%-30% more expensive than conventional ones. A number of fungal retardants(抑真菌剂)are in common use, including copper sulfate, riboflavin(核黄素), sodium dodecyl sulfate (SDS), and Roccall, a proprietary fungicidal cleaner used in a 2% solution. Remember, a fungicide will only protect the tray; there is no substitute for regular cleaning! Cleaning should be carried out regularly using 10% Roccall or an equivalent nontoxic antifungal cleaner. When the incubator is in use, any spillage must be mopped up immediately and contaminated cultures removed as soon as they are detected.

Cold Stores

Refrigerators and cold rooms also tend to build up fungal contamination on the walls in a humid climate, due to condensation that forms every time the door is opened, admitting moist air. The moist air increases the risk of deposition of spores on stored bottles; hence, they should be swabbed with alcohol before being placed in the hood. The cold store should be cleared, and the walls and shelving should be washed down with disinfectant(消毒剂)every few months.

Imported Cell Lines and Biopsies

Any biological material brought into the laboratory runs the risk of being contaminated. All such material should be maintained in quarantine until it is shown to be clear of contamination, at which point it can join other stocks in general use.

TYPES OF MICROBIAL CONTAMINATION

Bacteria, virus, yeasts, fungi, molds, and mycoplasmas(支原体)all appear as contaminants in tissue culture, and if protozoology is carried on in the same laboratory, or primary culture are derived from aquatic organisms, some protozoa can infect cell lines. Usually, the species or type of infection is not important, unless it becomes a frequent occurrence. It is only necessary to note the general kind of contaminant (e. g. , bacterial rods(杆菌) or cocci(球菌), yeast, etc.), how it was detected, the location where the culture was last handled, and the operator's name. If a particular type of infection recurs frequently, it may be beneficial to identify it in order to find its origin.

Monitoring Contamination

Potential sources of contamination are listed in Table 28.1, along with the precautions that should be taken to avoid them. Even in the best laboratories, however, contaminations do arise, so the following procedure is recommended:

(1) Check for contamination by eye and with a microscope at each handling of a culture. Check for mycoplasma every month.

(2) If it is suspected, but not obvious, that a culture is contaminated, check the culture with a microscope, preferably by phase contrast. If it is confirmed that the culture is contaminated, discard it and all suspected media and stock solutions.

(3) If the same kind of contamination has occurred before, check stock solutions for contamination (a) by incubation alone or in nutrient broth or (b) by plating out the solution on nutrient agar. If (a) and (b) prove negative, but contamination is still suspected, incubate 100 mL of solution, filter it through a 0.2 μm filter, and plate out filter on nutrient agar with an uninoculated control.

(4) If the contamination is widespread, multispecific, and repeated, check the laboratory's sterilization procedures (e. g. , the temperatures of ovens and autoclaves, particularly in the center of the load, the durations of the sterilization cycle, the packaging, the storage practices, and the integrity of the aseptic room and the filters on the laminar-flow hood filters).

(5) Do not attempt to decontaminate cultures unless they are irreplaceable.

Characteristic features of visible microbial contamination are as follows:

(1) A sudden change in pH, usually a decrease with most bacterial infections, very little change with yeast until the contamination is heavy, and sometimes an increase in pH with fungal contamination.

(2) Cloudiness in the medium, sometimes with a slight film or scum(浮渣) on the surface or spots on the growth surface that dissipate(消散) when the flask is moved.

(3) Under a low-power microscope ($\times 100$), spaces between cells will appear gran-

ular and may shimmer (发微光) with bacterial contamination (Fig. 28.1a). Yeasts appear as separate round or ovoid (卵形) particles that may bud off smaller particles (Fig. 28.1b). Fungi produce thin filamentous mycelia (菌丝体) (Fig. 28.1c) and, sometimes, denser clumps of spores. With toxic infection, some deterioration of the cells will be apparent.

(4) Under high-power microscopy ($\times 400$), it may be possible to resolve individual bacteria and distinguish between rods (杆菌) or cocci (球菌). At this magnification, the shimmering that is visible in some infections will be seen to be due to mobility of the bacteria.

(5) With a slide preparation, the morphology of the bacteria can be resolved at $\times$ 1,000, but this is not usually necessary. Microbial infection may be confused with precipitates of media constituents (particularly protein) or with cell debris, but can be distinguished by their regular morphology. Precipitates may be crystalline or globular and irregular and are not usually as uniform in size. If you are in doubt, plate out a sample of medium on nutrient agar.

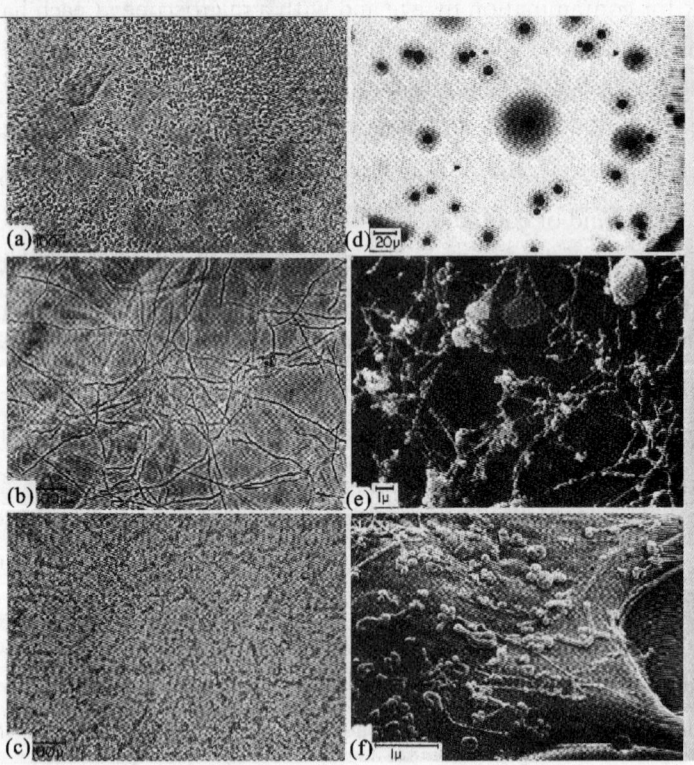

Fig. 28.1 **Contamination.** Examples of microorganisms found to contaminate cell cultures. (a) Yeast. (b) Mold. (c) Bacteria. (d) Mycoplasma colonies growing on special nutrient agar. (e,f) Scanning electron micrograph of mycoplasma growing on the surface of cultured cells (By R. Ian Freshney).

Mycoplasma

Mycoplasmal infections (Fig. 28.1d-f) cannot be detected by the naked eye other than through signs of deterioration in the culture. The culture must be tested specially by fluorescent staining, PCR, ELISA assay, immunostaining, autoradiography, or microbiological assay. Fluorescent staining of DNA by Hoechst 33258 is the easiest and most reliable method and reveals mycoplasmal infections as a fine particulate or filamentous staining over the cytoplasm at × 500 magnification. The nuclei of the cultured cells are also brightly stained by this method and thereby act as a positive control for the staining procedure. Most other microbial contaminations will also show up with fluorescence staining, so low levels of contamination or particularly small organisms such as micrococci（微球菌）can also be detected.

It is important to appreciate the fact that mycoplasmas do not always reveal their presence by means of macroscopic（肉眼可见）alterations of the cells or media. Many mycoplasma contaminants, particularly in continuous cell lines, grow slowly and do not destroy host cells. However, they can alter the metabolism of the culture in many different ways. Because mycoplasmas take up thymidine from the medium, infected cultures show abnormal labeling with $[^3H]$-thymidine. Immunological studies can also be totally frustrated by mycoplasmal contamination, as attempts to produce antibodies against the cell surface may raise antimycoplasma antibodies. Mycoplasmas can alter cell behavior and metabolism in many other ways, so there is an absolute requirement for routine, periodic assays to detect possible covert contamination of all cell cultures, particularly continuous cell lines.

Viral Contamination

Incoming cell lines, natural products, such as serum in media and enzymes such as trypsin, used for subculture, are all potential sources of viral contamination. Screening with a panel of antibodies by immunostaining or ELISA assays is probably the best way of detecting viral infection. Alternatively, one may use PCR with the appropriate viral primers.

ERADICATION OF CONTAMINATION

Eradication of Bacteria, Fungi, and Yeasts

Decontamination should be attempted only in extreme situations, under quarantine （隔离）, and with expert supervision. Wash the culture several times in a high concentration of antibiotics by rinsing the monolayer or by centrifugation of cells in suspension. Then grow the culture for three subcultures with, and three without, antibiotics. Test for contamination after each subculture.

Eradication of Mycoplasma

If mycoplasma is detected in a culture, the first and overriding rule, as with other forms of contamination, is that the culture should be discarded for autoclaving or incineration(焚化). In exceptional cases, one may attempt to decontaminate the culture. Decontamination should be done, however, only by an experienced operator, and the work must be carried out under conditions of quarantine. Several agents are active against mycoplasma, including kanamycin(卡那霉素), gentamycin(庆大霉素), tylosin(泰乐菌素), polyanethol sulfonate(聚茴香脑磺酸), and 5-bromouracil(5-溴尿嘧啶) in combination with Hoechst 33258 and UV light. Coculturing with macrophages, animal passage, and cytotoxic antibodies can also be effective in some cases. However, the most successful agents have been tylosin, Mycoplasma Removal Agent, ciprofloxacin (环丙沙星), and BM-Cycline (Boehringer-Mannheim 公司). It is far safer to discard infected cultures.

Eradication of Viral Contamination

There are no reliable methods for eliminating viruses from a culture at present; disposal or tolerance is the only options.

Eradication of Persistent Contamination

Many laboratories have suffered from periods of contamination which seems to be refractory(难治疗) to all the remedies suggested in Table 28.1. There is no easy resolution to this problem, other than to follow the previous recommendations in a logical and analytical fashion, paying particular attention to changes in technique, new staff, new suppliers, new equipment, and insufficient maintenance of laminar-flow hoods or other equipment. Typically, an increase in the contamination rate stems from deterioration in aseptic technique, an increased spore count in the atmosphere, poorly-maintained incubators, a contaminated cold room or refrigerator, or a minor, intermittent fault in a sterilizing oven or autoclave.

The constant use of antibiotics also favors the development of chronic contamination. Many organisms are inhibited, but not killed, by antibiotics. They will, therefore, persist in the culture, undetected for most of the time, but periodically surfacing when conditions change or when there are intrinsic host-parasite-type population fluctuations. It is essential that your cultures be maintained in antibiotic-free conditions for at least part of the time, and preferably all the time; otherwise cryptic contaminations will persist, their origins will be difficult to determine, and eliminating them will be impossible. If strict practices are maintained, contamination may not be eliminated entirely, but it will be detected early.

CROSS-CONTAMINATION

During the history of tissue culture, a number of cell strains have evolved with very short doubling times and high plating efficiencies. Although these properties make such cell lines valuable experimental material, they also make them potentially hazardous for cross-infecting other cell lines. The extensive cross-contamination of many cell lines with HeLa and other rapidly growing cell lines is now clearly established, but many operators are still unaware of the seriousness of the risk.

The following practices help avoid cross-contamination:

1. Obtain cell lines from a reputable cell bank that has performed appropriate characterization of the cells, or perform the necessary characterization yourself as soon as possible.

2. Do not have media bottles or culture flasks with more than one cell line open simultaneously.

3. Handle rapidly growing lines, such as HeLa, on their own and after other cultures.

4. Never use the same pipette for different cell lines.

5. Never use the same bottle of medium, trypsin, etc., for different cell lines.

6. Do not put a pipette back into a bottle of medium, trypsin, etc., after it has been in a culture flask containing cells.

7. Add medium and any other reagents to the flask first, and then add the cells last.

8. Do not use unplugged pipettes, or pipettors without plugged tips, for routine maintenance. Check the characteristics of the culture regularly, and suspect any sudden change in morphology, growth rate, etc.

IN SUMMARY

1. Check living cultures regularly for contamination by using normal and phase-contrast microscopy and for mycoplasmas by employing fluorescent staining of fixed preparations.

2. Do not maintain all cultures routinely in antibiotics: Grow at least one set of cultures of each cell line without antibiotics for a minimum of two weeks at a time, and preferably continuously, in order to allow cryptic contaminations to become overt.

3. Do not attempt to decontaminate a culture unless it is irreplaceable, and then do so only under strict quarantine.

4. Quarantine all new lines that come into your laboratory until you are sure that they are uncontaminated.

5. Do not share media or other solutions among cell lines or among operators, and check cell line characteristics periodically to guard against cross-contamination.

6. New cell lines should be characterized, preferably by DNA fingerprinting, as soon after isolation as possible.

It cannot be overemphasized that cross-contaminations can and do occur. It is essential that the preceding precautions be taken and that cell strain characteristics be checked regularly.

QUESTIONS FOR DISCUSSION

1. What are the resources of contamination in cell culture?

2. How many types of microbial contaminations often happen in cell culture? How to eradicate them?

3. What is cross contamination? How to eradicate it?

知识要点

无菌操作和避免污染是细胞培养成功的关键。培养器具和培养液的灭菌不彻底、培养箱和超净工作台不干净、实验操作不恰当等等，都可能导致微生物污染。常见的污染微生物包括细菌、病毒、酵母、真菌、霉菌和支原体，水生动物细胞培养时还常常发生原生动物的污染。通过普通光学显微镜可观察到细菌、酵母、真菌、霉菌和原生动物的污染。细菌污染严重时可导致培养液变浑浊和变黄。酵母污染时，可通过观察是否有出芽生殖来将其与细菌和细胞区分开来。真菌和霉菌污染时，显微镜下可见到树枝状菌丝；严重时，大的丝状菌斑肉眼可见。支原体污染时，不易察觉，需要借助DNA荧光染料或支原体培养基。病毒污染严重时可观察到蚀斑。培养细胞一旦被微生物污染，最好丢掉；只有极其珍贵的细胞，才有必要通过抗生素处理等方法进行挽救。实验室同时培养多个细胞系时，还要避免细胞系之间的交叉污染。

CHAPTER 29   CRYOPRESERVATION OF CULTURED CELLS(培养细胞的低温冻存)

NEED FOR CRYOPRESERVATION

The adoption of newly developed cell lines or imported cells lines into regular use implies an investment in time and resources increases, often exponentially, with continued use. The cell line becomes a valuable resource, replacement of which would be expensive and time-consuming. It is, therefore, essential to protect this considerable investment by preserving the cell line.

Cell lines in continuous culture are prone to variation, due to selection in early-passage culture, senescence in finite cell lines, and genetic instability in continuous cell lines. In addition, even the best-run laboratory is prone to equipment failure and contamination. Cross-contamination also continues to occur at an alarming frequency. There are many reasons, therefore, for freezing down a stock of cells; these reasons can be summarized as follows: (1) Genotypic drift due to genetic instability; (2) Senescence (衰老); (3) Transformation; (4) Phenotypic instability due to selection and dedifferentiation; (5) Contamination by microorganisms; (6) Cross-contamination by other cell lines; (7) Incubator failure; (8) Saving time and materials maintaining lines not in immediate use; (9) Need for distribution to other users.

CRYOPRESERVATION

Selection of Cell Line

A cell line is selected with the required properties. If it is a finite cell line, it is grown to around the fifth population doubling in order to create a sufficient bulk of cells for freezing. Continuous cell lines should be cloned and an appropriate clone selected and grown up to sufficient bulk to freeze. Prior to freezing, the cells should be maintained under standardized culture conditions, characterized, and checked for contamination, particularly cross-contamination.

Storage

Storage in liquid nitrogen is currently the most satisfactory method of preserving cultured cells. The cell suspension, preferably at a high concentration, should be frozen slowly, at $-1°C$ per min, in the presence of a preservative (保护剂) such as glycerol (甘油) or dimethyl sulfoxide (二甲基亚砜, DMSO). The frozen cells are transferred rapidly to liquid nitrogen when they are at or below $-70°C$; it is critical that they do not warm up above $-50°C$, when they will start to deteriorate. The ampules (小玻璃瓶) or plastic tubes for cryopreservation are then stored immersed in liquid nitrogen or in the gas phase above the liquid.

It must be realized that the cooling rate is proportional to the difference in temperature between the ampules and the ambient air. If the ampules are placed in a freezer at $-70°C$, they will cool rapidly to around $-50°C$, but the cooling rate falls off significantly after that. Hence, the time that the ampules spend in the $-70°C$ freezer needs to be longer than the amount of time projected by a $1°C$ cooling rate. It is safer to leave the ampules at $-70°C$ overnight before transferring them to liquid nitrogen. Furthermore, when removed from the insulated tube or container, they will heat up at a rate of $-10°C/min$, so the transfer to liquid nitrogen must take significantly less than two minutes.

When required, the cells are thawed rapidly and reseeded at a relatively high concentration to optimize recovery. If liquid nitrogen storage is not available, the cells may be stored in a conventional freezer. The temperature in this freezer should be as low as possible; little deterioration has been found at $-196°C$, but significant deterioration (5-10% per annum) may occur at $-70°C$.

Freezing Cells

Outline

Grow the culture to late log phase, prepare a high-cell-density suspension, and freeze it slowly with a preservative.

Protocol

1. Check the culture for the following: (a) healthy growth; (b) freedom from contamination; (c) specific characteristics.

2. Grow the culture up to the late log phase and, if you are using a monolayer, trysinize and count the cells. If you are using a suspension, count and centrifuge the cells.

3. Dilute one of the preservatives in growth medium to make freezing medium: (a) Add dimethyl sulfoxide (DMSO) to 5%-10% or (b) Add glycerol to 10%-15%.

4. Resuspend the cells in freezing medium at approximately $1 \times 10^6 - 1 \times 10^7$ cells/mL. It is not advisable to place ampules on ice in an attempt to minimize deterioration of the cells. A delay of up to 30 min at room temperature is not harmful when using DMSO and is beneficial when using glycerol.

5. Dispense the cell suspensions into pre-labeled ampules and seal the ampules, or

plastic tubes.

6. Freeze the ampules or tubes by one of the following methods: (a) Lay the ampules or tubes on cotton wool in a polystyrene foam box with a wall thickness of ~15 mm. This box, plus the cotton wool, should provide sufficient insulation such that the ampules will cool at 1℃/min when the box is placed at −70℃ or −90℃ in a regular deep freeze or insulated container with solid CO_2. (b) Place the cells in a Nalgene freezing container (Fig. 29.1). (c) Use a controlled-rate freezer programmed to freeze at 1℃/min (Fig. 29.2), with accelerated freezing through the eutectic point (共晶点).

7. When the ampules or tubes have reached −70℃ (a minimum of 4-6 h after placing them at −70℃ if starting from 20℃ ambient, but preferably overnight), transfer them to a liquid N_2 freezer. This transfer must be done quickly, as the ampules will reheat at −10℃/min, and the cells will deteriorate rapidly if the temperature rises above −50℃.

It is possible to check for leakage by placing glass ampules in a dish of 1% methylene blue(亚甲基蓝) in 70% alcohol at 4℃ for 10 min before freezing. If the ampules are not properly sealed, the methylene blue will be drawn into the ampule, and the ampule should be discarded.

Cryofreezers(冷冻罐)

There are four main types of liquid-nitrogen storage systems (see Fig. 19.5), based on whether the storage vessel is wide-necked or narrow-necked, and whether storage is in the vapor or liquid phase. Wide-necked freezers are chosen for ease of access and maximum capacity, and narrow-necked freezers for economy (since they have a slow evaporation rate). Storage in the vapor phase eliminates the risk of explosion with sealed ampules, while storage in the liquid phase means that the container can be filled and the liquid nitrogen will therefore last longer.

Fig. 29.1 Nalgene freezing container. Plastic holder with fluid-filled base. The specific heat of the coolant (冷却剂) of alcohol or isopropyl alcohol in the base insulates the container and gives a cooling rate of −1℃/min in the ampules.

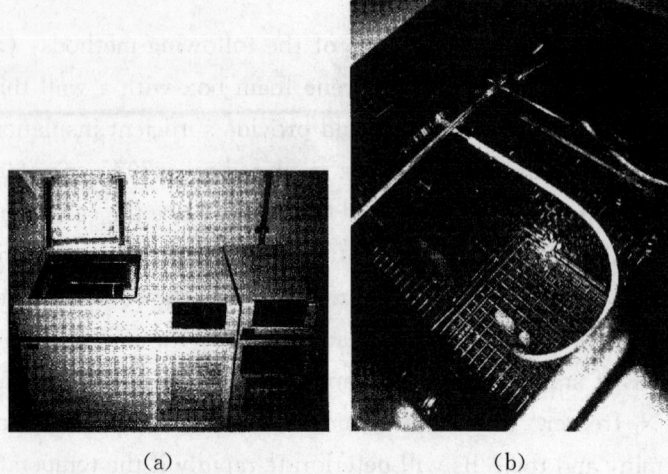

Fig. 29.2 Programmable Freezer(程序降温仪). Ampules are placed in an insulated chamber, and the cooling rate is regulated by injecting liquid nitrogen into the chamber at a rate determined by a sensor on the rack with the ampules and a preset program in the console unit (Planer Biomed). (a) Control unit and freezing chamber (lid open). (b) Close-up of a freezing chamber with four ampules, one with a probe in it. (By R. Ian Freshney)

Thawing Frozen Cells

Outline

Thaw the cells rapidly, dilute them slowly, and reseed them at a high cell density.

Protocol

1. Retrieve the ampule or tube from the freezer, check that it is the correct one, and place it in water bath at 37℃.

2. When the ampule or tube is thawed, check the label to confirm the identity of the cells; then swab the ampule thoroughly with 70% alcohol, and open it.

3. Transfer the contents of the ampule to a culture flask.

4. Add medium slowly to the cell suspension: 10 mL over about 2 min added dropwise at the start, and then a little faster, gradually diluting the cells and preservative. This gradual process is particularly important with DMSO, with which sudden dilution can cause severe osmotic damage and reduce cell survival by half.

5. The dregs(沉渣) in the ampule may be stained with naphthalene black(萘黑) or trypan blue(苔盼蓝) to determine their viability.

The number of cells frozen should be sufficient to allow for 1:10 or 1:20 dilution on thawing to dilute out the preservative but still keep the cell concentration higher than at normal passage. This dilutes the preservative from 10% to 0.5%, at which concentration it is less likely to be toxic. Residual preservative may be diluted out as soon as the cells start to grow (for suspension cultures) or the medium changed as soon as the cells have attached (for monolayers).

CELL BANKS

Several cell banks exist for the secure storage and distribution of validated cell lines. Since many cell lines may come under patent restrictions, particularly hybridomas (杂交瘤)and other genetically modified cell lines, it has also been necessary to provide patent repositories with limited access. As a general rule, it is preferable to obtain your initial seed stock from a reputable cell bank, where the necessary characterization and quality control will have been done. Furthermore, it is highly recommended that you submit valuable cultures to a cell bank in addition to maintaining your own frozen stock, as the former will protect you against loss of your own lines and allow their distribution to others. If you feel that your cells should not be distributed, then they can be banked with that restriction placed on them.

TRANSPORTING CELLS

Cultures may be transferred from one laboratory to another as frozen ampules or as living cultures. In either case,

(1) Advise the recipient as to when the cells are to be shipped;

(2) Fax or e-mail instructions on the following: (a) what to do on receipt, (b) medium or serum required, (c) any special supplements and (d) subculture regimen;

(3) tape the data sheet for the cells and a copy of the instructions to the outside of the box.

Frozen Ampules

Ship frozen ampules in solid carbon dioxide, in a thick-walled polystyrene foam container. The carrier must be informed when cells are shipped in solid CO_2. Usually, cells will remain frozen for up to 3 days if properly packed, but if they thaw slowly, their viability will decline rapidly.

Living Cultures

Alternatively, cells may be shipped as a growing culture. The cells should be at the mid to late log phase; confluent or post-confluent cultures will exhaust the medium more rapidly and may tend to detach in transit. The flask should be filled to the top with medium, taped securely around the neck with a stretch-type waterproof adhesive tape, and sealed in a small polythene bag. Place a label on the package that says "fragile" and, in large letters, the following instructions: DO NOT FREEZE!

On receipt, most of the medium is removed, leaving only the normal amount for culture-e. g., 5 mL for a 25-cm^2 flask. The culture can be weaned onto new medium when it is ready for the first feed, but keep the original shipping medium in case there are any problems of adaptation to the new stock.

QUESTIONS FOR DISCUSSION

1. What is the rationale for freezing cell line in continuous culture?
2. How to freeze cells?
3. How to thaw frozen cells?
4. How to increase the survival percentage of thawed cells?

知识要点

体外长期传代培养动物细胞耗时、费力和费材料,且容易发生微生物污染和遗传物质变异,因此对原代培养细胞和传代10次以内的二倍体细胞及时冻存具有重要意义。动物细胞一般冷冻保存于液氮(-196℃)中,并采用添加二甲基亚砜或甘油等冷冻保护剂缓慢冷冻和快速复苏的方法,尽可能减少冻存后细胞内形成冰晶的大小和数量,以提高冻存细胞的复苏率。玻璃化冻存是采用添加高浓度冷冻保护剂快速冷冻的冻存方法,尽量使细胞内不形成冰晶或形成的冰晶极小,以提高冻存细胞的复苏率,但技术上尚不成熟。

CHAPTER 30 CYTOTOXOCITY(细胞毒性)

New drugs, cosmetics, food additives, and so on go through extensive cytotoxicity testing before they are released for use by the public. The introduction of specialized cell lines and interactive organotypic cultures(器官型培养物), and the continued use of long-established cultures, provide us useful tools for this purpose. Toxicity is a complex event *in vivo*, where there may be direct cellular damage, as with a cytotoxic anticancer drug, physiological effects, such as membrane transport in the kidney or neurotoxicity (神经毒性)in the brain, inflammatory effects, both at the site of application and at other sites, and other systemic effects. Currently, it is difficult to monitor systemic and physiological effects *in vitro*, *so* most assays determine effects at the cellular level, or *cytotoxicity*. Definitions of cytotoxicity vary, depending on the nature of the study and whether cells are killed or simply have their metabolism altered.

All of these assays oversimplify the events that they measure and are employed because they are cheap, easily quantified, and reproducible. However, it has become increasingly apparent that they are inadequate for modern drug development, which requires greater emphasis on molecular target specificity and precise metabolic regulation. Gross tests of cytotoxicity are still required, but there is a growing need to supplement them with more subtle tests of metabolic perturbation, such as the induction of an inflammatory or allergic response.

It is not within the scope of this text to define all of the requirements of a cytotoxicity assay, many of which may be quite specialized, so instead I will concentrate on those aspects that influence cell growth or survival. Cell growth is generally taken to be the regenerative potential of cells, as measured by clonal growth (e. g. , in a plating efficiency assay), net change in population size (e. g. , in a growth curve), or a change in cell mass (total protein or DNA or metabolic activity (e. g. , DNA, RNA, or protein synthesis; MTT reduction).

IN VITRO LIMITATIONS

It is important that *any in vitro* measurement can be interpreted in terms of the *in vivo* response of the cells, or at least that the differences which exist between *in vitro* and *in vivo* measurements are clearly understood.

Pharmacokinetics(药物动力学)

The measurement of toxicity *in vitro is* generally a cellular event. It would be very difficult to re-create the complex pharmacokinetics of drug exposure *in vitro*, and between *in vitro* and *in vivo* experiments, there usually are significant differences in exposure time and concentration of the drug, rate of change of the concentration, metabolism, tissue penetration, clearance, and excretion.

Metabolism

Many nontoxic substances become toxic after being metabolized by the liver; in addition, many substances that are toxic *in vitro* may be detoxified by liver enzymes. For testing *in vitro* to be accepted as an alternative to animal testing, it must be demonstrated that potential toxins reach the cells *in vitro* in the same form as they would *in vivo*.

Tissue and Systemic Responses

The nature of the response must also be considered carefully. A toxic response *in vitro* may be measured by changes in cell survival or metabolism, while the major problem *in vivo* may be a tissue response (e.g., an inflammatory reaction, fibrosis(纤维化), kidney transport) or a systemic response (e.g., pyrexia(发热), vascular dilatation(扩张)). For *in vitro* testing to be more effective, models of these responses must be constructed, perhaps utilizing organotypic cultures reassembled from several different cell types and maintained in the appropriate hormonal milieu(环境).

NATURE OF THE CYTOTOXICITY ASSAY

The choice of assay will depend on the agent under study, the nature of the response, and the particular target cell. Assays can be divided into five major classes:

1. Viability: An immediate or short-term response, such as an alteration in membrane permeability or a perturbation of a particular metabolic pathway.

2. Survival: The long-term retention of self-renewal capacity (5-10 generations or more).

3. Metabolic: Assays, usually microtitration(微量滴定) based, of intermediate duration that can either measure a metabolic response (e.g., dehydrogenase(脱氢酶) activity; DNA, RNA, or protein synthesis) at the time of, or shortly after, exposure, or measure the same parameter two or three population doublings after exposure, when it is more likely to reflect cell growth potential and/or survival.

4. Transformation: Survival in an altered state (e.g., a state expressing genetic mutation(s) or malignant transformation).

5. Irritancy(刺激): A response analogous to inflammation, allergy, or irritation (刺激) *in vivo*; as yet difficult to model *in vitro*, but may be possible to assay by monitoring cytokine release in organotypic cultures.

VIABILITY

Viability assays are used to measure the proportion of viable cells following a potentially traumatic (外伤) procedure, such as primary disaggregation, cell separation, or freezing and thawing. Most viability tests rely on a breakdown in membrane integrity that is determined by the uptake of a dye to which the cell is normally impermeable (e. g., trypan blue, erythrosine (赤藓红), or naphthalene black (萘黑)) or the release of a dye normally taken up and retained by viable cells (e. g., diacetyl fluorescein (荧光素二乙酸酯) or neutral red (中性红)). However, this effect is immediate and does not always predict ultimate survival. Furthermore, dye exclusion tends to overestimate viability-e. g., 90% of cells thawed from liquid nitrogen may exclude trypan blue, but only 60% prove to be capable of attachment 24 h later.

SURVIVAL

While short-term tests are convenient and usually are quick and easy to perform, they reveal only cells that are dead (i. e., permeable) at the time of the assay. Frequently, however, cells that have been subjected to toxic influences (e. g., irradiation, antineoplastic drugs (抗肿瘤药)) show an effect several hours, or even days, later. The nature of the tests required to measure viability in these cases is necessarily different, since by the time the measurement is made, the dead cells may have disappeared. Therefore, long-term tests are used to demonstrate survival rather than short-term toxicity, which may be reversible. Survival implies the retention of regenerative capacity and is usually measured by plating efficiency. Plating efficiency measures survival by demonstrating proliferative capacity for several cell generations. Briefly, treat the cells with experimental agent at a range of concentrations for 24 h. Then trypsinize the cells, seed them at a low cell density, and incubate them for 1-3 weeks. Stain the cells for 10 min in 1% crystal violet, and count the number of colonies with >50 cells (>5 generations). Growth curve analyses can also be done by cell counting, but are feasible only with relatively small numbers of samples, as they become cumbersome (麻烦) in a large screen.

Some agents to be tested have low solubilities in aqueous media, and it may be necessary to use an organic solvent to dissolve them. Ethanol, propylene glycol (丙二醇), and dimethyl sulfoxide (二甲基亚砜) have been used for this purpose, but may themselves be toxic to cells. Therefore, you should use the minimum concentration of solvent to obtain a solution. The agent may be made up at a high concentration in, for example, 100% ethanol; then it should be diluted gradually with BSS and finally diluted into medium. The final concentration of solvent should be $< 0.5\%$, and *a solvent control* must be included (i. e., a control with the same final concentration of solvent but without the agent being tested).

Survival curve is a graph of the relative plating efficiency (the plating efficiency as a fraction of the control) against the log or linear drug concentration. (See Fig. 30.1) Determine the IC_{50} or IC_{90}, which is the concentration of compound promoting 50% or 90% inhibition of colony formation, respectively.

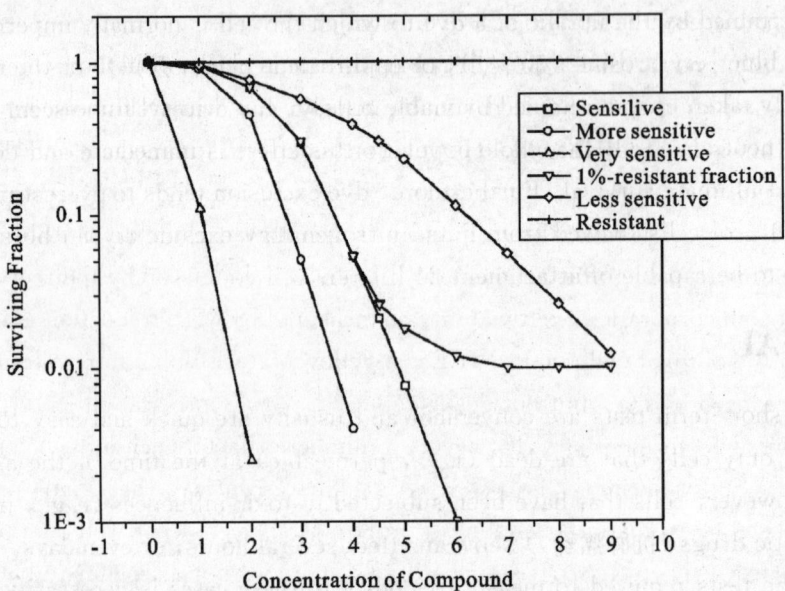

Fig. 30.1 **Interpretation of survival curves.** Semilog plot of cell survival against the concentration of cytotoxin. The slope increases with increasing sensitivity and decreases with reduced sensitivity until it becomes totally flat for complete resistance. Partial resistance can be expressed as the resistant fraction shown by the curve flattening out at the lower end. (By R. Ian Freshney)

METABOLIC ASSAYS

Plating efficiency tests are labor intensive and time consuming to set up and analyze, particularly when a large number of samples is involved, and the duration of each experiment may be anywhere from 2 to 4 weeks. Furthermore, some cell lines have poor plating efficiencies, particularly freshly isolated normal cells, so a number of alternatives have been devised for assaying cells at higher densities (e.g., in microtitration plates). None of these tests measures survival directly. Instead, the net increase in the number of cells (i.e., the growth curve), the increase in the total amount of protein or DNA, or the residual ability to synthesize protein or DNA is determined. Survival in these cases is defined as the retention of metabolic or proliferative ability by the cell population as a whole some time after removal of the toxic influence. However, such assays cannot discriminate between a reduction in metabolic or proliferative activity per cell and a reduced number of cells, and therefore any novel or exceptional observation should be confirmed by clonogenic survival assay.

Microtitration Assays

The introduction of multiwell plates revolutionized the approach to replicate sampling in tissue culture. These plates are economical to use, lend themselves to automated handling, and can be of good optical quality. The most popular is the 96-well microtitration plate, each well having 28-32 mm^2 of growth area and the capacity for 0.1 or 0.2 mL medium and up to 1×10^5 cells. Microtitration offers a method whereby large numbers of samples may be handled simultaneously, but with relatively few cells per sample. With this method, the whole population is exposed to the agent, and viability is determined subsequently.

The end point of a microtitration assay is usually an estimate of the number of cells. While this result can be achieved directly by cell counts or by indirect methods, such as isotope incorporation, cell viability as measured by MTT reduction is now widely chosen as the optimal end point. MTT is a yellow water-soluble tetrazolium(四唑盐) dye that is reduced by live, but not dead, cells to a purple formazan(甲䐳) product that is insoluble in aqueous solutions. However, a number of factors can influence the reduction of MTT. Sulforhodamine(磺酰罗丹明) is a fluorescent dye that stains protein and can also be used to estimate the amount of protein (i.e., cells) per well on a plate reader with fluorescence detection. It stains all cells and does not discriminate between live and dead cells. Labeling with [^{3}H]-thymidine (DNA synthesis), [^{3}H]-uridine (RNA synthesis), or other isotopes can be substituted for MTT reduction. Quantitation is achieved by microtitration plate scintillation counting. In practice, it may not matter which criterion is used for determining viability or survival at the end of an assay; it is, rather, the design of the assay (e.g., drug exposure, recovery, cell density, growth rate, etc.) that is most important.

The investigation of cytotoxicity often involves the study of the interaction of different drugs; drug interaction is readily determined by microtitration systems, in which several different ratios of interacting drugs can be examined simultaneously. Analysis of drug interaction can be performed using an isobologram(等效剂量分析方法) to interpret the data. A rectilinear plot(直线图) implies an additive response, while a curvilinear plot (曲线图) implies synergy(协同) if the curve dips below the predicted line and antagonism(对抗) if it goes above.

MTT based cytotoxicity assay

Cells in the exponential phase of growth are exposed to a cytotoxic drug. The duration of exposure is usually determined as the time required for maximal damage to occur, but is also influenced by the stability of the drug. After removal of the drug, the cells are allowed to proliferate for two to three population-doubling times(倍增时间) in order to distinguish between cells that remain viable and are capable of proliferation and those that remain viable but cannot proliferate. The number of surviving cells is then determined indirectly by MTT dye reduction. The amount of MTT-formazan produced can be determined spectrophotometrically once the MTT-formazan has been dissolved in a suitable solvent. Briefly, incubate monolayer cultures in microtitration plates in a range of drug concentrations. (Fig. 30.2) Remove the drug, and feed the plates daily for two to three population-doubling times; then feed the plates again, and add MTT to each well. Incubate the plates in the dark for 4 h, and then remove the medium and MTT. Dissolve the water-insoluble MTT-formazan crystals in DMSO, add a buffer to adjust the final pH, and record the absorbance in an ELISA plate reader. In the end, plot a graph of the absorbance (y-axis) against the concentration of drug (x-axis). Alternatively, the data can then be converted to a percentage-inhibition curve (Fig. 30.3), to normalize a series of curves. The IC_{50} concentration is determined as the drug concentration that is required to reduce the absorbance to half that of the control.

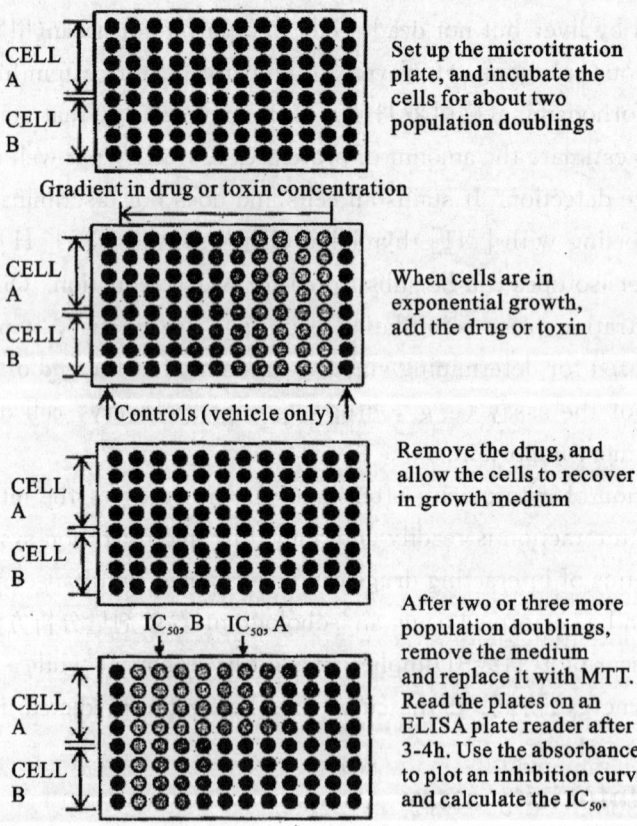

Fig. 30.2 Microtitration assay. Stages in the assay of two different cell lines exposed to a range of concentrations of the same drug and then allowed to recover before the estimation of survival by the MTT reaction. (By R. Ian Freshney)

Sulforhodamine B based cytotoxicity assay

Drug screening for the identification of new anticancer drugs can be a tedious and often inefficient method of discovering new active compounds. Method based on the determination of the amount of total protein using sulforhodamine B(磺酰罗丹明 B) is quicker and easier than the MTT assay. But it should be remembered that nonviable, and certainly nonreplicating, cells will still stain, so the assay should be confirmed when activity is detected, using a more reliable indicator, such as clonogenicity or MTT reduction.

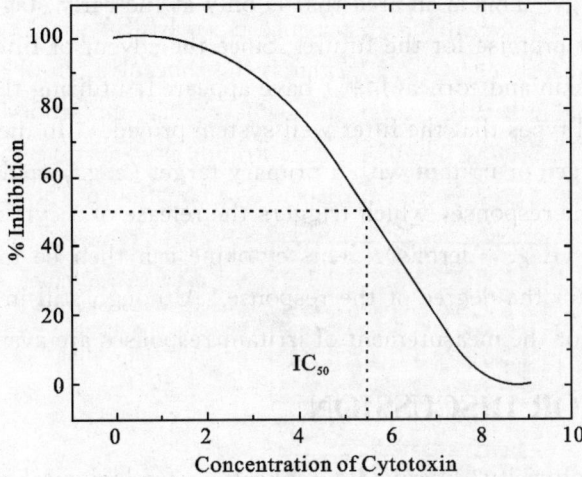

Fig. 30. 3 Percentage-inhibition curve. Percentage inhibition ([absorbance of test wells/absorbance of control wells] × 100) plotted against the concentration of cytotoxin. Typically, a sigmoid curve is obtained, and, ideally, the IC_{50} will lie in the center of the inflexion of the curve (although is often not the case). (By R. Ian Freshney)

TRANSFORMATION

Commonly used *in vitro* assays for transformation induced by agents include anchorage independence, reduced density limitation of cell proliferation, and evidence of mutagenesis. The potential for *in vitro* testing for carcinogenesis is considerable, but this is one area in which *in vivo* testing is far from adequate; the models are poor, and the tests often take weeks, or even months, to perform. The development of a satisfactory *in vitro* test is hampered (1) by the lack of a universally acceptable criterion for malignant transformation *in vitro* and (2) by the inherent stability of human cells used as targets. The most generally accepted tests so far assume that carcinogenesis, in most cases, is related to mutagenesis. This assumption is the basis of the Ames tests(彗星实验). The demonstration of increased oncogene expression or amplification, or the presence of increased or altered oncogene products, may provide reliable criteria, in some cases functionally related to the carcinogen. Likewise, deletions or mutations in the suppressor genes of p53, Rb, p16, and L-CAM (E-cadherin) genes would cover a high proportion of malignant transformation events.

INFLAMMATION

There is an increasing need for tissue culture testing to reveal the inflammatory responses that are likely to be induced by pharmaceuticals and cosmetics with topical application or by xenobiotics that may be inhaled or ingested and may be responsible for many forms of allergy. This is an area that is only at the early stages of development, but that bears great promise for the future. Since the advent of filter-well technology, several models for skin and cornea(角膜) have appeared, utilizing the facility for coculture of different cell types that the filter well system provides. In these systems, the interaction of an allergen or irritant with a primary target (e. g. , epidermis) is presumed to initiate a paracrine response, which triggers the release of a cytokine from a second, stromal component (e. g. , dermis). This cytokine can then be measured by ELISA technology to monitor the degree of the response. Although still in the early stages of development, kits for the measurement of irritant responses are available.

QUESTIONS FOR DISCUSSION

1. What is cytotoxicity? Decribe its implications for biological research area.
2. The nature of in vitro cytotoxicity assay can be classified into five classes-viability, survival, metabolic, transformation and irritacy. How to examine them?
3. How to carry out a MTT-based cytotoxicity assay?

知识要点

细胞毒性实验以其操作简单、快速和实验重复性好等优点而被广泛应用于新药、化妆品和食品添加剂等的毒性检测上。但是体外毒理学的实验结果并不能完全代替活体实验，因为肝脏的解毒作用、炎症和过敏反应等是体外细胞培养系统所缺乏的。体外毒理学的检测终点可分为5类：细胞活性、细胞存活率、代谢改变、转化和应激反应。MTT分析是常用的细胞毒性检测方法，借助多孔培养板，可快速得到药物的细胞生长抑制曲线，从而计算出药物的半数抑制浓度IC_{50}值。

CHAPTER 31　EMBRYONIC STEM CELLS(胚胎干细胞)

The first Embryonic stem (ES) cells were isolated from the inner cell mass (ICM, 内细胞团) of mouse blastocysts (胚泡) in 1981 by Evans and Kaufman, and Martin devised methods to grow them indefinitely. These cells are pluripotent because they can form chimera when reintroduced into mouse blastocysts and contribute to the formation of all tissues, including the germ line. That is, ES cells can self-renew indefinitely in vitro while maintaining the ability to differentiate into advanced derivatives of all three germ layers, features very useful for understanding the differentiation and function of tissues, for drug screen and toxicity testing, and for cellular transplantation therapies. In 1998, human ES cells were isolated successfully by Thomson et al. and inferred that stem cell technology may eventually benefit human disease therapy. First, stem cells, both embryonic and adult, hold the key for regenerative medicine, which may be considered the third therapeutic modality(方式)after drug therapy and surgery. Second, stem cells, especially ES cells, are ideal models for basic research in fields such as signal transduction, development, and epigenetics(表观遗传学). Last, stem cells could be useful tools for drug screening and safety assessments. Despite the excitement associated with stem cell research, we are still in the early phase of our exploration toward a molecular understanding of stem cells in normal development, diseases, and regenerations.

ORIGINS OF MAMMALIAN EMBRYONIC STEM CELLS

Mammalian development starts from a single cell (fertilized egg) that can give rise to all cells required for a new life, but through subsequent differentiation events, developmental potential becomes increasingly restricted. As the one-cell embryo divides, it forms a morula, a "mulberry"-like cluster of undifferentiated cells. The first differentiation event occurs when the outer layer of cells of the morula differentiates to the trophectoderm(滋养外胚层), forming the blastocyst(胚泡). The cells inside the blastocyst (inner cell mass, or ICM) give rise to all cells of the adult body, while the trophectoderm gives rise to the outer layer of the placenta(胎盘). That is, ICM cells are pluripotent stem cells in the early embryo and provide us a source of ES cells.

If the ICM cells are grafted to a permissive ectopic site, such as the testis(睾丸)or kidney capsule(肾小囊) of a syngeneic(同源)or immunocompromised(免疫抑制)

mouse, they will generate large multidifferentiated tumors known as teratocarcinomas （畸胎癌）. Teratocarcinomas are malignant germ cell tumors that comprise an undifferentiated embryonal carcinoma (EC) component and a differentiated component that can include all three germ layers.

EC cells are capable of both unlimited self-renewal and multilineage differentiation, but most EC cells show some restrictions in differentiation potential and in their ability to integrate normally into embryogenesis.

DERIVATION OF EMBRYONIC STEM CELLS

Studies with EC cells laid the intellectual and experimental groundwork for the establishment of "true" embryo stem cell cultures. A seminal point was the realization that pluripotency was best sustained in a coculture system. Martin and Evans observed that in primary cultures of teratocarcinoma, EC cells tended to thrive in proximity to differentiated cell types but to expand poorly in isolation. This prompted investigation of the potential of established cell lines to support EC cell propagation. Coculture with mitotically inactivated embryonic fibroblasts was found not only to allow the efficient establishment of EC cultures, but also to result in stem cells with high differentiation capacity. It was reasoned that the fibroblasts were providing some critical nutrient or trophic (营养) factor support, hence they were described as "feeder" cells.

In 1981, the derivation of pluripotent cell lines from mouse blastocysts was first reported (Evans and Kaufman 1981; Martin 1981). The protocols for ES cell derivation are relatively simple(Fig. 31.1). Embyros at the blastocyst stage are plated, either intact or following surgical isolation of the ICM, onto a feeder layer. Conventional tissue culture medium is supplemented with 2-mercaptoethanol(2-巯基乙醇)and 10%-20% fetal calf serum. After several days of culture, epiblast（上胚层或初级外胚层，由内细胞团中央的非极性细胞分化而来的多能性细胞）outgrowths are disaggregated and replated onto fresh feeders. Various types of differentiated colonies arise along with colonies of undifferentiated morphology. The latter are individually dissociated and replated. If secondary colonies of undifferentiated cells arise, these can generally be expanded further, and continuous ES cell lines can be established. These cells are formally called ES cells only if they meet the following two criteria: they display ES markers like high activity of alkaline phosphatase, over-expression of SSEA-1 (specifically expressed in mouse) or TRA-1-60 (specifically expressed in human); they are immortalized and can self-renew indefinitely *in vitro*.

DERIVATION OF EMBRYONIC GERM (EG) CELLS

In addition to experimental induction from explanted embryos, teratocarcinomas can originate spontaneously from germ cells. Testicular (睾丸) teratocarcinoma is particularly prevalent in strain 129 mice. In 1992 pluripotent stem cells (embryonic germ cells or EG cells) were successfully derived from mouse primordial germ cells(PGCs, 原始生殖细胞)directly *in vitro*. In contrast to mouse ES cells, the initial derivation of mouse EG cells requires a combination of stem cell factors of the steel cell growth factor (SCF, 青灰细胞生长因子), the leukemia inhibitory factor (LIF, 白血病抑制因子), and basic fibroblast growth factor (FGF) in the presence of a feeder layer. In culture, EG cells are morphologically indistinguishable from mouse ES cells and express typical ES cell markers such as SSEA-1 and Oct4. And similar to ES cells, upon blastocyst injection, they can contribute extensively to chimeric mice including germ cells. Unlike ES cells, however, EG cells retain some features of the original PGCs, including genome-wide demethylation, erasure of genomic imprints, and reactivation of X-chromosomes, the degree of which likely reflects the developmental stages of the PGCs from which they are derived. The derivation of human EG cells was reported in 1998, but in spite of efforts by several groups, their long-term proliferative potential appears to be limited.

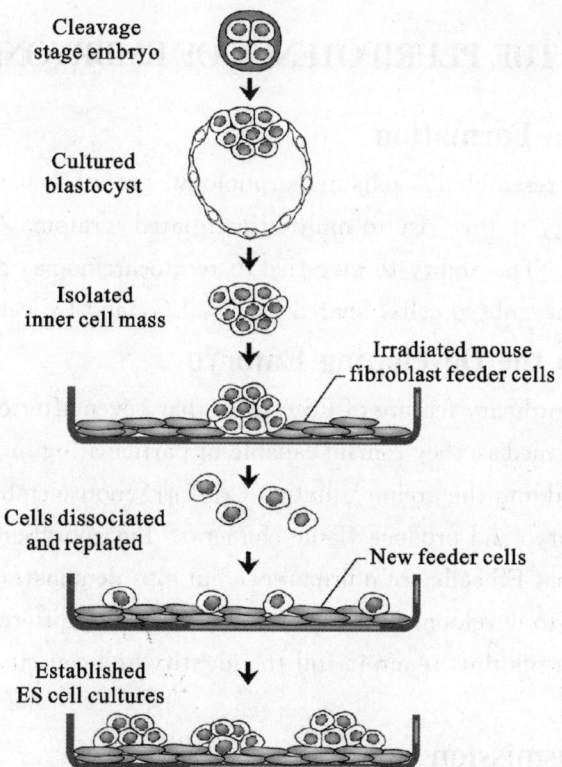

Fig. 31.1 Derivation of mammalian ES cell lines. Mammalian blastocysts were grown from cleavage-stage embryos produced by *in vitro* fertilization. ICM cells were separated from trophectoderm by surgery, plated onto a fibroblast feeder substratum in medium containing fetal calf serum. Colonies were sequentially expanded and cloned. (By J. S. Odorico, et al., 2001)

FACTORS INFLUENCING ES CELL DERIVATION

Establishing an ES cell culture entails the liberation of pluripotent epiblast cells from their fated differentiation. Prior induction of diapause (implantation delay,延迟着床) appears to enhance the efficiency of ES cell generation. This may be attributable to an increase in epiblast cell numbers during diapause, although this is relatively modest. Perhaps more likely is that the arrest of normal development pre-configures the epiblast cells for continued self-renewal by activating dependency on cytokine signaling for maintenance of pluripotency. The process by which a state of continuous self-renewal is arrived at is not automatic, however, and is poorly understood. Usually only a minority of embryos give rise to ES cells, suggesting that some epigenetic event is rate-limiting. Furthermore, the isolation of ES cell lines from some strains of mice has generally proven very problematic. Thus, there is a strong genetic component to ES cell derivation. Interestingly, this is not reflected in the propensity(倾向)of embryos to give rise to teratocarcinomas, which does not exhibit significant strain dependency.

CRITERIA FOR THE PLURIPOTENCY OF EMBRYONIC STEM CELLS

Teratocarcinoma Formation

ES cells closely resemble EC cells in morphology, growth behavior, marker expression, and the capacity to give rise to multidifferentiated teratomas(畸胎瘤)and teratocarcinomas(畸胎癌). The ability to give rise to teratocarcinomas clonally is a defining feature of pluripotent embryo cells, shared by ES, EG, and EC cells.

Integration into the Developing Embryo

The most extraordinary feature of ES cells is that, even after extended propagation *in vitro* in synthetic media, they remain capable of participating in normal embryogenesis. When introduced into the preimplantation (着床前) mouse embryo, ES cells can integrate into the embryo and produce viable chimeras. Incorporation into embryogenesis not only confirms that ES cells are pluripotent, but also demonstrates that they can respond appropriately to developmental cues for proliferation, differentiation, migration, and patterning. ES cells thus retain in full the identity and capacity of resident epiblast cells.

Germ-line Transmission

A key property of ES cells is that they maintain a euploid karyotype. This is crucial because a balanced diploid chromosome complement is permissive for meoisis. Thus, unlike EC cells, if ES cells colonize the germ cell lineage in a chimera, they are capable of progression to functional gametes. The landmark of deriving mice from cultured stem cells was reported by the Evans laboratory in 1984. Retention of germ-line competence depends absolutely on adherence to a rigorous tissue culture regime, with avoidance of any untoward selective pressures such as overgrowth or nutrient deprivation. Of course, random mutational events will always occur in the culture and epigenetic modifications may also arise; for example, alterations in imprinting status, so it is advisable to use low-passage stocks and/or to isolate new subclones periodically for transgenic work.

ES Cell-derived Fetuses

ES cells are not in themselves capable of generating a blastocyst and should therefore not be described as totipotent. The issue of whether ES cells are self-sufficient for generation of the fetal component of the conceptus (孕体) has been addressed by Nagy et al. (1991, 1993), who introduced ES cells into tetraploid recipient embryos. In tetraploid embryos, extra embryonic lineages are produced normally but fetal lineages develop poorly. Consequently, in chimeras between tetraploid and diploid embryos, the fetus becomes almost exclusively colonized by the diploid cells. ES cells show a similar propensity to dominate the tetraploid contribution to the fetus, and such fetuses can develop to term. Thus, it can be argued that ES cells alone are competent to generate the entire fetus. However, up to date, tetraploid complement technology is still a golden criterion for the pluripotency of ES cells.

Although there may be few or possibly no tetraploid cells persisting in the animal at birth, a resident tetraploid ICM compartment is present initially. Tetraploid embryos can also be preparaed by electrically inducing cell fusion of 2-cell embryos. Although liveborn offspring may be obtained from ES cell tetraploid chimeras, many embryos die in utero, and those that do persist usually die shortly after birth, in contrast to the situation with ICM chimeras. This is likely attributable to cryptic epigenetic or possibly mutational changes that have arisen during derivation or propagation of the ES cells. Such changes may be masked in diploid chimeras. Consequently, ES cells that give good somatic and germ-line colonization in diploid chimeras vary greatly in performance in the tetraploid setting. Thus, a note of caution is required in any assertion that an ES cell is unaltered from an epiblast cell in situ.

MAINTENANCE OF ES CELL PLURIPOTENCY

Indefinitely Self-renewal

ES cells multiply by symmetrical cell division. They can routinely be expanded to give relatively homogeneous and undifferentiated populations (Fig. 31.2), judged by morphology, marker expression, efficient generation of equipotent(等效)subclones, and reproducibly broad colonization of chimeras from a few cells. This expansion can be continued over several weeks, and very large ($10^9 \sim 10^{10}$) populations of substantially pure stem cells can be generated. In fact, ES cells appear to be immortal and show no evidence of either crisis or senescence, in contrast to other primary cultures. The symmetric amplification of ES cells contrasts with most other stem cells ex vivo and, in conjunction with the facility for genetic manipulation, provides a tractable system for experimental characterization of self-renewal.

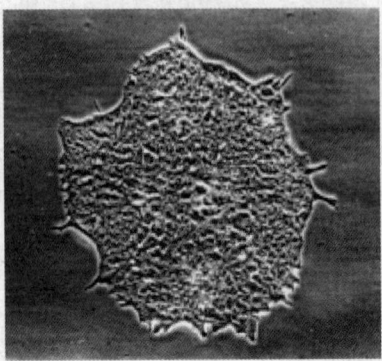

Fig. 31.2 Colony of self-renewing ES cells. (By A. Smith)

Maintenance of Oct-3/4 level

Oct-3/4 is a POU family transcriptional regulator restricted to early embryos, germ-line cells, and undifferentiated EC, EG, and ES cells. In vivo, zygotic expression of *Oct*-3/4 is essential for the initial development of pluripotential capacity in the ICM. In ES cells, continuous function of *Oct*-3/4 is necessary to maintain pluripotency. If *Oct*-3/4 expression is acutely eliminated in ES cells, self-renewal ceases and an unorthodox (非正统) differentiation process is triggered. Instead of forming the normal ES cell derivatives of endoderm and mesoderm, the cells differentiate into trophoblast (Fig. 31.3). ES cells do not normally form trophoblast either in vitro, in teratomas, or in chimeras. It appears that this developmental restriction may be necessary for manifestation of pluripotency and is imposed directly by *Oct*-3/4. In other words, *Oct*-3/4 acts in part as a lock that prevents default differentiation into trophoblast. *Oct*-3/4 also contributes positively to pluripotency by directing expression of multiple target genes. Although *Oct*-3/4 seems to be a pivotal player in the determination of pluripotent cell fate, maintenance of *Oct*-3/4 expression is not in itself sufficient to sustain the pluripotent phenotype, extrinsic signals are also needed.

Self-renewal by Cytokine Stimulation

In monoculture using media supplemented with serum alone, ES cells can neither be derived nor maintained. As discussed above, ES cells were originally isolated by coculture with a feeder layer. Subsequently it was discovered that the feeders can be substituted by conditioned medium preparations, indicating that their key function is to provide trophic stimulation. In fact, a purified cytokine, leukemia inhibitory factor (LIF), is sufficient to sustain ES cell self-renewal. This effect is exclusive to LIF and a small group of related cytokines that act via the gp130 receptor. LIF is expressed by feeder cells, and this expression is elevated in the presence of ES cells. LIF does not act via inducing expression of *Oct*-3/4 because transgenic expression of *Oct*-3/4 does not remove the requirement for LIF. On withdrawal of LIF (or feeders), proliferation continues, but differentiation is induced and ES cells do not persist beyond a few days (Fig31.3).

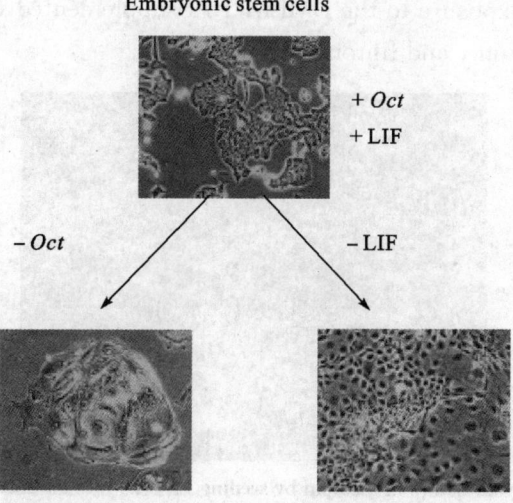

Fig. 31.3　Alternative ES cell fates induced by repression of *Oct*-3/4 or withdrawal of LIF. (By A. Smith)

IN VITRO DIFFERENTIATION OF ES CELLS

A major aspiration at the outset of EC and ES cell research was to elucidate the decision-making processes in lineage commitment and cell type differentiation of pluripotent cells. From their differentiation in teratomas and chimeras, ES cells clearly have the capacity to produce every type of fetal and adult cell. Understanding and controlling cell fate determination remains a major challenge, however.

Differentiation in Embryoid Bodies

Embryoid bodies are aggregates of ES cells after directly seeded upon non-tissue culture treated plates or flasks, preventing cells from adhering to a surface to form the typical colony growth. Upon aggregation, differentiation is initiated and occurrs in a three dimensional manner. First it simply appears as a ball of cells, then a hollow ball after a few days, and next forms internal structures such as a yolk sac, and cardiomyocytes. Heart muscle cells which beat in a rhythmic pattern and neurons also commonly appear as part of the embryoid body. Now it is possible to bias the differentiation for or against certain cell types by addition of retinoic acid(视黄酸). However, the final cultures are always a heterogeneous mixture of various cell types.

Differentiation in Monolayer Culture

ES cells differentiate readily in monolayer culture when deprived of LIF or feeder support (Fig. 31.4). Various differentiated morphologies emerge, and markers of mesoderm and endoderm become expressed. Use of defined media will likely be required to realize the full potential of this approach. Neural differentiation of mouse ES cells in monolayer culture can be achieved by a 2-day culture in serum-containing DMEM followed by continuous exposure to the DMEM/F12 supplemented with a mixture of insulin, transferrin, selenium, and fibronectin.

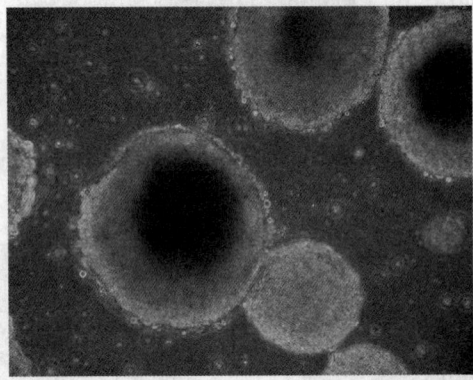

Fig. 31.4 Suspended embyoid bodies in medium by seeding ES cells into an uncoated substrate without feeder cells.

MECHANISM OF DIFFERENTIATION

In the absence of LIF, undifferentiated ES cells are rapidly depleted from the cultures. Although there is selective activation of a small number of specialized genes, lineage commitment (谱系提交) is in essence (在本质上) a restriction of global (整体) gene expression potential. This entails (意味着) the heritable repression of the majority of non-housekeeping genes. How such epigenetic (表观遗传学) mechanisms operate and how they may be erased during nuclear transfer or dedifferentiation has yet to be determined. Important insights could be obtained, however, by studying how chromatin architecture is modified during ES cell differentiation.

The chromatin organization in a pluripotent cell nucleus must be permissive for activation of lineage-specific gene transcription. A noteworthy observation from gene trapping (基因捕获) studies is that many developmentally regulated genes are already transcriptionally active at low levels in undifferentiated ES cells. It is also possible to detect allegedly(据称)tissue-restricted transcripts in ES cells by reverse transcription PCR. This is reminiscent(回忆)of the "lineage priming" (谱系预激) concept proposed for hematopoietic stem cells; But if stem cells express lineage-specific genes, how is the undifferentiated pluripotent state maintained? There are at least three possible and nonexclusive explanations:

(1) The level of expression of differentiation genes may not be functionally significant, but may simply reflect random transcription occurring through open chromatin.

(2) Commitment may require coordinated expression of a battery of genes, individual expression of which has no consequence.

(3) Stem-cell-specific transcriptional determinants may specifically antagonize the action of lineage commitment genes.

Self-renewal of ES cells appears to rely on the interplay of conflicting intracellular signals and transcriptional determinants. This is so finely balanced that the alternative outcome of differentiation can readily be triggered. Thus, some level of "spontaneous" differentiation is usually evident in ES cell cultures. It will be interesting to discover whether other types of stem cells are regulated in a similar manner such that they are constantly "poised" to differentiate.

GENOME MANIPULATION IN ES CELLS

Insertional Mutagenesis and Gene Trapping

DNA can be introduced into ES cells by conventional infection or transfection protocols. Their capacity for clonogenic expansion then allows independent integrants to be expanded and transgenic mice to be generated. Random insertion of viral vectors into the ES cell genome has been employed to mutate and tag genes in phenotype-driven screens. Gene trapping is a refinement (改进) of this approach that facilitates isolation of a disrupted gene and can allow a degree of preselection for desired categories of target gene based on expression pattern or subcellular localization of the gene product. This technique has been widely used as a gene discovery tool in mice and further pursued as a method for annotated mutagenesis of the entire mouse genome.

Targeted Gene Modification

The major use of ES cell genetic modification to date, however, has been for the directed modification of nominated genes, known as gene targeting (基因打靶). Pioneering work in the mid-1980s established that transfected DNA could be integrated into designated loci in the ES cell genome via homologous recombination (同源重组). In 1989 the first incidence of germ-line transmission of a targeted allele was reported, demonstrating that the manipulations and drug selections involved in isolating homologous recombinant clones did not in themselves compromise (损害) ES cell pluripotency. There are now well-established procedures for introducing a range of different types of modifications, such as deletion, point mutation, reporter insertion, or coding sequence replacement, into the mouse genome. Conditional mutations can be created by incorporation of site-specific recombinase technology. In such cases, short recognition sequences for a recombinase such as Cre or Flp are targeted by homologous recombination to flank the gene segment of interest. This interval can then be deleted in a stage-or tissue-specific fashion by appropriate transgenic expression of the recombinase.

Chromosome Engineering

The use of site-specific recombination can be extended to the engineering of long-range modifications in the ES cell and thence(因此)the mouse genome. Deletions, inversions, duplications, or translocations can be generated according to the respective orientation and cis or trans-localization of the recombinase recognition sequences. This is a powerful method for interrogating(审问)the genome, increasingly so with the amassing(积累)of sequence information and gene localization data. Autonomous chromosomal elements have also been introduced into ES cells via cell fusion. These minichromosomes(微小染色体)can be maintained stably in ES cells and chimeras, and in some cases, can be transmitted through the germ line. This creates the foundations of a system for genetic dissection of centromere function in mammalian mitosis and meiosis.

PLURIPOTENT EMBRYO CELLS FROM OTHER SPECIES

Derivation of permanent stem cell lines that fulfill the criteria of epiblast origin, sustained symmetrical self-renewal, pluripotency, integration into fetal development, and germ-line colonization has to date only been validated in mice. Germ-line colonization from cultured cells has been reported in chickens and medaka fish, but only after short-term culture. Chimeras have been reported in rabbits, pigs, and cattle, but in no case has germ-line colonization been corroborated. These data may suggest that the situation in the mouse is the exception rather than the rule. The ES cell phenotype represents a ground state for mouse epiblast or primordial germ cells in teratocarcinomas or ex vivo(体外). However, although diploid stem cell cultures can readily be derived from rat ICMs, rather than exhibiting multilineage differentiation, these cells appear restricted to extraembryonic development. It may be significant in this regard that ectopically grafted rat embryos do not produce teratocarcinomas and that rat epiblast appears to retain the ability to produce hypoblast(下胚层)even into egg cylinder stages. Therefore, the possibility should be considered that the ES cell phenomenon is specific to inbred laboratory mice and that the ground state in other species may differ and perhaps even be a more primitive "pre-pluripotent" cell.

Stem cell cultures have also been established from human blastocysts. These cells can generate teratomas in immunocompromised(免疫抑制)mice and show some capacity for multilineage differentiation in vitro. Therefore, they could represent human equivalents of ES cells. However, the critical functional tests of chimera contribution and gamete production obviously should not be undertaken for ethical reasons. It is noteworthy that these human cells differentiate into trophoblast, indicating that they may not represent exactly the same developmental stage as mouse ES cells. Furthermore, they are difficult to expand and seemingly do not respond to LIF. Intriguingly, human EG-like cells derived from fetal primordial germ cells, in contrast, appear to be dependent on LIF for continued propagation. The molecular characterization of these human cells and comparison against mouse ES and EG cells is now a pressing issue, particularly in light of the desire to develop human pluripotent cells for regenerative therapies.

FISH ES CELLS

ES cells build up a bridge between *in vitro* and *in vivo* genetic modifications. They can also be a source of pluripotent donor nuclei suitable for nuclear transfer procedures. Transgenic animals derived from targeted-gene inactivation in ES cells open the door for selectively improving productivity in farmed species, as well as identifying functions of genes in model species for basic biomedical research. In addition, ES cells can be used as tools for studying processes of differentiation of pluripotent cells into various lineages. Moreover, ES cells may be a method to preserve biodiversity in species for which embryo or gamete cryopreservation is not possible. The attractive prospect of ES cells has triggered the isolation of fish ES cells in order to produce transgenic fish in a precise and efficient way.

In spite of the discouraging situation encountered in other animal species, fish are especially attractive for developing ES cell technology for two reasons: (1) Piscine species（鱼类）are of considerable interest for both basic studies in molecular, cellular, and developmental biology as well as for commercial interest. As model vertebrate organisms, zebrafish and medaka are competitive with mouse for the analysis of gene functions relevant to humans. In the aquaculture activity, there is increasing interest for incorporating new fish species for market diversification（多样化）;（2）Fish have several technical advantages over other vertebrates such as high fecundity（繁殖力）, large transparent embryos, and rapid development. These features simplify manipulation and allow phenotypic observations of markers during early development, especially in small aquarium fish such as zebrafish and medaka whose generation times are relatively short, about 2 to 3 months. Conventional approaches to create transgenic fish involve direct introduction of transgenes into germ cells or embryos or fertilized eggs. The efficiency of the process is poor because integration of foreign DNA is exceedingly low, often less than 10^6. In general, integration is random, which often results in position effects; plasmid DNA is often cointegrated, which can silence the eukaryotic promoters; and late integration produces mosaic fish that may or may not include the germ line. Hence, there is an urgent need for site-directed integration of a transgene into fish genomes. ES cell-mediated gene transfer is a promising approach for producing site-mutated transgenic fish with enhanced growth rates or disease resistance, as well as for analyzing functions of fish genes. ES cells along with other cellular based strategies, such as primordial germ cells and nuclear transfer, allow selecting the desired transformation events before transferring the transgene to the whole animal.

The first attempts to isolate and culture fish ES cells were concentrated in zebrafish and medaka. Most of the work done in zebrafish has been carried out by the Collodi group, which initially obtained undifferentiated and stable cultures for more than 40 doublings, by using the Buffalo rat liver (BRL) line as feeder. Later, MBE (mid-blastula embryo) cell cultures were able to differentiate into neurons and astrocytes with either BRL or zebrafish embryo fibroblasts (ZEF) as feeders. A significant achievement was the generation of germline chimeras from primary MBE cells cultured less than 9 days, either supported by the rainbow trout spleen cell line (RTS) or in the presence of RTS conditioning medium. Adult chimeric fish revealed that MBE cells contributed extensively to the chimera, including gonads. The germline competence in zebrafish MBE cells was checked by the expression of the pluripotent stem cell marker POU2, a homologue of the mammalian *Oct*-4, which interestingly was maintained only under feeder conditions.

In medaka, Wakamatsu et al. (1994), also adopted the feeder layer technique by using primary cultures from blastula and gastrula medaka embryos. They formulated a rich medium that allowed establishment of a pluripotent cell line (OLES1). The line manifested stable growth, ES-like morphology, high alkaline phosphatase (AP) activity, and the potential to be induced by retinoic acid to differentiate into several cell types. A significant step was the work carried out over a decade by Hong et al. in medaka. A key contribution was the achievement of undifferentiated long-term embryonic cell cultures under feeder-free conditions. From this work several stable medaka ES cell lines were derived, one of which (MES1) was extensively characterized. *In vitro* this line showed all the features of mouse ES cells and differentiated under defined conditions into melanin-synthesizing pigment cells, contracting muscle cells, nerve cells, and fibroblasts.

Derivation of Fish ES Cells

The procedure for developing fish ES cells and their application in the production of transgenic fish is diagrammed in Figure 31.5. Fish embryos often develop outside the mother, which is a big advantage for initiating embryonic cultures from fish compared to mouse. The starting point is the mid-blastula embryo (Figs. 31.5-A and 31.6-A). Blastocytes, which remain undifferentiated, are easier to handle than cells from younger embryos and are competent for generating germ line chimera offspring. To obtain blastocytes, eggs are dechorionized by digestion with hatching enzyme in medaka, with pronase in zebrafish, or mechanically with fine forceps (Fig. 31.6-B) in gilthead seabream

（黄鳍鲷），red seabream（真鲷）and sea perch（鲈鱼）. After washing and removal of the yolk, blastocytes are seeded either on a layer of feeder cells or in gelatine-coated culture wells (Figs. 31. 5-B and 31. 6-C). Once the blastocytes are attached to the bottom, they acquire a typical ES-like form (Figs. 31. 5-C and 31. 6-D).

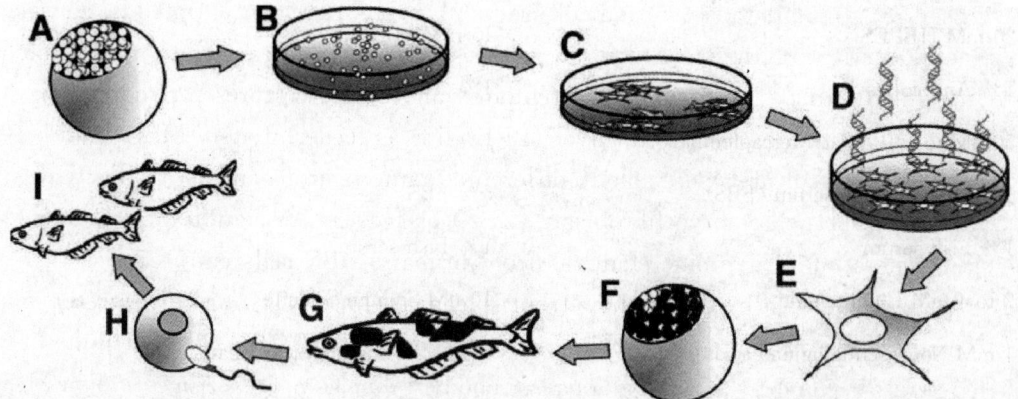

Fig. 31. 5 Diagram depicting steps for developing fish ES cells and their application in transgenesis. A. Mid-blastula embryo. B. Blastocysts primary culture. C. Clonal cell culture. D. Transfection with a gene-targeting construction. E. Selected homologous recombinant ES-cells. F. Chimera formation. G. Adult chimeric fish. H. Egg fertilization by ES-cell-derived sperm. I. Transgenic fish heterozygous for the gene-target event. (By M. C. Alvarez et al. , 2007)

To establish conditions that supported the growth while avoiding spontaneous differentiation of fish ES cells, much works were done in zebrafish and medaka and showed the necessity of culturing in a very rich medium that contained fish embryo extract and bovine fibroblast growth factor (bFGF) as a mitogen (Table 31. 1). To avoid differentiation, strategies used in mice, such as feeder cells, conditioning media, or leukemia inhibitory factor (LIF), were adopted both in zebrafish and medaka. Feeder-free conditions were established for medaka and successfully applied to gilthead seabream（黄鳍鲷），red seabream（真鲷）and sea perch（鲈鱼）using species-specific embryo extract and fish serum components. For these species, LIF had no effect on preventing either differentiation or cell growth. The independence of fish ES cells from feeder layers and/or inhibitory differentiation factors is an important advantage in the manipulation of fish ES cells compared to those from non-piscine species.

Characterization of Pluripotency of Fish ES Cells *In Vitro*

The actual proof of pluripotency is the ability of transplanted ES cells to enter the germ line of a chimera. However, some *in vitro* tests can give some insights on pluripotency, including (a) morphology, the cells must be small, round, or polygonal and contain large nuclei (Fig. 31. 6-D); (b) karyotype, the cells must be euploid to contribute to chimera germ line; and (c) growth rate, the cells must show rapid growth and be able to form colonies. Following satisfaction of these conditions, further markers of

Table 31.1 Standard medium composition for fish ES-like cells cultured with feeder layer and feeder free. (By M. C. Alvarez et al., 2007)

Feeder-free cultures	Feeder layer culture
DMEM/L-15 + 4.5 g/L glucose	L-15/DMEM/Ham's 12 (50:35:15)
20 mM HEPES	15 mM HEPES
1% Antibiotics	1% Antibiotics
50 μM to 100 μM 2-Mercaptoethanol	
15% Fetal bovine serum (FBS)	5% FBS
1% Fish serum	1% Fish serum
2 to 4 mM L-Glutamine	10 nM Sodium Selenite
1 mM Nonessential amino acids	0.18 g/L Sodium bicarbonate
1 mM Sodium pyruvate	10 μg/mL Bovine insulin
2 to 8 nM Sodium selenite	50 ng/mL EGF
2 to 10 ng/mL of bFGF	50 ng/mL bFGF
Embryo extract (0.5 embryos/mL)	Embryo extract (0.5 embryos/mL)

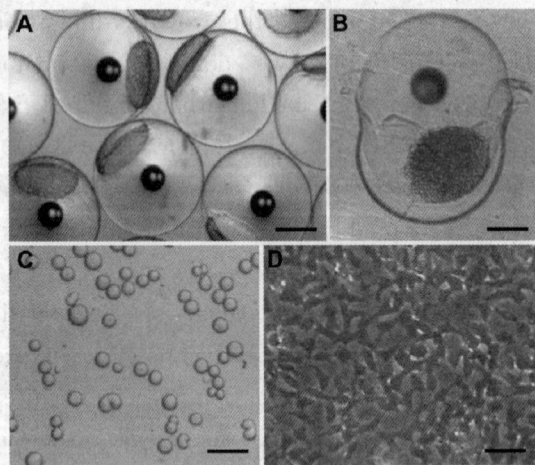

Fig. 31.6 Development of ES cell cultures in the gilthead seabream. A. Mid-blastula embryos (bar = 500 μm). B. Partially dechorionized embryo (bar = 400 μm). C. Primary culture (bar = 80 μm). D. ES-like cells at passage 10 (bar = 70 μm). (By M. C. Alvarez et al., 2007)

pluripotency can be tested. Among them, the AP activity is a valuable marker for putative ES cells in fish, with a level similar to those of parental blastula cells (Fig. 31.7-A). Another marker of totipotency is telomerase activity. High telomerase activity has been associated with undifferentiated cells and the ability of a cell culture to proliferate indefinitely. In fish ES cells, the elevation of telomerase activity has been found in the

SaBE-1c line of gilthead seabream. The most compelling *in vitro* test for ES cell pluripotency is their ability to respond to induction by differentiating factors. In the presence of the inducer all-trans retinoic acid(全反式视黄酸), most cells rapidly lose their ES-like morphology, and after a few days, they transform into epithelial, fibroblastic, or neuronal-like cells (Fig. 31. 7-B).

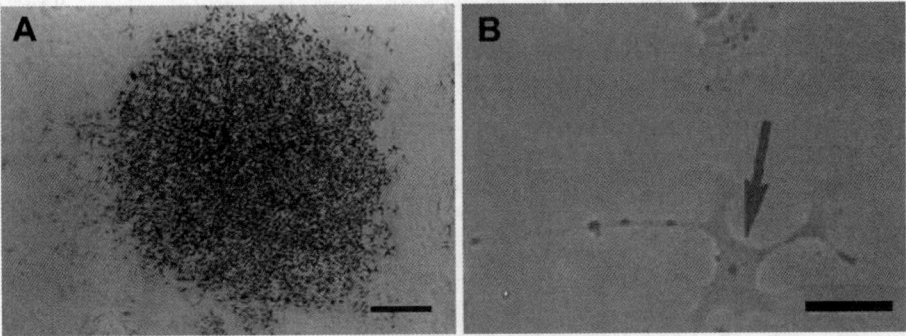

Fig. 31.7 Traits of pluripotency. A. SaBE-1c colony, derived from the blastula embryos of gilthead seabream (*Sparus aurata*), showing intense alkaline phosphatase staining (bar = 100 μm). B. Neuron-like cell obtained after retinoic acid treatment of medaka ES cells (MES-1) (bar = 30 μm). (By M. C. Alvarez et al., 2007)

In Vivo Characterization of Pluripotency-Chimera Production

To be useful in transgenics, ES cells must be able to produce chimeras following injection into recipient embryos, to have the opportunity of colonizing the germinal line (Fig. 31. 5-F). This is the ultimate test for pluripotency of ES cells. The procedure in fish is based on the microinjection technique developed for noncultivated blastomeres in medaka. Cells suspended in specific medium are injected into mid-blastula, dechorionated embryos (approximately 1000 cells), which are arranged on agarose ramps, by using a micro-transplantator equipped with borosilicate glass needles (彩页 Fig. 31. 8-A). The efficiency of chimera production has been thoroughly evaluated in medaka, by injecting either wild-type cells into albino recipient embryos (彩页 Fig. 31. 8-B) or GFP-transfected cells into normal embryos (彩页 Figs. 31. 8-C, D and E). Injection of approximately 100 cells was optimal; there was an inverse correlation(负相关)between duration of cells in cultivation and chimera rate. Cells from 27 to 66 passages produced 90% of chimeric fry and donor cells appeared in all three embryonic layers. The degree of chimerism was 2% to 10%, which is significantly lower than for mouse ES cells. In zebrafish, Collodis group reported a rate of 37% chimeras from 2-day-old MBE cell cultures and 15% from 14-day-old MBE cell cultures. In zebrafish, germ chimeras have been produced with mid blastula cells of up to 6 weeks in culture.

APPENDIX

ES Cell Culture

Media

High glucose DMEM (not supplemented with pyruvate and glutamine), 20% Heat-inactivated Fetal calf serum, 1× L-glutamine, 1× Penicillin/streptomycin, 1× Non-essential amino acids, 1× ribonucleosides, 1/100 volume β-mercaptoethanol (β-ME) stock (stock is 7 μL β-ME in 10 mL DMEM), 1:1000 dilution of LIF (add 500 μL cell conditioned media to 500 mL media)

Plating ES cells

Generally, ES cells are split 1:6-1:10. This depends on how "confluent" the plate is. Since ES cells grow as discrete colonies, you should NEVER get a plate that is confluent in ES cells the way fibroblasts are. To split, simply add 1 mL of the appropriate dilution of cells to a 100 cm plate containing a feeder layer, and then add 4 mL of ES cell media.

Splitting ES cells

When ES cells are "confluent" and ready to split, feed them with fresh media an hour or two before splitting. Aspirate media and wash cells once briefly with 1 trypsin-EDTA, aspirate. Then, add 0.5 mL pre-warmed (37℃) trypsin to a 6-well plate, or 2 mL to a 100 cm dish and put at 37℃ for 5-6 min. When cells are loosened add 0.5 mL ES cell media to the well with a 5 mL pipette. Attach a sterile yellow tip to the end of a 10 mL sterile plastic pipette and pipette cells up and down several times. ES cells are sticky and this is necessary to achieve a single-cell suspension to prevent differentiation after plating. Aspirate and resuspend cells in ES cell media, plate onto fresh feeder cells (remember to be careful about the split ratio. ES cells will grow slowly, or not at all, if they are not dense enough. Unless the plate that you are splitting from is extremely dense, split cells no more than 1:10)

QUESTIONS FOR DISCUSSION

1. What are embryonic stem (ES) cells? How to establish and maintain mouse ES cell lines?
2. What are teratocarcinoma and embryonic carcinoma (EC) cells?
3. What are embryonic germ (EG) cells?
4. What factors may influence the derivation of ES cells?
5. How to characterize and maintain the pluripotency of ES cells?
6. How to derive fish ES cells? List some applications of fish ES cells in aquaculture.

7. How to subculture ES cells?

知识要点

分离哺乳动物胚泡的内细胞团（或原始生殖细胞），在体外进行抑制分化培养，可成功获得具有无限自我更新能力和多分化潜能的胚胎干细胞（或胚胎生殖干细胞，ES 细胞）。通过延迟胚胎着床可提高胚胎干细胞的获得率。将体外培养的胚胎干细胞移植到同源鼠或免疫抑制鼠的睾丸或肾小球，可形成多胚层分化的畸胎癌（瘤），这是检测 ES 细胞多能性的常用指标。将胚胎干细胞移植到胚泡的内细胞团，ES 细胞可参与胚胎发育形成生殖系嵌合体，并传递给子代，这是 ES 细胞多能性的最有说服力的实验证据。ES 细胞多能性的维持依赖于细胞内 $Oct3/4$ 基因的表达水平以及饲养层细胞或细胞因子 LIF 的支持，否则 ES 细胞很快分化，失去多能性。基因打靶技术提高了体外修饰和改造 ES 细胞遗传物质的效率，大大推动了 ES 细胞在细胞治疗、基因转移和核移植上的应用。分离鱼类的中期囊胚细胞，进行体外抑制分化培养，可以建立鱼类的胚胎干细胞。鱼类 ES 细胞具有与小鼠 ES 细胞相似的形态特征和多分化潜能。

CHAPTER 32   NUCLEAR REPROGRAMMING AND INDUCED PLURIPOTENT STEM CELLS(细胞核重编程与诱导性多能干细胞)

NUCLEAR REPROGRAMMING IN CELLS

As a fertilized egg develops into an adult organism, specialized cells are formed by a one-way process, and they become increasingly, and normally irreversibly, committed to their fate. A skin cell does not naturally turn into, or give rise to, a brain cell, nor does an intestine cell generate a heart cell. Nevertheless, there are certain experimental procedures that enable just these kinds of changes to take place. They entail nuclear reprogramming, a term that describes a switch in nuclear gene expression of one kind of cell to that of an embryo or other cell type. This process is of interest for three reasons. First, identifying how reprogramming takes place can help us understand how cell differentiation and specialized gene expression are normally maintained. Second, nuclear reprogramming represents a first major step in cell-replacement therapy, in which defective cells are replaced by normal cells of the same or a related kind but derived from a different cell type. Eventually, it may be possible to derive replacement heart, pancreas, or other types of cells from the skin of the same individual, thereby avoiding the need for immunosuppression(免疫抑制). Third, nuclear reprogramming enables the culture of lines of cells from diseased tissues, and hence allows us to analyze the nature of the disease and to screen for therapeutic drugs.

Up to date, nuclear reprogramming in cells can be initiated by three kinds of experimental techniques including nuclear transfer(核移植)to eggs and oocytes(卵母细胞), cell fusion and molecular reprogramming(分子重编程)by pluripotent factors(彩页 Figs 32.1 and 32.5).

NUCLEAR TRANSFER TO EGGS AND OOCYTES

The earliest evidence for the experimental reversal of cell differentiation came from the transplantation of a viable cell nucleus into an enucleated frog egg. Briggs and King (1952) first succeeded in producing normal swimming tadpoles of Rana pipiens(豹蛙)by transplanting the nuclei of embryo (blastula) cells. They found, however, that the transfer of nuclei from slightly older (gastrula) embryos resulted only in abnormal development and concluded that cell differentiation was likely to involve irreversible nuclear changes. Soon after this, similar experiments were carried out with eggs of the South African frog Xenopus laevis(爪蟾). In due course, it was found that even when Xenopus nuclei were transplanted from fully differentiated cells, in this case from the intestinal epithelium of feeding tadpoles, entirely normal and fertile male and female frogs were obtained. These results led to the conclusion that the process of cell differentiation can be fully reversed and does not require irreversible nuclear changes; it involves changes in nuclear gene expression but not in gene content. Therefore, although cells become stably and functionally very different from each other during development, their genome stays the same in all cells (with the exception of antibody-producing cells) and therefore retains the potential to form any cell type.

The next major advance in this field came with the production of a normal adult sheep (Dolly) by transplanting the nuclei of cultured mammary gland cells derived from an adult sheep to enucleated(去核)sheep eggs. This and later work showed that it is possible to completely reverse the process of mammalian cell differentiation using nuclei from an adult mammal, and this suggests that this same procedure might work with humans. An important step in this direction has recently been taken by the generation of monkey embryonic stem (ES) cells from the nuclei of adult monkey cells. These proliferation-and differentiation-competent cells were derived from blastocysts(胚泡)grown after transplanting nuclei from adult monkey cells to enucleated monkey eggs. It is therefore likely that human eggs contain the components required to reverse the differentiation of adult human somatic cells.

The gold standard for the completeness of reprogramming by eggs has been described as the formation of a fertile adult animal containing functional cells of every kind (termed totipotency). In the case of somatic cell nuclear transfer, it is important to determine the efficiency of obtaining a particular differentiated cell type by using the transplanted nucleus of an entirely unrelated cell type. It has been shown that the success of nuclear reprogramming decreases as donor cells become more differentiated. The frog experiments include the results of serial nuclear transfers (transplanting nuclei from a nuclear transplant embryo to another set of enucleated eggs) and grafts (transplanting nuclei from a nuclear transplant embryo to host embryos reared from fertilized eggs) to produce the conclusion that about 30% of intestinal epithelium cell nuclei can generate functional muscle and nerve cells. In mammals, the frequency that a normal adult is obtained from the nucleus of a specialized cell is usually 1 to 2%, as compared with about 30% from embryo nuclei.

An appeal of using eggs to reprogram nuclei is that eggs have the natural ability to reprogram highly specialized sperm nuclei with 100% efficiency. Another advantage of this procedure is that it does not require a permanent genetic change to the transplanted nucleus or to the resulting reprogrammed cells. Therefore, it is important to discover the mechanisms involved and ask, how is successful reprogramming achieved, and what makes the process frequently unsuccessful even when eggs are used?

The mechanism of nuclear reprogramming by eggs (in second meiotic metaphase) has been explored by the use of oocytes (female germline cells in first meiotic prophase and immediate progenitors of eggs). Multiple mammalian somatic nuclei transplanted to the germinal vesicle of an oocyte are directly reprogrammed to transcribe stem-cell marker genes, including *Oct*4, Nanog, and *Sox*2 (彩页 Fig. 32.1B). Nuclear reprogramming by oocytes does not yield new cells but, in contrast to eggs, takes place without cell division and does not need protein synthesis. Mechanisms accompanying this reprogramming include (1) a massive volume increase of 30 times in transferred nuclei and chromatin decondensation (彩页 Fig. 32.2), due in part to an oocyte histone chaperone nucleoplasmin; (2) the removal of differentiation marks, such as DNA methylation and histone modifications; and (3) chromatin protein exchange, especially of the oocyte-specific linker histone H1 by the oocyte-specific histone variants B4 or H1foo (彩页 Fig. 32.3). The general principle here seems to be that, during their formation, oocytes (and hence eggs) acquire very high concentrations of certain proteins that are responsible for the above effects. If egg proteins can be exchanged in seconds or minutes for those in transplanted somatic nuclei (as suggested by most fluorescence recovery after photobleaching experiments), complete reprogramming should always take place.

This concept of rapid exchange does not, however, agree with the fact that eggs are often unsuccessful in fully reprogramming somatic nuclei. If the rapid exchange of chromosomal proteins referred to above applies to all those components of an egg that normally reprogram sperm nuclei after fertilization, there would be time in frogs, and even more in mammals, for transplanted somatic nuclei to be fully reprogrammed before the first egg division (24 hours in mammals). This often does not happen. One reason may be that transplanted nuclei carry an epigenetic memory of their gene expression in their donor cells. For example, nuclei taken from muscle cells sometimes continue to strongly express muscle genes in neural and other non-muscle cells of an embryo obtained by nuclear transfer. This may be caused by the incorporation of an abundant egg histone variant (H3.3) into the chromatin of daughters of transplanted nuclei. The incorporation of the H3.3 histone is thought to prevent reprogramming and so to preserve a memory of previous gene expression.

CELL FUSION

It is possible to fuse two somatic cells and to use a cell-division inhibitor to ensure that the two nuclei remain separate (Fig. 32.1C). In these heterokaryons, the dominant cell, usually the larger and more actively dividing partner, imposes its own pattern of gene expression on the other partner. Examples include the fusion of an erythrocyte with a growing cultured cell or of a human liver cell with a multinucleate muscle cell. If enucleated cytoplasms of one kind of somatic cell (cytoplasts) are fused to another cell, they also impose gene expression of their original cell type on the incoming nucleus. However, these fused cells do not proliferate well, and therefore are not likely to be of therapeutic value.

Some important conclusions can be drawn from these experiments. One is that reprogrammed gene expression is commonly preceded by nuclear swelling and chromatin decondensation, such as in nuclear transfers to eggs and oocytes (Fig. 32.2). Another is that new gene expression does not depend on the extinction of donor cell specific gene expression, nor on cell division; therefore, neither of these is a necessary part of reprogramming. The third conclusion is that differentiated cells (as well as embryo cells) contain regulatory molecules that can redirect gene expression in the nuclei of other cells. When the recipient cell is very large, such as an egg or myotube (100 or so muscle cells fused into one large syncytial cell), it is understandable that its own programming molecules can override a much smaller supply of regulatory molecules introduced by the incoming nucleus or cell (彩页 Fig. 32.1). These molecules probably have a role in normal (non-nuclear transfer) conditions by ensuring that cells and their daughters do not escape from their lineage or change cell type; in other words, cells seem to continually self-reprogram themselves and their daughters to remain in the same lineage.

MOLECULAR REPROGRAMMING BY PLURIPOTENT FACTORS

A spectacular advance in this field came when Takahashi and Yamanaka (2006) discovered that viral transfection of four genes (*Oct 3/4*, *Sox2*, *c-Myc* and *KLF4*) into an adult mouse fibroblast population can lead to the appearance of some cells with the characteristics of ES cells. After further selection for the expression of Nanog, in addition to the first four genes, the resulting stem cells were shown to enter all cell lineages when transplanted to immunotolerant host embryos; hence, they are pluripotent and termed induced pluripotent cells, or iPS cells. iPS cells from human somatic cells require the same set of factors used in mice (above) or the combination of *Oct4*, *Sox3*, *Nanog* and *Lin28* (彩页 Figs. 32.4). These procedures have now been confirmed and extended. iPS cells have been obtained from differentiated stomach and liver cells and can be obtained even if Myc, which can induce cancer, is omitted. The resulting stem cells do not appear to be substantially different from ES cells and may eventually provide a suitable source of different cell types for patient-specific cell replacement therapy in humans and of disease-specific cell lines to test potential therapeutic agents, but only after methods are developed to eliminate the concern of genome integration by the associated viral vectors. Recent work provides a step in this direction by showing that stable viral integration is not required to generate iPS cells when nonintegrating adenoviruses or plasmids, and even protein transporters are used.

Pluripotency

Pluripotency is the central property of all ES cells. It refers to the ability to generate any type of cells in the body. Developmentally, zygotes are totipotent, capable of giving rise to a whole animal, including all cell types. After several cell divisions, zygotes differentiate into blastocysts, from which the inner cell mass can be isolated and cultured into ES cells. Thus, ES cells are developmentally arrested at the pluripotent stage and can be propagated indefinitely in vitro. Although less potent than totipotent zygotes, ES cells have been experimentally proven to be able to contribute to all cell types except the trophectoderm in mouse. However, ES cells are generally described as being pluripotent, not totipotent, reflecting the fact that no animal has been generated by ES cells alone.

Pluripotency is maintained by a process called self-renewal. Self-renewal allows ES cells to duplicate themselves without losing the ability to differentiate, thus maintaining pluripotency. This can be achieved through both symmetric and asymmetric cell divisions. In vitro, ES cells undergo self-renewal through symmetric divisions. In vivo, tissue stem cells tend to self-renew through asymmetric divisions to generate one exact copy and another one for differentiation. Experimentally, ES cells have to be grown under special conditions to be kept in a pluripotent state. Mouse ES cells should be cultured on top of a feeder layer of cells, presumably supplying unknown factors to the ES cells. In addition, LIF or other cytokines are routinely added to prevent ES cells from undergoing spontaneous differentiation, a phenomenon encountered on a daily basis during ES cell culturing. Human ES cells appear to have different requirement for cytokines. Instead of LIF, human ES cells require both bone morphogenetic protein and fibroblast growth factors to prevent differentiation. The removal of feeders or cytokines leads to spontaneous differentiation and the loss of pluripotency.

Differentiation is the process during which pluripotency is expressed. During differentiation, stem cells commit to one cell lineage while losing the ability to commit to the rest of the cell lineages. Reprogramming is the process that converts differentiated cells back to pluripotent ones, effectively the reversal of differentiation. Experimentally, reprogramming has been achieved through somatic cell nuclear transfer or cloning. More recently, iPS cell technology has accomplished the same feat via the introduction of pluripotency factors, including *Oct*4 and *Sox*2, into somatic cells. Taken together, self-renewal, differentiation, and reprogramming can be viewed as three different aspects, i. e. maintenance, expression, and acquisition, of pluripotency (彩页 Fig. 32.4).

Control of Stem Cell Pluripotency by Transcription Factors

*Oct*4 was the first gene to be identified as a master regulator of pluripotency.

Nichols et al. (1998) demonstrated that *Oct*4-deficient embryos develop to the blastocyst stage but that the inner cell mass cells are not pluripotent. In fact, *Oct*4 was originally discovered as a member of the murine octamer-binding protein family that interacts specifically with the octamer motif, a transcription regulatory element found in the promoter and enhancer regions of many genes. The expression profile of *Oct*4 suggests that it may regulate cell fate during early developmental control. Biochemically, *Oct*4 has been shown to be a DNA-binding protein with a bipartite(两个)POU/homeodomain encoded by a 324-amino acid open reading frame. *Oct*4 relies on two transactivation(反式激活) domains flanking the DNA-binding domain to exert its transcription activities. *Oct*4 protein is synthesized in the cytosol and transported into the nuclei via a typical nuclear localization signal. The nuclear localization signal of *Oct*4 is required for its transcription activity, and its ablation leads to the generation of a dominant-negative form of *Oct*4, which is capable of inducing ES cell differentiation by interfering with wild-type *Oct*4 activity.

In ES cells, *Oct*4 appears to regulate cell fate in a dosage-dependent fashion. Using a conditional expression and repression system, Niwa et al. (2000) demonstrated that the level of *Oct*4 activity specifies three distinct fates of ES cells: (1) a < 2-fold increase in expression turns ES cells into primitive endoderm and mesoderm; (2) repression of oct4 induces the formation of trophectoderm; and (3) only an optimal amount of *Oct*3/4 can sustain stem cell self-renewal. These results suggest that ES cells must possess a network of regulators to keep *Oct*4 expression at the optimal level to ensure pluripotency.

How many transcription factors are involved in the regulation of *Oct*4 expression? This was the question asked by several groups in light of the observation that *Oct*4 must be maintained in a narrow range of expression levels to ensure stem cell pluripotency. The discovery of Nanog offered a clear candidate for *Oct*4 regulation. Named after Tir Nan Og, Nanog was discovered based on its ability to sustain stem cell self-renewal in the absence of LIF. Nanog possesses two potent transactivators and behaves as a strong activator of the *Oct*4 promoter, thus participating in the regulation of *Oct*4 expression in ES cells (彩页 Fig. 32.4B).

*Sox*2 often partners with *Oct*4 to regulate gene expression. Gene knock-out experiment demonstrated that *Sox*2 is required for epiblast and extraembryonic ectoderm formation, suggesting that *Sox*2 and *Oct*4 cooperatively specify the fate of pluripotent stem cells at implantation. Recent results demonstrated that *Sox*2 is necessary for regulating multiple transcription factors that affect *Oct*4 expression, thus stabilizing ES cells in a pluripotent state by maintaining the requisite level of *Oct*4 expression. In a word, *Oct*4, *Sox*2, and Nanog regulate overlapping targets. They collaborate to form regulatory circuitry consisting of autoregulatory and feed-forward loops that contribute to pluripotency and self-renewal.

Reprogramming by Transcription Factors

One of the goals in dissecting the molecular networks that control pluripotency is to regain pluripotency lost during development and differentiation. This would entail reversing the well programmed process of development from a fertilized egg to a grown adult. In higher mammals, it was thought that the differentiation process is irreversible until the successful cloning of Dolly. The cloning experiment demonstrated that somatic cells can be reprogrammed back to the totipotent zygotic state by the cellular factors of unfertilized eggs. It could have taken a considerable amount of time and effort to identify those unknown factors responsible for reprogramming.

Takahashi and Yamanaka (2006) stunned the scientific community with their study showing molecular reprogramming of mouse somatic cells into induced pluripotent stem (iPS) cells using just four factors: *Oct*4, *Sox*2, *Klf*4, and *c-Myc*. They leapfrogged (跳过) this hurdle through a candidate gene approach. To enhance their chance of success, a selection marker driven by fbx15, a gene known to be specifically expressed in ES cells, was engineered into the recipient cells by gene targeting. Remarkably, ES-like colonies were recovered after ~ 2 weeks. Eventually, only four genes, *Oct*4, *Sox*2, *Klf*4, and *c-Myc*, were deemed sufficient to reprogram fibroblasts into ES-like cells. These ES-like cells were later coined as iPS cells for induced pluripotent stem cells to differentiate them from blastocyst-derived pluripotent ES cells. Mouse iPS cells are indistinguishable from ES cells in morphology, proliferation, gene expression, and teratoma formation. Furthermore, when transplanted into blastocysts, mouse iPS cells can give rise to adult chimeras, which are competent for germline transmission. These results are proof of principle that pluripotent stem cells can be generated from somatic cells by the combination of a small number of factors.

Later, Takahashi, Yamanaka, and their colleagues (Takahashi et al., 2007) translate their remarkable findings from mouse to human (彩页 Fig. 32.5). They selected adult human dermal fibroblasts and two other human fibroblast populations (from synovial tissue and neonatal foreskin, 滑膜组织和新生儿包皮) from different human donors as their reprogramming target cell populations. They then transduced the human fibroblast cultures with retroviral vectors carrying transgenes for the human versions of *Oct*4, *Sox*2, *Klf*4, and *c-Myc*, and cultured the cells under human ES cell culture conditions. Thirty days after transduction, the culture plates were covered with human ES cell-like iPS colonies (among other colonies), which could be further propagated and expanded. The retroviral vectors enabled silencing of all four transgenes after human iPS formation (as found in the mouse system) indicating that the iPS cells are fully reprogrammed and no longer depend on transgene expression. Unlike the mouse study, human iPS cells were generated without any genetic selection procedures. Given the lower mitotic index of human ES cells, it is not surprising that the generation of human iPS cells takes notably longer than in the mouse system.

The recapture of pluripotency lost during differentiation by these four magic factors established a new paradigm for our understanding of pluripotency. This elegant iPS approach opened a new era for stem cell and regenerative medicine. iPS-mediated reprogramming of somatic cells removes the ethic as well as technical hurdles associated with therapeutic cloning, the use of human eggs for the generation of patient-specific pluripotent cells. Indeed, patient-specific iPS cells have been reported at an accelerated pace in the literature. These iPS cells may become important models for us to understand the mechanisms associated with a particular disease. However, given the use of viral delivery and four potent oncogenes in the iPS process, these patient-specific iPS cells are not safe for therapeutic purposes. Efforts are under way in many laboratories to identify small molecules that can functionally substitute for these four reprogramming factors. It has been found that a mixture of chemical regulators may function to reprogram somatic cells. The chemical approach, or ciPS for chemical iPS, to reprogramming may eventually yield clinical grade pluripotent stem cells for therapies and regenerative medicine.

Mechanical steps involved in the reprogramming of somatic cells into pluripotent ones by *Oct4/Sox2/Klf4/Myc*

Mechanically, iPS reprogramming involves the following key steps (彩页 Fig. 32. 6). First, the recombinant viruses carrying *Oct4*, *Sox2*, *Klf4*, and *c-Myc* enter the somatic cells and integrate into the host genomes. Following transcription driven by the viral promoters, all four proteins are produced in the cytosol and then imported back to the nuclei to activate the first wave of genes whose promoters are accessible to them. These first responders must then engage the epigenetic(表观遗传学) machinery to remodel the chromatins through the histone modification system and the DNA methylation system. Through this process, genes critical for pluripotency must be switched on by transcription factors and kept on through chromatin remodeling. Conversely, genes responsible for differentiation must be turned off by the transcription machinery and kept silent through epigenetic mechanisms. One remarkable feature of the iPS process is the silencing of the integrated viral genomes carrying the reprogramming initiators *Oct4*, *Sox2*, *Klf4*, and *c-Myc*. As such, iPS cells function indistinguishably from ES cells derived from blastocysts.

MECHANISMS FOR NUCLEAR REPROGRAMMING

Protein-DNA Interactions. Two basic characteristics of cell differentiation influence our understanding of nuclear reprogramming. One is that every cell seems to express those genes whose products determine its state of differentiation, a conclusion especially clear from cell-fusion experiments. Thus, a muscle cell will maintain by autoactivation a high enough content of MyoD, for example, to continually program itself to be a muscle cell. The larger the cell, and/or the more embryonic it is, the greater abundance it will have of self-reprogramming molecules. Therefore, eggs will be particularly effective without added factors; a second characteristic of all nuclear reprogramming experiments is that the experimental resetting of gene expression becomes increasingly difficult as cells become more differentiated. The differentiated state becomes more firmly established as cells embark（着手）on their terminal pathways and shut down inappropriate lineages. To understand the basis of this is a major challenge in this field, and much informative work has already been done on DNA and histone modifications.

A general hypothesis is the idea of "fleeting access（短暂的访问）." The structure of chromatin is dynamic, varying between heterochromatin (condensed) and euchromatin (extended) forms. Any cellular process that uses DNA as a template, such as transcription and replication, is regulated by the structure of chromatin. In embryonic cells, most genes (and in differentiated cells, the active genes) will be in a decondensed configuration with relatively short dwell times for histone protein complexes. According to this view, the probability of reprogramming taking place in nuclear transfer, cell fusion, iPS, and lineage-switching experiments would depend on the statistical access frequency of gene regulatory regions, together with the duration, and the concentration of transcription or other regulatory factors. Large cells such as eggs or myotubes（肌管）with a high content of factors would be especially successful at reprogramming, as would any cell with an experimentally enhanced content of factors. A major advance in the future will be to understand why the nuclei of differentiated cells are reprogrammed so much less well than those of embryonic cells. This will probably require an explanation of chromatin decondensation.

Will the mechanism of reprogramming be the same in nuclear transfer to eggs, iPS experiments, and lineage switching? Probably not. The concept of fleeting access will be the same, but the actual reprogramming molecules will be different. We already know that eggs have very high concentrations of certain molecules such as nucleoplasmin and histones B4 and H3.3. The eventual identification of egg-reprogramming molecules may well be able to enhance the efficiency of the iPS and lineage-switching routes for adult cells.

PROSPECT

The future value of reprogrammed cells is of two kinds. One is to create long-lasting cell lines from patients with genetic diseases, in order to test potentially useful drugs or other treatments. The other is to provide replacement cells for patients. To be therapeutically beneficial, replacement cells will probably need (i) to be provided in sufficient numbers; (ii) to carry out their function, even though they are not normally integrated into host tissues; and (iii) to be able to produce the correct amount of their product.

A human adult has about 10^{15} cells, and the liver contains about 10^{14} cells. To create this number of cells starting from a 10^{-4} success rate of deriving iPS cells from skin would require an enormous number of cell divisions in culture, although the prolonged culture of ES-like cells provides a valuable amplification step. However, many parts of the human body need a far smaller number of cells to improve function. An example is the human eye retina, in which only 10^5 cells could be of therapeutic benefit.

Will introduced cells be useful even if not "properly" integrated into the host? Most organs consist of a complex arrangement of several different cell types. The pancreas(胰脏), for example, contains exocrine(外分泌)cells, ductal cells, and at least four kinds of hormone-secreting cells in the endocrine islet(内分泌胰岛). Replacement endocrine cells can provide useful therapeutic benefit even if not incorporated into the normal complex pancreas cell configuration. In some cases, introduced cells can have functionally beneficial effects, even if indirectly. It is not yet clear whether introduced cells will be correctly regulated to produce the desired amount of product.

Looking ahead, alternative routes to cell replacement may emerge. One is to avoid the need to transfect genes into cells if the right combinations of small molecules that can easily enter cells can be found. It may also be increasingly fruitful to find populations of naturally dividing cells in adult organs so that these cells in their naturally less-specialized state can be expanded and differentiated in culture before implantation. A future objective, in our view, is to aim for unipotency and oligopotency (the generation of only one or a few cell types) rather than pluripotency (the potential to differentiate into any of the three germ layers) and certainly not totipotency (the potential to differentiate into all embryonic and extra-embryonic cell types). Likewise, we would much prefer to be able to create new cells by switching normal cells from a closely related lineage than by going back to totipotency and then narrowing down the differentiation options from a wide range. For replacement therapy, totipotency and germline transmission are not desirable criteria or objectives. An oligopotent state with limited differentiation potential is likely to be much safer and more useful from a therapeutic point of view.

Appendix

Molecular Reprogramming of Human Foreskin Fibroblasts (HFF)

Retrovirus production

Thaw the packaging cells 293T and culture the cells to 80% confluence in 10 cm culture dish. Five dishes are needed for the four plasmids of *pMXs-Oct4*, *pMXs-Sox2*, *pMXs-cMyc*, *pMXs-Klf4* and *pMXs-GFP*, respectively. In 2 mL eppendorf tube containing 2 mL Opti-MEM without serum, dilute the plasmids at a ratio of pVSV-G: pGal-pol: pMXs-Oct4/Sox2/cMyc/Klf4/GFP = 3/5/8, and mix gently. Add Lipofectamine Plus into the DNA solution at an amount of 1 μL per μg plasmid DNA, mix gently and incubate at room temperature for 5 min. Then add Lipofectamine LTX at an amount of 2 μL per μg plasmid DNA into the mixture of Opti-MEM-plasmid DNA-Lipofectamine Plus, mix gently and incubate at room temperature for 30 min. After replacement of the medium by adding 10 mL Opti-MEM without antibiotics, add all the above mixture into the 293T cells in 10 cm culture dish drop by drop. The transfected 293T cells are incubated at 37℃, 5% CO_2 incubator overnight, then changed with 15 mL fresh DMEM supplemented with 1%-10% FBS (dependent on the cell type you will infect). Collect the medium from 293T dish every 24 hours post transfection. Filter the collected medium through a 0.45 μm filter first and then consentrate viruses using superfilter centrifuge tube (超滤离心管) with a cut off of 10 kDa, at 6000 g for 20-30 min.

Measurement of virus titers

Virus titers are based on the expression of transgene such as GFP. Plate 293T cells and add dilution of viral supernants and 8 μg/mL polybrene (聚凝胺). After 24 hours incubation, count the percentage of GFP positive by flurescent microscope and calculate the transduction units (TU). If 10% cells are GFP positive with 0.02 mL viral supernant, then the titer is 10% × target cell numbers in total (2×10^5 for example)/ 0.02 mL =1×10^6 TU/mL.

Infection of HFF cells

Prepare 2 mL mixture of collected retrovirus medium containing equal titers of pMXs-Oct4, pMXs-Sox2, pMXs-cMyc and pMXs-Klf4. Add polybrene to a final concentration of 4μg/mL, mix gently. Discard the medium of prepared HFF cells in 6-well plate. Add the 2 mL mixed infection medium and incubate overnight. From the next day on, change the medium daily (antibiotics can be added from this day).

Isolation of HFF iPS cells

At the 6th day post infection, passage the HFF cells onto feeder and change ES medium next day. Then monitor and pick up the iPS clones daily. To passage the iPS clones, add collagenase V (1 mg/mL) and incubate in the incubator until the edges of all colonies are slightly rolled up while all feeders detached. Aspirate the supernatant, then scrape off the iPS clone using a bending glass needle and transferred to a new well.

QUESTIONS FOR DISCUSSION

1. What is nuclear reprogramming in cells? What methods can initiate and realize nuclear reprogramming? And what are the possible mechanisms for reprogramming?

2. What is induced pluripotent stem (iPS) cells? List some applications of iPS cells in clinical research.

3. How to derive iPS cells by molecular reprogramming? And what are the mechanical steps involved in the reprogramming of somatic cells into pluripotent cells by $Oct4$, $Sox2$, $Klf4$ and $c\text{-}Myc$?

知识要点

细胞核的重编程是指细胞核的基因表达状态从一种细胞类型转变成另外一种类型，如将体细胞逆转成未分化的胚胎细胞。核移植实验其实就是卵胞质对供体细胞核的重编程过程。细胞融合实验中，分裂活跃的细胞核对静止核也存在细胞核的重编程作用。分子重编程技术是通过在靶细胞中过表达多能性转录因子来实现对细胞核的重编程。诱导性多能干细胞(iPS)技术的成功是干细胞研究领域中里程碑式的突破性进展。通过在体细胞中过表达 4 个转录因子 $Oct4$, $Sox2$, $Klf4$ and $c\text{-}Myc$，已成功获得小鼠、大鼠、人和猪的 iPS 细胞。细胞核的分子重编程是一个表观遗传学过程，涉及染色质的去致密化、组蛋白的乙酰化和 DNA 的去甲基化等修饰，以改变基因表达状态。

CHAPTER 33   ORGAN CULTURE, HISTOTYPIC CULTURE AND ORGANOTYPIC CULTURE(器官培养、组织型培养和器官型培养)

In cell culture, nutritional and hormonal supplementation are in themselves inadequate to recreate full structural and functional competence in a given cell population. The vital missing factor is cell interaction and the signalling capacity that it entails. Interacting populations of cells have a mutual effect on their respective phenotypes, and the resultant phenotypic changes lead to new interactions. Cell interaction is therefore not a single event, but, instead, a cascade of events. Epithelium differentiates in response to matrix constituents that are often determined jointly by the epithelium on one side and connective tissue on the other, as may be the case with the interaction between epidermis and dermis *in vitro*. A primitive neural crest cell(神经嵴细胞) may become a neuron, an endocrine cell, or a teratoma(畸胎瘤), depending on its ultimate location, its interaction with adjacent cells, and its response, mediated by neighboring cells, to hormonal stimuli. Thus, if you want to learn something of the integrated function, or dysfunction, of whole organs, a histotypic or organotypic model will be required.

There are two major ways to examine the cell interactions. One is to accept the cellular distribution within the tissue, explant it, and maintain it as an organ culture. The second is to purify and propagate individual cell lineages, study them alone under conditions of homologous cell interaction, recombine them, and study their mutual interactions. These approaches have given rise to three main types of technique: (1) organ culture, in which whole organs, or representative parts, are maintained as small fragments in culture and retain their intrinsic distribution, numerical and spatial, of participating cells; (2) histotypic culture, in which propagated cell lines are grown alone to high density in a three-dimensional matrix; and (3) organotypic culture, in which cells of different lineages are recombined in experimentally determined ratios and spatial relationships to re-create a component of the organ under study.

Organ culture seeks to retain the original structural relationship of cells of the same or different types, and hence their interactive function, in order to study the effect of exogenous stimuli on further development. This relationship may be preserved by explanting the tissue intact or recreated by separating the constituents and recombining them. Organotypic culture represents the synthetic approach, whereby a three-dimensional, high-density culture is regenerated from isolated (and, preferably, purified and characterized) lineages of cells that are then recombined, after which their interaction is studied, and, in particular, their response to exogenous stimuli is characterized. The exogenous stimuli may be regulatory hormones, nutritional conditions, or xenobiotics. In each case, the response is likely to be different from the responses of a pure cell type in isolation, grown at a low cell density.

ORGAN CULTURE

Gas and Nutrient Exchange

A major deficiency in tissue architecture in organ culture is the absence of a vascular system, limiting the size (by diffusion) and potentially the polarity of the cells within the organ culture. When cells are cultured as a solid mass of tissue, gaseous diffusion and the exchange of nutrients and metabolites becomes limiting. The dimensions of individual cells cultured in suspension or as a monolayer are such that diffusion is rapid, but aggregates of cells beyond about 250 μm in diameter (5,000 cells) start to become limited by diffusion, and at or above 1.0 mm in diameter (-2.5×10^5 cells) central necrosis is often apparent. To alleviate this problem, organ cultures are usually placed at the interface between the liquid and gaseous phases, to facilitate gas exchange while retaining access to nutrients. Most systems achieve this by positioning the explant on a raft or gel exposed to the air, but explants anchored to a solid substrate can also be aerated by rocking the culture, exposing it alternately to a liquid medium and a gas phase, or by using a roller bottle or tube.

Anchorage to a solid substrate can lead to the development of an outgrowth from the explant and resultant alterations in geometry, although this effect can be minimized by using a nonwettable surface. One of the advantages of culture at the gas-liquid interface is that the explant retains a spherical geometry if the liquid is maintained at the correct level. If the liquid is too deep, gas exchange is impaired; if it is too shallow, surface tension will tend to flatten the explant and promote outgrowth.

Increased permeation of oxygen can also be achieved by using increasing O_2 concentrations up to pure oxygen or by using hyperbaric oxygen (高压氧). As increasing the O_2 tension will not facilitate CO_2 release or nutrient-metabolite exchange, the benefits of increased oxygen may be overridden by other limiting factors.

Structural Integrity

Structural integrity, above other considerations, is the main reason for adopting organ culture as an *in vitro* technique in preference to cell culture. While cell culture utilizes cells dissociated by mechanical or enzymic techniques or spontaneous migration, organ culture deliberately maintains the cellular associations found in the tissue. It was discovered that certain elements of phenotypic expression were found only if cells were maintained in close association, and associated cells do exchange signals via junctional communications (缝隙连接, gap junctions), via paracrine factors, and via cell adhesion molecules. Signaling between cells is most striking during organogenesis, but is probably also required for the maintenance of fully mature tissues. Therefore, maintenance of the structural integrity of the original tissue may preserve the correct homologous and heterologous cellular interactions present in the original tissue and maintain the correct configuration of the extracellular matrix.

Growth and Differentiation

Differentiated cells no longer proliferate. It is also possible that cessation of growth may in itself contribute to the induction of differentiation. Because of density limitation of cell proliferation and the physical restrictions imposed by organ culture geometry, most organ cultures do not grow, or, if they do, proliferation is limited to the outer cell layers.

Limitations of Organ Culture

Analysis of organ cultures depends largely on histological techniques and they do not lend themselves readily to biochemical and molecular analyses. Organ cultures are also more difficult to prepare than replicate cultures from a passaged cell line and do not have the advantage of a characterized reference stock to which they may be related. Organ cultures cannot be propagated, and hence each experiment requires recourse(求助) to the original donor tissue. Preparation is labor intensive, and as a result, the yield of usable tissue is often too low to be of value in biochemical or molecular assays. Furthermore, as the population of reacting cells may be a minor component of the culture, it is difficult to analyze the biochemical nature of the response and attribute it to the correct cell type, other than by autoradiographic, histochemical, or immunocytochemical techniques, which tend to be more qualitative than quantitative.

Organ cultures are useful in the demonstration of processes such as embryonic induction, where it is important to maintain the integrity of whole tissue. However, they are slow to prepare and present problems of reproducibility between samples. Growth is limited by diffusion, mitosis occurs only around the periphery, while the centers of explants frequently become necrotic.

HISTOTYPIC CULTURE

Various attempts have been made to regenerate tissue-like architecture from dispersed monolayer cultures. Green and Thomas [1978] showed that human epidermal keratinocytes will form friction ridges (指纹) if they are kept for several weeks without transfer, and Folkman and Haudenschild [1980] were able to demonstrate the formation of capillary tubules in cultures of vascular endothelial cells cultured in the presence of endothelial growth factor and medium conditioned by tumor cells.

Gel and Sponge Techniques

Both normal and malignant cells can penetrate cellulose sponge. Collagen coating of the sponge may facilitate occupation, and Gelfoam (a gelatin sponge matrix used in reconstructive surgery) may be used in place of cellulose.

Hollow Fibers(中空纤维)

Since medium supply and gas exchange become limiting at high cell densities, Knazek et al. [1972] developed a perfusion chamber from a bed of plastic capillary fibers, now available commercially. The fibers are gas and nutrient permeable and support cell growth on their outer surfaces. Medium, saturated with 5% CO_2 in air, is pumped through the centers of the capillaries, and cells are added to the outer chamber surrounding the bundle of fibers. The cells attach and grow on the outside of the capillary fibers, fed by diffusion from the perfusate (灌流液), and can reach tissue-like cell densities. Different plastics and ultrafiltration properties give molecular weight cut-off points at 10, 50, or 100 kDa, regulating the diffusion of macromolecules. It is claimed that cells in this type of high-density culture behave as they would *in vivo*. However there are considerable technical difficulties in setting up the chambers and they are costly. Furthermore, sampling cells from these chambers and determining the cell concentration are difficult. Overall, however, hollow fibers appear to present an ideal system for studying the synthesis and release of biopharmaceuticals and are now being exploited on a semi-industrial scale.

Spheroids

When dissociated cells are cultured in a gyratory shaker (回旋摇床), they may re-associate into clusters. Dispersed cells from embryonic tissues will sort during reaggregation in a highly specific fashion. For example, Muller cells of chick embryo retina re-aggregated with neuronal cells from the retina were inducible for glutamine synthetase, but those reaggregated with neurons from other parts of the brain were not. Cells in these heterotypic aggregates appear to be capable of sorting themselves into groups and forming tissue-like structures. Homotypic reaggregation also occurs fairly readily, and spheroids generated in gyratory shakers or by growth on agar have been used as models for chemotherapy *in vitro* and for the characterization of malignant invasion. As with organ cultures, the growth of spheroids is limited by diffusion, and a steady state may be reached in which cell proliferation in the outer layers is balanced by central necrosis.

Immobilization(固定化) of Living Cells in Alginate(海藻酸)

The technique of encapsulating living cells within alginate beads has been widely used in experimental research-e. g. , encapsulation of hybridoma cells for monoclonal antibody production. Alginate is derived from brown seaweeds(褐藻) and consists mainly of two types of monosaccharides(单糖): L-guluronic acid (古洛糖醛酸, G) and D-mannuronic acid (甘露糖醛酸, M). It is composed of alternating molecules of M and G, and divalent cations bind strongly between separate G blocks and initiate the formation of an extended alginate gel network. Alginate gels can be formed into beads by dripping the alginate solution into a buffer containing divalent cations, such as Ca^{2+}. Mechanical strength, volume, stability, and porosity correlate with the G content such that alginate beads with a high G content have the largest pore sizes, ranging between 5 and 200 nm. Pores of such sizes allow free diffusion of macromolecules out of, as well as into, the alginate. At present, numerous cell types can be genetically engineered to produce specific proteins of choice. By encapsulating such cells in alginate, a valuable vehicle is obtained for delivering specific recombinant proteins to the organism. Thus, such alginate "bioreactors" may have an important therapeutic potential for the treatment of a number of diseases, in which the alginate may prevent the encapsulated cells from being destroyed by the immune system.

ORGANOTYPIC CULTURE

The advent of filter-well technology, boosted by its commercial availability, has produced a rapid expansion in the study of organotypic culture methods. It gives the opportunity for the formation of both high-density polarized cultures and heterotypic combinations of cell types. Skin equivalents have been generated by coculturing dermis with epidermis, with an intervening layer of collagen, or with dermal fibroblasts incorporated into the collagen, and models for paracrine control of growth and differentiation have been developed with cells from lung, prostate, and breast. The opportunity for heterotypic cell interaction has also opened up numerous opportunities for studying inflammation and irritation *in vitro* and for creating other models for tissue interaction with increased *in vivo* relevance.

Filter-Well Inserts. The opportunities provided by filter-well inserts for cell interaction, stratification, and polarization have made them a popular culture system in many areas (Fig. 33. 1). Filter-well inserts have been used to generate stratified epidermis and polarized intestinal and kidney epithelium. Others have used them to study invasion by granulocytes or malignant cells. One of the major advantages of filter-well inserts is that they allow the recombination of cells at very high, tissue-like densities, with ready access to medium and gas exchange, but in a multireplicate form. Filters can be obtained precoated with collagen, laminin, fibronectin, or Matrigel(基质胶).

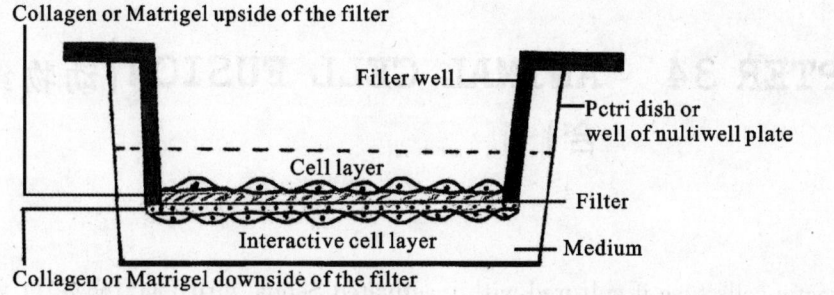

Fig. 33.1 Schematic diagram of a filter-well insert. The filter-well is inserted into a Petri dish or the well of multiwell plate. Monolayer is grown on matrix on top of a filter. Interactive cell layer added to the underside of a filter with matrix coating.

QUESTIONS FOR DISCUSSION

1. What is an organ culture? How to maintain the culture conditions of an organ culture? What are the limitations for organ culture?
2. What is histotypic culture? List some techniques to maintain a hisotypic culture.
3. What is an organtypic culture?

知识要点

研究细胞与细胞之间的相互作用及其对表型的影响,是不能依靠细胞培养技术来解决的,但是可以通过器官培养,或将体外培养的相同类型细胞(组织型培养)或不同类型细胞(器官型培养)重新组合到一起的方法来进行研究。器官培养中,保证氧气和营养物质的供给,维持器官的结构和功能,抑制细胞的迁移,是必须要解决的关键问题。组织型培养方法有多种,中空纤维和海藻胶包埋是行之有效的方法。插入式滤膜井技术是常用的器官型培养方法,操作简单易行。

CHAPTER 34 ANIMAL CELL FUSION(动物细胞融合)

Somatic cells fuse if cultured with inactivated Sendai virus(仙台病毒)or with polyethylene glycol (聚乙二醇,PEG). A proportion of the cells that fuse progress to nuclear fusion, and a proportion of these cells progress through mitosis, such that both sets of chromosomes replicate together and a hybrid is formed. In some interspecific hybrids-e. g., human-mouse one set of chromosomes (the human) is gradually lost(Fig. 34.1). Thus, genetic recombination is possible *in vitro*, and, in some cases, segregation is possible as well. Genetic recombination experiments can also be carried out with isolated nuclei. Nuclei can be isolated by centrifuging cytochalasin B(细胞松弛素 B)-treated cells and fusing the extracted nuclei to recipient whole cells or enucleated cytoplasts in the presence of PEG (Fig. 34.2(a)).

Chromosomes may be isolated from metaphase cells as micronuclei (微核)after prolonged colcemid (秋水仙碱) treatment, followed by cytochalasin B treatment and centrifugation. Incubation of these micronuclei with whole host cells in the presence of PEG results in fusion with the host cells and, ultimately, their incorporation into the nucleus (Fig. 34.2(b)). Incorporation of a selectable marker (e.g. HyTK) into individual chromosomes of the donor cell allows the selection of resistant clones containing the marked chromosome with hygromycin (潮霉素)after incorporation into the host cell.

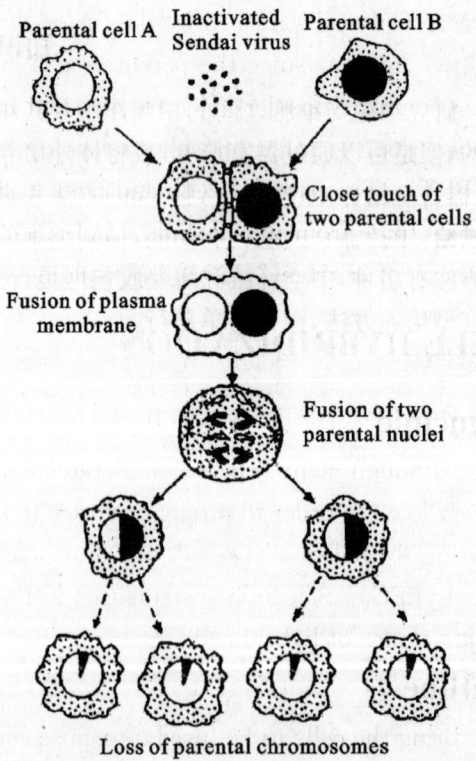

Fig. 34.1 Cell fusion by Sandai virus.

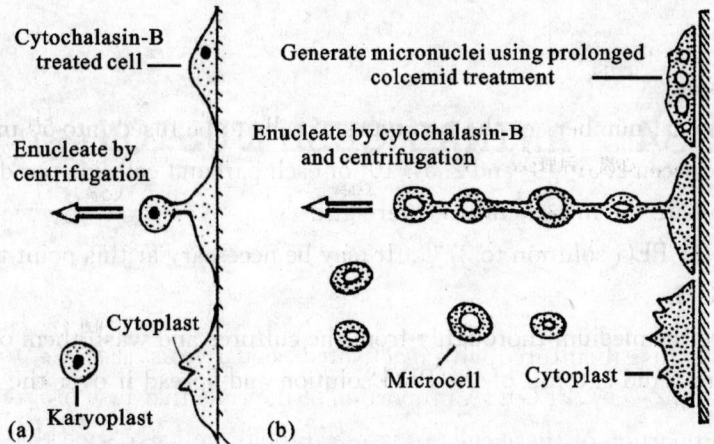

Fig. 34.2 Isolation of karyoplast(核体,小型细胞)(a) and microcell (微细胞)(b) by cytochalasin-B and/or colcemid treatment and centrifugation.

Since the proportion of viable hybrids is low, selective media are required to favor the survival of the hybrids at the expense of the parental cells. Two mutant parental cell types of TK^- (deficient in thymidine kinase) and $HGPRT^-$ (deficient in hypoxanthine guanine phosphoribosyl transferase) are used, and the selection is carried out in HAT medium containing hypoxanthine(次黄嘌呤), aminopterin(氨基喋呤), and thymidine (胸苷). Only cells formed by the fusion of two different parental cells (heterokaryons, 异核体) survive, since the parental cells and fusion products of the same parental cell type (homokaryons,同核体) are deficient in either thymidine kinase(胸苷激酶,TK)or hypoxanthine guanine phosphoribosyl transferase(次黄嘌呤鸟嘌呤磷酸核糖转移酶, HGPRT). The parental cells and homokaryons cannot, therefore, utilize thymidine or hypoxanthine from the medium, and since aminopterin(氨基喋呤)blocks endogenous synthesis of purines(嘌呤)and pyrimidines(嘧啶), they are unable to synthesize DNA.

CELL HYBRIDIZATION

Principle

Although many cell lines undergo spontaneous fusion, the frequency of such events is very low. In order to produce hybrids in significant numbers, cells are treated with either inactivated Sendai virus or, more commonly, the chemical fusogen polyethylene glycol (PEG). Selection systems that kill parental cells but not hybrids are then used to isolate clones of hybrid cells.

Outline

Bring the cells to be fused into close contact, either in suspension or in monolayers. Treat the cells with PEG, briefly to minimize cell killing. Usually, the cells are given a 24-h period to recover before selection for hybrids.

Protocol

Monolayer Fusion

Inoculate equal numbers of the two types of cells to be fused into 50-mm tissue culture dishes. Between 2.5×10^5 and 2.5×10^6 of each parental cell line per dish is usually sufficient. Incubate the mixed culture overnight.

2. Warm the PEG solution to 37℃. It may be necessary at this point to readjust its pH, using NaOH.

3. Remove the medium thoroughly from the cultures and wash them once with serum-free medium. Add 3.0 mL of the PEG solution and spread it over the monolayer of cells.

4. Remove the PEG solution after exactly 1.0 min, and rinse the monolayer three times with 10-ml of serum-free medium before returning the cells to complete medium.

5. Culture the cells overnight before adding selection medium.

Suspension Fusion

1. Centrifuge a mixture of 4×10^6 cells of each of the two parental cell lines at 150 g for 5 min at room temperature. Carry out centrifugation and subsequent fusion in 30-ml plastic universal containers or centrifuge tubes.

2. Resuspend the pellet in 15 mL of serum-free medium, and centrifuge again.

3. Aspirate all the media, and resuspend the cells in 1 mL of PEG solution by gently pipetting.

4. After 1.0 min, dilute the suspension with 9 mL of serum-free medium, and transfer half of the suspension to each of two universal containers or centrifuge tubes containing a further 15 mL of serum-free medium.

5. Centrifuge the suspensions at 150 g for 5 min. Remove the supernatant, and resuspend the cells in complete medium.

6. After overnight incubation, clone the cells in selection medium.

A large number of variations of the PEG fusion technique have been reported. While the procedure described here works well with a range of mouse, hamster, and human cells in interspecific and intraspecific fusions, it is unlikely to be optimal for all cell lines. Inclusion of 10% DMSO in the PEG solution has the advantage of reducing its viscosity and has been reported to improve fusion. Also, the molecular weight of the PEG used need not be 1,000 D. Preparations with molecular weights from 400 to 6,000 D have been successfully used to produce hybrids. Although now largely superseded (取代) by PEG as a fusogen, for reasons of convenience, Sendai virus fusion remains a reliable method.

SELECTION OF HYBRID CLONES

The method of selection used in any particular instance depends on the species of origin of the two parental cell lines, the growth properties of the cell lines, and whether selectable genetic markers are present in either or both cell lines. Hybrids are most frequently selected using the HAT system: 10^{-4} M hypoxanthine, 6×10^{-7} M aminopterin (氨基喋呤), and 1.6×10^{-6} M thymidine. This system can be used to isolate hybrids made between pairs of mutant cell lines deficient in the enzymes thymidine kinase (TK$^-$) and hypoxanthine guanosine phosphoribosyl transferase (HGPRT$^-$), respectively. TK$^-$ cells are selected by exposure to BUdR and HGPRT$^-$ cells by exposure to thioguanine(硫鸟嘌呤). When only one parent cell line carries such a mutation, HAT selection can still be applied if the other cell line does not grow, or grows poorly in culture (e.g., lymphocytes, senescing primary cultures).

Differential sensitivity to the cardiac glycoside ouabain(毒毛旋花甙) is an important factor in the selection of hybrids between rodent cells and cells from a number of other species, including human. Rodent cells are resistant to concentrations of this antimetabolite up to 2.0 mM, while human cells are killed at 10^{-5} M ouabain. The hybrids are much more resistant to ouabain than the human parental cells. If a rodent cell line that is HGPRT deficient is fused to unmarked human cells, then the hybrids can be selected in medium containing HAT and low concentrations of ouabain.

Although many other selection systems have been reported, only complementation of auxotrophy(营养互补) has been widely used. It must be stressed that, whichever method is used to isolate clones of putative hybrid cells, confirmation of the hybrid nature of the cells must be obtained. This is usually done using cytogenetic or biochemical techniques.

PRODUCTION OF MONOCLONAL ANTIBODIES

Principle

Monoclonal antibodies have become indispensable tools in research, diagnostics, and therapeutics. They have gradually replaced polyclonal antibodies since hybridoma technology(杂交瘤技术) was first introduced by Kohler and Milstein in 1975. Hybridomas are produced by fusing a nonsecreting myeloma(骨髓瘤) cell with an antibody-producing B-lymphocyte in the presence of polyethylene glycol. (Fig. 34.3) The myeloma cell is deficient in an enzyme hypoxanthine-guanine phosphoribosyl transferase (HGPRT) or thymidine kinase (TK) necessary for DNA synthesis and cannot survive in selection medium containing hypoxanthine, aminopterin, and thymidine. Any unfused B-lymphocytes from the spleen cannot survive in culture for more than a few days. Any B-cell-myeloma hybrids should contain the genetic information from both parent cells and are thus able to survive in the HAT selection medium. They can be cultured indefinitely and will produce unlimited quantities of antibody. Supernatants from surviving hybridomas are screened for antibody by ELISA. Those hybridomas selected are then subcloned to ensure that they are producing antibody that is specific for a single epitope. Antibody production can be scaled up *in vivo* as ascites in mice or *in vitro* as a suspension culture. Hybridomas also grow very well in various hollow-fiber and fermentation culture systems.

Outline

Using polyethylene glycol (PEG), fuse spleen cells from an immunized mouse with myeloma cells. Select hybrid colonies (hybridomas) in HAT medium. Screen the supernatants by ELISA 10-14 d after fusion, and expand, freeze, and subclone the desired hybridomas, to ensure monoclonality.

Protocol

Immunization

1. Bleed the mice on day 0 prior to the initial injection, and check the serum for background antigen reactivity.

2. Immunize the mice (A/J or Balb/c) with antigen emulsified in Freund's adjuvant (弗氏佐剂) or mixed with a 1/10 volume of alum (明矾) and vortexed. Give three injections of antigen intraperitoneally (腹腔), according to the following schedule:

Day	Amount of antigen	Adjuvant
0	50 μg	Alum or complete Freund's adjuvant
14	25 μg	Alum or incomplete Freund's adjuvant
28	25 μg	Alum or PBSA

3. Bleed the mice on day 35 and measure the serum titer (滴度) of antibody by an ELISA assay.

4. Dilute the serum serially 1∶4 after a 1∶30 dilution, and up to 1∶30720.

5. Select mice with the highest ratio of serum titers to antigen for fusion.

6. Give the selected mice a final boost of 10 μg of antigen i. v. or 25 μg of antigen i. p. 3 d prior to fusion.

Myeloma

1. It is convenient to perform fusions on a Thursday, with the mice receiving a final boost of antigen on a Monday.

2. Maintain the P3.653 myeloma cell line in MEM + 10% fetal calf serum + 8-azaguanine (8-氮鸟嘌呤).

3. Dilute the P3.653 cells to 3.5×10^5 cells/mL each day for the three days prior to

fusion.

T-Cell Depletion

1. Bleed mice with appropriate serum titers, and sacrifice them by cervical dislocation(颈椎脱位). Aseptically remove the spleens, and place them in a sterile Petri dish with 5 mL of sterile PBSA. Gently tease the spleens with two 23G needles on 1-ml syringes. Teasing spleens roughly will result in a high concentration of fibroblasts.

2. Transfer the cells to a 15-ml conical tube(锥形管), and allow clumps to settle. Transfer spleen cells (without clumps) to a 50-ml conical tube, and, following a 1:100 dilution, count the cells with a hemocytometer(血球计数板).

3. Spin the cells at 1,000 rpm for 8 min. To lyse the red blood cells, resuspend the resultant pellet in 0.84% NH_4Cl (10 mL/spleen), and incubate the suspension at 4℃ for 15 min. Underlayer the cell suspension with 14 mL of horse serum, and spin the solution at 1,500 rpm for 8 min.

4. Resuspend the resultant pellet in 50 mL of TCD buffer (Hanks' balanced salt solution + 10 mM HEPES + 0.3% BSA + 0.16 M NH_4Cl), and spin the suspension at 1,000 rpm for 8 min.

5. For T-cell depletion, resuspend the resultant cell pellet in Antimouse Thy 1.2 antibody at a final concentration of 1×10^7 cells/mL. Incubate the suspension at 4℃ for 45 min, and then spin it at 1,000 rpm for 8 min. Resuspend the resultant pellet in rabbit complement (reconstituted in 1 mL of cold UPW, diluted 1:12 in TCD buffer, and filter sterilized). Incubate the suspension at 37℃ for 45 min, and then spin it at 1,000 rpm for 8 min. Count the cells by trypan blue exclusion on a hemocytometer. B-cell recovery should be 30%-50%.

Fusion

1. Mix the myeloma and B-cells in a 50-ml centrifuge tube. One fusion can be done on a maximum of 1.2×10^8 spleen cells. Mix the spleen cells with P3.653 myelomas at a ratio of 4:1; thus, the maximum number of P3.653 cells per fusion is 3×10^7 cells.

2. Centrifuge the suspension at 1,000 rpm for 8 min.

3. Break up the resultant pellet by tapping, and add 1 mL of PEG to the tube over 15 s.

4. Mix the suspension by gently swirling the tube for 75 s.

5. Add 1 mL of serum-free medium over 15 s, and gently swirl the tube for 45 s.

6. Add 2 mL of serum-free medium over 30 s, and swirl the tube for 90 s.

7. Add 4 mL of HAT medium over 30 s, and swirl the tube for 90 s.

8. Finally, add 8 mL of HAT medium over 30 s, and swirl tube for 90 s.

9. Add this volume (16 mL) to a sterile Nalgene bottle containing the calculated amount of HAT medium (125 mL if the maximum cell concentration has been used). 16

mL, containing 1.5×10^8 cells, from step 1 in this section of the protocol plus 125 mL of HAT medium in the bottle makes 141 mL. With the wash in the next step (step 10), the total volume is 150 mL and will result in a final concentration of 1×10^6 cells/mi.

10. Wash the 50-ml conical tube with 9 mL of HAT medium, and add this volume to the bottle.

11. Mix the contents of the bottle well, and transfer the cells to the sterile reservoir.

12. Using a 12-channel multipipettor, plate the cells at 200 μL/well into a sterile 96-well plate. The final concentration is then 2×10^5 cells/well.

Selection of Hybridomas

1. Feed the fusion plates 5 d after fusion, by aspirating most of the culture media from the wells and replacing it with 150-200 μL/well of fresh HAT medium (Basal medium, such as MEM or RPMI, + 10% FBS + 20% spleen-conditioned medium + 1 mM hypoxanthine + 4 μM aminopterin + 0.16 mM thymidine).

2. Feed the plates twice per week.

3. Screen the clones for selection of positive hybridomas (杂交瘤), usually two weeks after fusion, using ELISA.

4. After a further 48 h, retest those clones that tested positive in the previous step.

5. Expand the most productive hybridomas by culturing them in two wells of a 96-well plate in media containing 10% FBS and HT (containing hypoxanthine, thymidine, and glycine).

6. Retest the clones, expand the positive hybridomas to a 24-well plate, and wean them off HT medium, at which time 2 mL of culture supernatant should be harvested for screening. At this step, enough volume is harvested to perform several selection assays to ensure that the antibody is directed only at the antigen of interest.

7. Expand the hybridomas to be kept to 4 wells of a 24-well plate, and cryopreserve them.

8. Perform a second cryopreservation after expanding the hybridoma to a 75-cm² flask.

Subcloning

To ensure monoclonality, subclone hybridomas of interest. This can be done by serially diluting cells and plating the equivalent of 1 cell per 3 wells in a 96-well plate or by sorting with an automated cell deposition unit (ACDU) on a FACStarplus (Becton Dickinson) and plating at one cell per well. Subcloning can be done on top of a mouse spleen feeder layer plate. After subcloning, colonies can usually be seen at day 5 and must be checked visually for monoclonality. Plates are fed with fresh medium beginning on day 7. Screening for positive hybridomas is usually done between days 10-14. Those clones selected are then expanded and frozen in the same way as the parental hybridoma.

Antibody Production

Concentrated antibody from clones of interest can be produced *in vivo* as ascites (腹水) in IFA primed mice (Balb/c or nu/nu) or *in vitro* as a suspension culture. Several hollow-fiber cell culture systems are also available. When hybridomas are inoculated into a hollow-fiber system, the cells are maintained in a compartment of the bioreactor, while fresh media and waste from the cells are recirculated. High concentrations of antibody are produced in the cell compartment, and culture supernatant containing antibody can be harvested at multiple time-points.

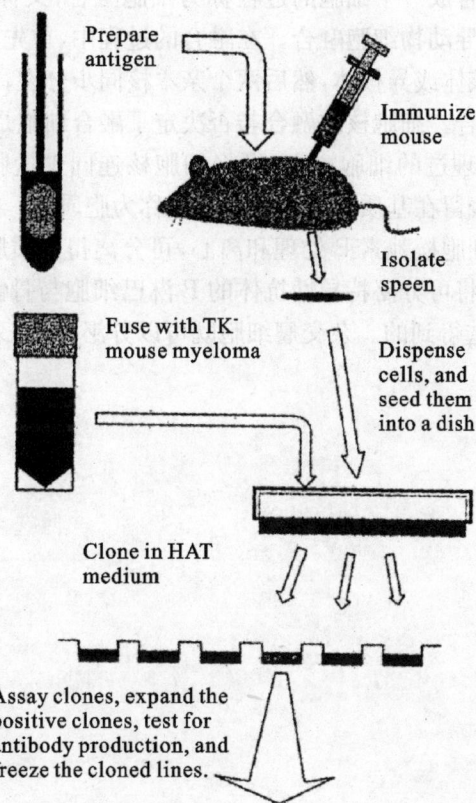

Fig. 34.1 Production of Hybridomas. Schematic diagram of the production of hybridoma clones capable of secreting monoclonal antibodies. (Modified from R. Ian Freshney)

Screening

Take care in developing the screening strategy to obtain a monoclonal antibody with the characteristics that you want. Hybridoma culture supernatants should be screened as early as feasible for desired reactivity patterns. After initial selection by ELISA for reactivity to the immunogen, the expanded culture supernatant should be tested in the application for which it was developed (e.g., Western blot, competitive immunoassay, flow cytometry, etc.).

QUESTIONS FOR DISCUSSION

1. What is cell fusion and cell hybridization?
2. How to get animal cells fused? And how to select hybrid cells from the fusion mixture successfully?
3. How to produce monoclonal antibody by hybridomas techniques?

知识要点

将两个不同细胞并合成一个细胞的过程称为细胞融合,又称为细胞杂交。灭活的仙台病毒和聚乙二醇可诱导动物细胞融合。在融合的过程中,首先是细胞与细胞相互靠近,随后质膜融合,形成同核体或异核体,然后两个亲本核同步分裂,形成一个纺锤体,从而实现两个亲本细胞核的融合。细胞核的融合与否决定了融合细胞是否能分裂和长期存活下来。将细胞松弛素 B 处理过的细胞离心,可将细胞核连同少量胞质从细胞中分离出来,形成外包质膜的核体;残留在基质上的无核的胞质称为胞质体。用秋水仙素处理细胞,使细胞微核化,然后通过细胞松弛素 B 处理和离心,可分离得到质膜包被的微核,称为微细胞。杂交瘤细胞是通过将可分泌特异性抗体的 B 淋巴细胞与骨髓瘤细胞融合,然后通过 HAT 选择性培养基筛选得到的。杂交瘤细胞既可以分泌单抗,又能够在体外无限增殖。

CHAPTER 35   ANIMAL CELL TRANSFECTION(动物细胞转染)

In order to study the function of individual genes, the sequence of interest can be cloned and then transferred into host cells by a variety of transfection technology, such as coprecipitation with calcium phosphate(磷酸钙共沉淀), electroporation(电穿孔), lipofection(脂质体转染), and retroviral infection(反转录病毒感染). Cloned DNA is often conveniently maintained as part of a bacterial plasmid. Many plasmids can attain a high *copy* number during bacterial growth, thus ensuring a plentiful stock of DNA for experimentation. Plasmid DNA is purified from the bacteria prior to use. Once the sequence of interest has been cloned, it can be further cloned into an expression plasmid containing prokaryotic and/or eukaryotic promoter sequences. By genetic manipulation, promoter sequences can also be linked to a reporter gene (e.g., β-gal or CAT) whose products can be readily assayed subsequent to transfection(转染). In this way, tissue-specific gene expression can be analyzed in detail. Oncogenes and tumor suppressor gene function can also be analyzed by similar manipulations.

Transfections may be *transient* or *stable*. Transient transfections are short term and used shortly after transfection, and the efficiency of transfection is determined by reporter gene assays. DNA used for stable transfection contains a selectable marker, such as *neo* or *hyg B*, that confers resistance to G418 [geneticin(遗传霉素), an analogue of neomycin(新霉素)] or hygromycin(潮霉素), respectively. Transfected cells are then selected by continued exposure to the selection agent, and resistant clones can be isolated. Three protocols for gene transfer into mammalian cells are presented below, two for stable transfection using calcium phosphate and electroporation and one for transient using lipofection. Selection protocols for stable transfection can be added to the lipofection protocol, provided that the construct used for transfection contains the appropriate selectable marker.

STABLE DNA TRANSFECTION BY COPRECIPITATION WITH CALCIUM PHOSPHATE

Principle

The calcium phosphate technique for introducing genes into mammalian cells was first described by Graham and Van der Eb [1973] and is still widely used. In this method, exogenous DNA is mixed with calcium chloride and is then added to a solution containing phosphate ions. A calcium-phosphate-DNA coprecipitate is formed, which is taken up by mammalian cells in culture, resulting in expression of the exogenous gene. This method can be used to introduce any DNA into mammalian cells for transient expression assays or long-term transformation.

Outline

Transfect cells with the appropriate DNA carrying a selectable marker (e. g., aminoglycoside phosphotransfera*se* (*aph*) gene, 氨基糖苷类磷酸转移酶基因), and apply selection to eliminate the cells that have not taken up and expressed the exogenous gene.

Protocol

1. Harvest exponentially growing cells by trypsinization.

2. Replate the cells at a density of 5×10^5 cells per flask (25-cm^2 growth area) in 5 mL of SF12 medium containing 15% fetal bovine serum.

3. Incubate the culture at 37℃ for 24 h.

4. Add 1 mL of 2 × HBS (HEPES-buffered saline: NaCl, 1.63 g, HEPES, 1.19 g, $Na_2HPO_4 \cdot 2H_2O$, 0.023 g, per 100 mL) to a vial.

5. Into a second, plastic vial, dilute plasmid DNA to a final concentration of 80 μg/mL in 0.5 mL of TEB (Tris-EDTA buffer: 0.1 mM EDTA, 1.0 mM Tris-HCl; pH 8.0).

6. Add 0.4 mL of TEB and 0.1 mL of 2.5 M $CaCl_2$ to the vial, and mix the solution.

7. Add this DNA solution slowly (over about 30 s), with continuous mixing, to the 1 mL of 2 × HBS in the first vial.

8. Mix the contents of the vial immediately by vortexing, and leave the solution at room temperature for 30 min. The DNA concentration at this stage is 20 μg/mL.

9. After the incubation, a fine precipitate will have formed.

10. Add 0.5 mL of this DNA-calcium phosphate suspension to each flask containing cells in 5 mL of growth medium.

11. Incubate the flasks at 37℃ for 24 h, to allow absorption of the DNA-calcium phosphate coprecipitate by the cells.

12. Preselection expression stage:

(a) Replace the medium in the flasks with fresh, prewarmed medium.

(b) Incubate the flasks at 37℃ for a further 24 h, to allow expression of the transferred gene(s) to occur.

13. Selection stage:

(a) Replace the medium with an appropriate selection medium-in this case, SF12 medium containing 15% serum and 200 μg/mL geneticin (Cultured cell lines differ in their sensitivity to geneticin, and the most suitable concentration of geneticin to use must be determined empirically).

(b) Renew the selection medium every 2-3 d for up to 2-3 weeks when the colonies are routinely counted.

(c) Pick colonies.

STABLE TRANSFECTION BY ELECTROPORATION

Principle

DNA can be introduced into cells by electroporation(电穿孔), when a high cell concentration is briefly exposed to a high-voltage electric field in the presence of the DNA to be transfected. Small holes are generated transiently in the cell membrane, and the DNA is allowed to enter the cell and, in some of the cells, becomes incorporated into the genome. Equipment for electroporation is available commercially. Most cells refractory to chemical methods of gene transfer are successfully transfected by electroporation.

Electroporation is usually performed at a constant capacitance(电容) setting and therefore a constant pulse duration, with various field strengths (500-1500 kV/cm) for pilot investigations. For most cells, the settings at which approximately 20-50% of the cells remain viable after electroporation are sufficient for DNA transfer. Electroporation is usually performed at room temperature, and the cells are subsequently kept on ice, to extend the period of time that the membrane pores remain open.

There is a linear relationship between DNA concentration, DNA uptake, and reporter gene expression. It is believed that linearized DNA is more efficient for the production of stable transfectants than is supercoiled(超螺旋)DNA, presumably due to the increased efficiency with which linear DNA integrates into the genome DNA. Electroporation results in the integration of DNA in low copy number, although the copy number introduced can be adjusted by altering the concentration of DNA in the cell suspension. Chemical methods of transfection usually result in the integration of large concatamers (串联体), which may inherently interfere with cell function and obscure investigations involving specific gene overexpression.

Suspension cells are more easily transfected by electroporation than are adherent cells, since adherent cells must be detached from the culture vessel. The drawbacks of electroporation include its requirement for more cells and DNA than chemical methods of gene transfer and its variability in optimal parameters between cell types.

Expression vectors

pSVTKGH, linearized plasmid. This vector contains the SV40 enhancer and the herpes virus thymidine kinase(疱疹病毒胸苷激酶) promoter sequences that drive the expression of human growth hormone (hGH). hGH is secreted directly into the tissue culture medium and is detected by radioimmunoassay.

pcDNA3, linearized plasmid. This expression vector contains the human cytomegalovirus（细胞肥大病毒）enhancer-promoter sequences upstream from its multicloning site(多克隆位点), and the polyadenylation signal(多腺苷酸信号) and transcription-termination sequences of the bovine growth hormone gene (bGH) downstream from the multicloning site. The pcDNA3 vector also contains the gene for neomycin resistance, alleviating the need to cotransfect an antibiotic resistance gene for the selection of stable transformants overexpressing a particular gene of interest. pcDNA3 may also be used as a selectable marker when cotransfected with reporter constructs, as presented in the protocol.

Protocol

1. Seed cells at 5×10^5 cells/mL in a 75 cm² culture flask, and incubate the flask at 37℃ in 5% CO_2 in growth medium supplemented with 50 IU/mL penicillin, 5μg/mL streptomycin, 10% heat-inactivated FBS.

2. During the late log phase, collect the cells by centrifugation at 4℃ at 380 g for 5 min.

3. Wash the resultant cell pellet in 10 mL of PBSA (Ca^{2+} and Mg^{2+} free PBS), and centrifuge at 4℃ at 380 g for 5 min.

4. Wash the resultant cell pellet in 5 mL of electroporation buffer, and count the cells, using a hemocytometer. Collect the cells by centrifugation at 4℃ at 380 g for 5 min.

5. Resuspend the cells at 1×10^6 cells/0.8 mL of electroporation buffer.

6. Transfer 0.8 mL of cells into prechilled electroporation cuvettes(槽).

7. Add 50 μg of linearized pSVTKGH, along with 5 μg of linearized pcDNA3, to the cell suspension.

8. Mix the DNA/cell suspension by holding the sides of the cuvette and flicking the bottom. Incubate the suspension on ice for 10 min.

9. Electroporate the suspension at 400 volts/500 μF, and record the duration of the shock. Remove the cuvette from the shocking chamber, and incubate it on ice for 10 min.

10. Transfer the electroporated cells into 10 mL of growth medium, rinse the cuvette with medium to remove all of the cells, and collect the cells by centrifugation.

11. Resuspend the cells in 20 mL of growth medium, and culture the suspension in a 75 cm² flask at 37℃ in 5% CO_2 to allow expression of the neomycin selectable marker gene.

12. After 24-48 h, collect the cells by centrifugation, and then resuspend the cells in 20 mL of selective medium.

13. Change the selective medium every 2-4 d for at least 2 weeks, to remove the debris of dead cells and to permit resistant cells to grow.

14. Assay for hGH expression in the cell supernatant.

TRANSIENT TRANSFECTION BY LIPOFECTION

Principle

The original method for cationic lipid-mediated DNA transfection into cultured cells was improved in 1993 by replacement of the monocationic(单价阳离子)lipid reagent with a polycationic(多价阳离子)one, Lipofectamine. The method is based on an ionic interaction of DNA and liposomes(脂质体)to form a complex, which can deliver functional DNA into cultured cells. Plasmid DNA is complexed(络合), but not encapsulated (封装), within unilamellar liposomes 600-1200 nm in size, formed by cationic lipids in water.

The advantages of cationic liposome-mediated transfection over other methods include generally higher efficiency; the ability to transfect successfully a wide variety of eukaryotic cell lines, many of which are refractive to other transfection procedures; and relatively low cell toxicity. Another advantage is that the basic procedure of DNA transfection can be adapted for transfection with RNA, synthetic oligonucleotides, proteins, and viruses. Finally, cationic liposomes can be used for the successful delivery of functional genes or viral genomes *in vivo*. Its disadvantage is the relatively high cost of reagents, which practically precludes large-scale use.

Protocol

Transfection of Adherent Cells

1. Seed approximately 1×10^5 to 3×10^5 cells per well in 6-well plates in 3 mL of growth medium.

2. Incubate the cells at 37℃ in a CO_2 incubator until the cells are 50% to 90% confluent. This step usually takes 18-24 h and should not take less then 16 h.

3. Before transfection, prepare the DNA and lipid solutions in sterile tubes:

(a) For the DNA solution, dilute 1-2 μg of DNA into 0.5 mL of serum-free medium.

(b) For the lipid solution, dilute 2 to 25 μL of cationic lipid reagent into 0.5 mL of serum-free medium.

(c) Combine the two solutions, mix the resultant solution gently, and incubate it at room temperature for 15-45 min, to allow the formation of DNA-lipid complexes.

4. Rinse the cells once with 2 mL of serum-free medium.

5. Overlay the DNA-lipid complex onto the cells. Antibacterial agents should be o-

mitted during transfection.

6. Incubate the cells with the complexes at 37℃ in a CO_2 incubator for 2-24 h; five or 6 hours are usually enough. Then replace the medium to remove the transfection mixture from the cells.

7. Assay the growth medium or cells for transient gene activity as appropriate.

Transfection of Suspension Cells

1. Prepare the transfection mixture in sterile tubes as follows:

(a) Dilute 2-5 μg of DNA in 0.5 mL of serum-free medium.

(b) Dilute 2-20 μL of cationic lipid reagent in 0.5 mL of serum-free medium.

(c) Combine the two solutions, mix the resultant solution gently, and incubate it at room temperature for 15-45 min, to allow the formation of DNA-lipid complexes.

2. Centrifuge a cell suspension containing approximately 1×10^6 to 2×10^6 cells, and aspirate the medium.

3. Resuspend the cells in the transfection mixture, and transfer the suspension to a 35-mm dish.

4. Incubate the cells in a CO_2 incubator for 4 to 6 h.

5. To each dish, add 0.5 mL of growth medium, supplemented with 30% serum. (A large amount of serum is necessary to protect cells in suspension, which are more sensitive to the toxic effects of liposome reagents.)

6. Incubate the dishes in a CO_2 incubator overnight.

7. Add 2 mL of complete growth medium to each dish, and incubate the dishes in a CO_2 incubator.

8. At 24 to 72 h after the start of transfection, assay the cells or medium for gene activity as appropriate.

OTHER DNA TRANSFER METHODS

Retroviral Infection

Retroviruses have a high efficiency of gene transfer, are able to incorporate larger DNA fragments than plasmids, and infect host cells spontaneously. The introduced gene becomes permanent, and the process of insertion is achieved by normal cellular processes, is not harmful to the host cell, and does not cause any other genetic alterations.

Baculovirus

Inserting genomic sequences into baculoviruses, which are then propagated in insect cells (such as Sf9 cells), also allows large sequences (>100 kbp) to be cloned. The proteins produced have post-translational modifications that are not available in prokaryotic systems, although there are differences in processing from mammalian cells. Baculoviruses are not transmissible to mammalian cells, and so cells are unlikely to carry any risk of contamination with mammalian viruses, provided that any mammalian-derived supplements (e. g., FBS) are thoroughly screened before use. Sf9 can be subcultured without trypsin.

Yeast Artificial Chromosomes

Yeast artificial chromosomes (YACs) also provide a genome that is capable of packaging larger sequences of DNA than bacterial plasmids, with downstream post-translational processing such as found in eukaryotic cells, although this latter aspect is likely to have significant differences from that in mammalian cells. Propagation in yeast also gives a high-yield and stable culture system and is less difficult to maintain than large-scale insect or mammalian cell cultures.

Mammalian Artificial Chromosomes

There has been considerable success in applying the principle of YACs to mammalian systems, in order to incorporate large mammalian sequences containing one or more structural and multiple regulatory genes into one construct. These constructs are known as mammalian artificial chromosomes (MACs) and are introduced into mammalian cells by monochromosomal transfer techniques. The technology has now progressed to the construction of human artificial chromosomes opening up a whole new era for genetic therapy.

QUESTIONS FOR DISCUSSION

1. What is gene transfection?
2. How to obtain stable transfected cells by coprecipitation with calcium phosphate?
3. How to obtain stable transfected cells by lipofection?
4. How to obtain stable transfected cells by electroporation?

知识要点

真核细胞的基因转染技术是研究基因表达调控和信号转导通路的重要手段,分为瞬时转染和稳定转染两类,其差别源于载体质粒上是否携带有筛选标记。磷酸钙共沉淀法是最早出现的最便宜的转染方法。其原理是通过细胞对 DNA-磷酸钙沉淀颗粒的胞饮作用,使外源 DNA 进入细胞内,实现基因的转移。电穿孔法是利用脉冲电场,在细胞膜上打孔,使外源 DNA 进入细胞内,实现基因的转移。电穿孔法操作简单,而且转染效率高,不过需要配备电转染仪。阳离子脂质体法是目前最常用的基因转染技术,已有多种商品化的脂质体可供选择,如 Lipofectamine 2000/LTX,FuGene HD 等。脂质体法操作简单,但转染剂的价格很高。其原理是利用 DNA-脂质体混合物与细胞膜的融合或胞饮作用,使外源 DNA 进入细胞内,实现基因的转移。此外,反转录病毒、棒状病毒以及酵母和哺乳动物人工染色体基因转移技术的应用也越来越多。

CHAPTER 36   TRANSGENIC ANIMAL(转基因动物)

In gene transfer, animals carrying new genes (integrating foreign DNA segments into their genome) are referred to as "transgenic," a term first coined by Gordon and Ruddle (1981). As such, transgenic animals were recognized as specific variants of species following the introduction and/or integration of a new gene or genes into the genome. Especially the excellent work of Richard Palmiter and Ralph Brinster (1983) on the production of growth hormone transgenic mice-"supermouse" (Fig. 36.1) greatly influenced this emerging field in a most compelling manner for both basic and applied sciences. Today, transgenic animals embody (具体表现) one of the most potent and exciting research tools in the biological sciences. Transgenic animals represent unique models that are custom tailored to address specific biological questions such as the regulation of gene expression as well as the regulation of cellular and physiological processes. Hence, the ability to introduce functional genes into animals provides a very powerful tool for dissecting complex biological processes and systems. Gene transfer is of particular value in those animal species, where long life cycles reduce the value of classical breeding practices for rapid genetic modification. Furthermore, classical genetic monitoring cannot engineer a specific genetic trait in a directed fashion.

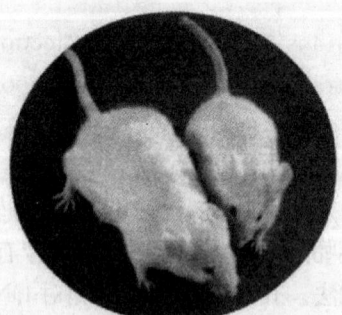

Fig. 36.1 Production of transgenic mice harboring a growth hormone (GH) fusion construct. Animals harboring the GH transgene and expressing the GH gene product grew at a rate 2-to 4-fold greater than control littermates, reaching a mature weight twice that of controls. This dramatic phenotype led the way for the exponential development of gene transfer technology. (By Palmiter et al. 1983)

During the development of gene transfer technology in whole animal, all the following techniques are of great importance, such as recombinant DNA techniques, and the capacity to artificially regulate or synchronize embryo development, the *in vitro* culture and transfer techniques of egg and embryos, and the skills in the production of chimeric (嵌合) animals, the transfer of inner cell mass cells and teratocarcinoma(畸胎瘤)cells, nuclear transfer and injections of nucleic acids into developing ova, etc.

New strategies for producing genetically engineered animals have been developed on and on. In this chapter we outline four currently used techniques to develop transgenic animals, including (1) producton of transgenic animal by DNA microinjection; (2) gene targeting in embryo stem cells by homologous recombination; (3) gene targeting in embryo stem cells by Cre-LoxP system; (4) retrovirus-mediated gene transfer.

PRODUCTION OF TRANSGENIC ANIMAL BY DNA MICROINJECTION

Over the last 20 years, DNA microinjection has become the most widely applied method for gene transfer in mammals. While relative efficiencies are debated and new methodologies explored, DNA microinjection has been the workhorse(主力) providing the bulk of useful animal models for study. DNA microinjection and embryonic stem (ES) cell transfer were the two most practiced methods. However, to identify characteristic function of dominant genes in a host of applications, following *in vitro* studies, DNA microinjection can be the first step in delineating possible research directions.

The steps involved in producing and evaluating transgenic mice derived by DNA microinjection are summarized in Fig. 36.2. In general, it takes between 5 and 12 months of dedicated effort to become proficient at the necessary techniques. One must begin with embryo handling and practice embryo transfer skills to develop good coordination. These skills will ultimately maximize all gene transfer efforts. Then, microinjection practice will lead to the production of the first transgenic mice, but continued commitment will be necessary in order to develop proficiency.

In general, microinjected DNA appears to integrate randomly into the mouse genome. In most cases the DNA integrates stably as a tandem head-to-tail repeat containing from 1 to 200 copies of the injected DNA. Multiple factors can influence the pattern of expression of the transgenic DNA, such as integration site and copy number. The tyrosinase minigene can be used to allow visual identification of transgenic mice by coat and ocular color. Interestingly, certain tyrosinase transgenic mice show that, even for a stable integration at a single site in the genome, there can be variations in the level and pattern of transgene expression.

To optimize experimental efficiencies, the following considerations are important. a. Linear DNA fragments integrate with greater efficiency than supercoiled DNA. The DNA fragment size or length does not generally affect integration frequency; b. A low-ionic-strength microinjection buffer should be prepared (10 mM Tris, pH 7.4, with

0.1-0.3 mM EDTA); c. The efficiency of producing transgenic mice (DNA integration and development of microinjected eggs to term) appears most efficient at a DNA concentration between 1.0 and 3.0 ng/l; d. Linear DNA fragments with blunt ends have the lowest chromosomal integration frequency, whereas dissimilar ends are more efficient; e. Injection of DNA into the male pronucleus is slightly more efficient than injection into the female pronucleus for producing transgenic mice; f. Nuclear injection of foreign DNA is dramatically more efficient than cytoplasmic injection.

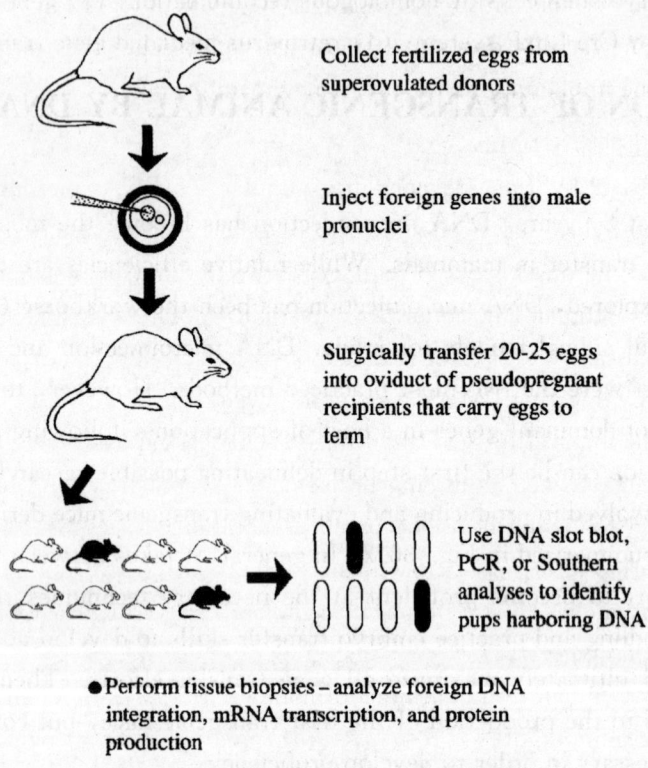

Fig. 36.2 Gene transfer in mice. The methodology employed in the production and subsequent evaluation of transgenic mice is depicted. (By H. G. Polites and C. A. Pinkert)

DNA Microinjection and Transgenic Mouse Production

DNA Preparation and Purification

DNA microinjection was one of the only methods where fragment size was not at issue. However, fragment size constraints were originally imposed by plasmid (12-kb), P1 bacteriophage and cosmid (粘粒) (45-kb) cloning vectors. Larger *stable* constructs can be established by yeast artificial chromosome (YAC) and bacterial artificial chromosome (BAC) vectors, as well as large chromosomal fragments (megabase lengths).

The given DNA fragment for microinjection should not contain nicks or strand breaks and be as pure as possible. Try to remove as much of the vector sequences as possible because numerous studies point to the relative severity and/or influence of vector DNA sequences on the function of foreign genes after chromosomal integration.

Integration frequencies for foreign genes are concentration dependent, but egg viability is inversely related to DNA concentration. At DNA concentrations of 1-to 2-ng/μL injection buffer, viability averages 20% and still allows on average 30% live-born transgenic mice. At higher concentrations (e. g., 6 ng/μL), integration frequencies can increase to 60%, but viability drops to 10% or less and reaches the point where there are insufficient uterine implantations for pregnancy maintenance.

Superovulation(超排卵)*of Mice*

Several different hybrids are currently popular for DNA microinjection, and the C57BL/6×SJL Fl hybrid has been shown to be efficient in the generation of transgenic mice. For superovulation, PMSG (Pregnant Mare's Serum Gonadotropin,孕马血清促性腺激素) or HCG (Human Chorionic Gonadotropin,人绒毛膜促性腺激素) are injected into the peritoneal cavity（腹腔）of female mouse (3-8 weeks of age) both at 5.0-7.5 units per mouse, by injecting PMSG at noon followed at 48 h with HCG for most strains or at a 26-to 48-hr interval for C57BL/6 mice. A biological assay related to the quality and quantity of eggs as well as mating performance（copulatory plug formation,阴道栓形成）is evaluated following hormone batch preparation. Excessive PMSG will lead to hormone refractoriness（抗药性）or will increase proportions of nonfertilized, crenated （皱缩）, or abnormal eggs. At a proper dosage, one can reasonably expect to obtain 20-30 eggs per female mouse.

Use of FSH (Follicle Stimulating Hormone,卵泡刺激激素) delivered via an implantable osmotic minipump(植入式微型渗透泵) and induction of ovulation with either HCG or LH (Luteinizing Hormone,黄体激素) can significantly improve egg quality and quantity in poorly responding strains. The FSH/pump protocol for superovulation is significantly more expensive and time consuming than the PMSG/HCG regimen. But the FSH osmotic pump method has been used successfully to induce superovulation in nonovulating transgenic females from a variety of lineages.

Harvesting Superovulated Eggs

1. Mating. After female egg donors receive HCG, they are placed in cages with male mice. The next morning, the females are checked for copulatory plugs（阴道栓）. Bred females are sacrificed for egg recovery.

2. Oviduct Dissection. Harvesting of the oviducts and eggs should be done within 8-9 h of the midpoint of the last dark cycle. This allows one to easily isolate the eggs as a mass from the oviduct, obviating the need to flush individual eggs. After donor fe-

males are euthanized (无痛处死), the oviducts (输卵管) are obtained through a midventral (腹中部) or dorsolateral (背外侧) approach. After cutting through the skin and body wall is complete, the fat pad (脂肪垫) by the ovary is located, and the uterotubal junction (where the uterus and oviduct join, 子宫角) is grasped with a pair of forceps. The uterus is then severed just below the forceps. A second cut between the oviduct and the ovary frees the oviduct, which is still grasped with forceps. The oviduct is then placed into a sterile Petri or tissue culture dish filled with bicarbonate buffer. The dissections from donor females continue as rapidly as possible to remove all the oviducts, but care is taken not to excessively manipulate oviducts, thus causing possible rupture and release or loss of egg masses.

3. Release of Eggs from the Oviduct. The oviducts are collected in a barrier hood then brought to the microinjection laboratory and transferred to a clean dish. Some laboratories will collect or place oviducts into a hyaluronidase-containing medium at this point. We postpone hyaluronidase (透明质酸酶) treatment until the eggs, contained in cumulus masses (卵丘细胞团), are liberated from oviducts in order to minimize enzyme exposure to eggs. The eggs remain until the cumulus cells (卵丘细胞) surrounding the eggs are digested and the zonae pellucidae (透明带) are free of any attached cells or debris. Practice is necessary to keep the exposure of eggs to hyaluronidase to a minimum (3-6 min in total).

To remove the eggs from the hyaluronidase-containing medium and for routine handling of eggs, a glass pipette (200-μm final inner diameter) is first filled by capillary action with medium. Then, two small air bubbles are drawn into the narrow bore to enhance fine control for egg manipulation (Fig. 36.3). The isolated eggs are carefully picked up, avoiding rough "vacuuming action" that may damage them. The eggs are placed in a new dish of hyaluronidase-free medium, minimizing the transfer of any remaining extraneous cells/debris. The eggs are then pipetted into a second wash and counted. Good-quality eggs are then selected and placed in a dish containing microdrops of medium until readied for microinjection.

In Vitro Culture of Superovulated Eggs

1. Equipment. A dissecting stereozoom microscope (实体解剖显微镜) (6.5-40×) with wide-field (16-20×) eyepieces and long-working distance objective(s) offers excellent resolution for the harvest and manipulation of eggs.

2. Atmosphere for culturing eggs. Eggs and media are maintained in a 5% O_2, 5% CO_2, and 90% N_2 atmosphere (by volume) that is controlled with a flowmeter (流量计). The mixture is passed through a gas humidifier, then into a small plastic incubator covering the microdrop dishes containing eggs.

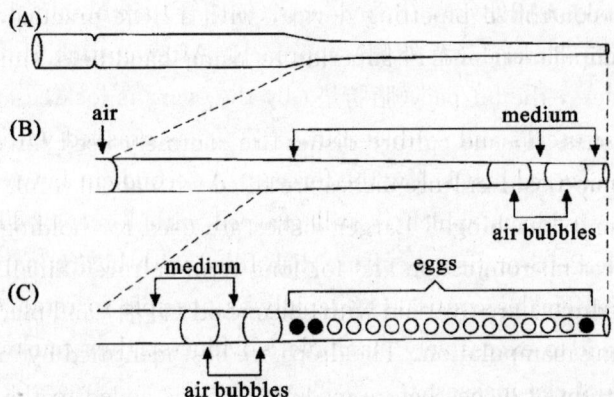

Fig. 36.3 Egg handling pipette. After heating the tapered portion of a 9-in. Pasteur pipette to a molten state, the pipette can be pulled to an appropriate diameter (90-200 μm, depending on particular needs). After cooling (∼3 sec), the two ends can again be pulled (without heat), breaking the glass evenly. Alternatively, microcapillary tubing may also be used. The diagrams illustrate the procedure from pulling a pipette (A) to loading the pipette (B and C represent enlargement of the distal end of the pipette). (B) Medium is drawn up the pipette by capillary action. (C) Following the air bubbles, eggs are then drawn into the pipette. (By H. G. Polites and C. A. Pinkert)

3. Microinjection Media. For microinjection, a HEPES-buffered medium such as modified BMOC-3 plus HEPES supplemented with cytochalasin B (5 μg/mL, 细胞松弛素 B) is used. If the egg membrane is dragged into the cytoplasm during injection, then the egg will likely lyse rapidly. Cytochalasin B stiffens (使变硬) the membranes during microinjection and helps prevent lysis of the egg. Alternatively, 7% (v/v) ethanol has a similar effect. Egg membranes can also be stiffened by lowering the slide temperature to 10℃ with a stage cooler. This will not alter egg survival or the percentage of transgenic mice produced, as long as the eggs are returned to 37.5℃ within approximately 45 min. The cooling effect takes up to 10 min before the egg membranes are noticeably stiffer (更硬); Therefore, the simpler route using cytochalasin B is preferred.

4. Glass Pipettes for Manipulating Eggs. Pipettes for manipulating eggs can be prepared from 9-in. Pasteur pipettes (Fig. 36.3). The taper (细端) of the Pasteur pipette is heated over a flame, and the tube is rapidly pulled apart when the glass becomes pliable (易弯). The smaller the area of tubing that is heated, the smaller the final tube diameter. The pulled end is pinched off to a length of 3 in., placed in a clean and/or sterile container, and sterilized. Small diameters (i.e., slightly greater than one egg [∼80-90 μm] in internal diameter) are appropriate for egg transfers; however, a larger inner diameter (200 μm) of the final taper facilitates a more rapid collection and transfer of eggs during the microinjection procedure. Once the glass is prepared, the wide end of the glass pipette is attached with latex tubing (乳胶管) to a capillary adapter. Using larger bore (中空) latex tubing, the adapter is then attached to a plastic mouthpiece.

Use of this mouth-controlled pipetting device, with a little practice, greatly enhances control of egg manipulation and is superior to using hand-held, micrometer-type devices.

5. Culture Dishes. Tissue culture dishes are routinely used for egg culture procedures. Small, 35-mm tissue culture dishes are filled with about 3 mL of media and used for egg collection and "washing." Larger dishes are used for holding large numbers of eggs before and after microinjection and for long-term culture. Small volumes of media (10-50 μL in microdrops) are overlaid with silicone oil (硅油) and placed in the dishes to maintain eggs during manipulation. The drops are best identified by marking quadrants/sections on the bottom of dishes before medium or oil is added to the dishes, similar to labeling quadrants on Petri dishes before pouring plates. Silicone oil is used routinely and readily minimizes diffusion and evaporation of the microdrops. After microdrop dishes are prepared, they are placed in the temperature-and gas-controlled environment. A minimum of 10 min is required to equilibrate conditions before use.

Egg Development and Microinjection Timing

Media conditions and strains have been chosen to obviate the two-cell block in egg development. This allows one to test the quality of all media components and manipulations by culturing the eggs *in vitro* for several days. In optimal *in vitro* culturing conditions, we have obtained 95% or greater development of C57BL/6 SJL hybrid-derived zygotes to the blastocyst stage in 4 days, with 70% then being capable of growing out inner cell masses within 3 days. Any suboptimal components will lower these efficiencies or delay development.

During any manipulation with bicarbonate buffer, attention is paid to the color of the medium and egg morphology to ensure pH balance. As the CO_2 diffuses, egg viability decreases rapidly. If eggs are harvested early before pronuclei are well formed, the microdrop dish can be stored in a modular incubator chamber and placed in a standard temperature-controlled incubator in order to maintain temperature, humidity, and atmospheric conditions relatively inexpensively. The optimal time windows for microinjection range from the time when the male pronuclei at the periphery of the egg membrane are identifiable until the male and female pronuclei merge, just prior to the first cleavage division. The time of this window varies between strains. Typically, the injection window is 3-4 h, but it is affected by *in vitro* culturing methods and care during routine handling.

Microinjection of DNA into the pronuclei of fertilized eggs

1. Preparation of Injection and Holding Pipettes (see Fig. 36.4).

(1) Holding Pipette (see Fig 36.5). The holding pipette can be inserted directly into the pipette holder, after the system has been checked for air bubbles. A few turns on

the Hamilton syringe will push oil all the way to the tip of the pipette. The holding pipette is then oriented above the microscope lens so that the bend in the pipette is parallel to the slide.

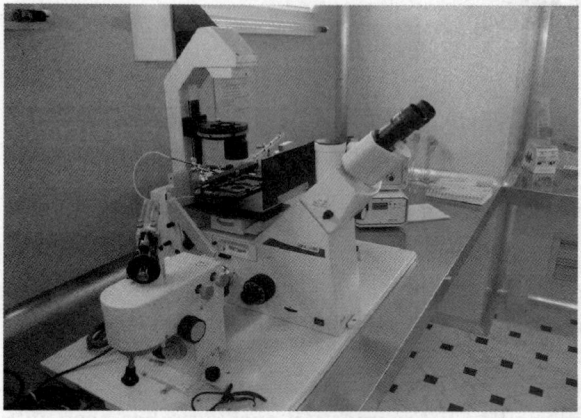 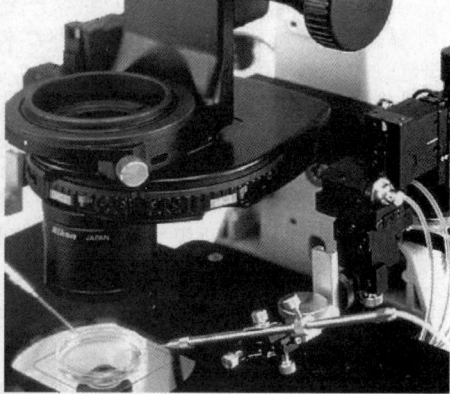

Fig. 36.4 Micromanipulator assemblies used for DNA microinjection. Microinjection systems for controlling microinjection of DNA and holding eggs are grouped into either air-driven systems or oil-driven hydraulic systems(油驱动的液压系统). The air-driven systems can also be attached to electronically controlled delivery systems that allow preprogrammed regulation of fluid delivery volume, time, and pressure. Oil-driven injection system is very inexpensive compared to other automated injector systems. The syringes are loaded with either silicone or mineral oil. The tubing, the microinjection needle holder, and the needle itself are filled with fluid to dampen(使潮湿) control and allow continuous flow of DNA without repeated syringe adjustments.

(2) Injection Pipette (see Fig 36.5). The injection pipette is first filled with oil. While filling, keep the injection needle vertical and fill from the tip back, making sure to trap only one air pocket in the tip of the pipette. Once the injection needle is filled and the injection system is cleared of air bubbles, turn the microsyringe micrometer to push some oil out the tip of the injection pipette holder. Insert the back of the injection needle into this oil drop and down into the pipette holder. The microsyringe micrometer is then turned to increase the pressure on the air trapped in the tip of the injection needle, which will keep it in place until it is all pushed through the opening at the tip.

(3) Aligning the Injection Needle. The injection needle and holder are then placed on the micromanipulator and adjusted to orient it parallel to the slide. As with the holding pipette, the injection pipette is first oriented so that it is perpendicular(垂直) to the microscope. Next, under low power, adjust the angle of the injection pipette so that the full length of the bent tip section is in focus. This adjustment is sometimes time consuming but is important because it ensures that the pipette enters the egg at a straight angle and avoids shearing the egg as one pushes the needle into the pronucleus.

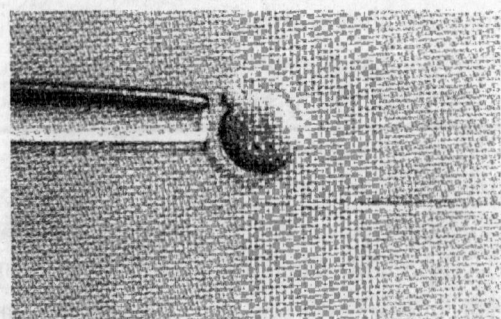

Fig. 36.5 Injection and holding pipettes. With an egg in place (-70-m diameter), the relative size of the injection and holding pipettes becomes evident. The holding pipette may range from 15 to 50 μm in diameter depending on style (not polished to very polished), and the injection pipette tip is ~0.75 μm in diameter. Generally, the injection pipette is aligned with the pronucleus before insertion into the egg. (By H. G. Polites and C. A. Pinkert)

(4) Loading DNA into the Injection Needle. If the injection needle tip has been sufficiently beveled, then oil should be flowing out of the tip by the time orientation adjustments are complete. The size of the oil beads running down the pipette will indicate the diameter of the tip. The micrometer assembly is dialed back to reduce the flow, and the injection pipette is lowered into the DNA drop. The quality of the tip can be observed at this point and the needle replaced if any problems are found. DNA is drawn up into the injection pipette as far as possible, and filling is allowed to continue while the eggs are being selected and prepared for microinjection. Once the injection pipette is loaded and the eggs are ready for microinjection, switch the stopcock value to open the microsyringe micrometer and reverse oil flow to start the DNA/oil meniscus (弧面) moving as slowly as possible down the pipette.

2. Preparing Eggs for Microinjection. Once the DNA is readied in the injection pipette, the eggs to be injected are selected from the holding dish. The eggs are first transferred to a drop of injection medium (containing cytochalasin B) in a dish under oil. Only eggs with visible pronuclei and normal appearance are selected and transferred to the drop of medium on the microinjection slide. There are many arrangements for the eggs on the slide, but we prefer to keep them organized in a line above and below the microinjection area (where needles are located).

3. Visualizing and Injecting the Egg Pronuclei.

(1) Orientation. The optimal orientation for microinjection of pronuclei is to have the male pronucleus closest to the injection needle (Fig. 36.6). The height control of the holding pipette is raised and lowered until one is sure the center of the pronucleus is in focus. The injection needle tip should also be in focus prior to penetrating the egg.

(2) Egg Membrane Penetration with the Injection Needle. Because the DNA aliquot is constantly running out the tip of the injection needle, one has to work efficiently to

penetrate the zona pellucida membrane and pierce the pronuclear membrane to prevent the buffer and DNA from "pushing" the membranes.

(3) Filling the Pronuclei with DNA Solution. The injection pipette should fill the pronucleus with DNA until the expansion of the pronucleus stops (an approximate 50%-100% volume expansion). One must simultaneously observe the swelling of the egg to ensure that not too much DNA is delivered. The injection pipette does not always penetrate the pronucleus perfectly, and leaking of DNA out of the pronucleus will frequently occur. In this case, the pronucleus will not expand as rapidly, and the whole egg will enlarge. The egg will lyse if too much DNA is delivered into the cytoplasm.

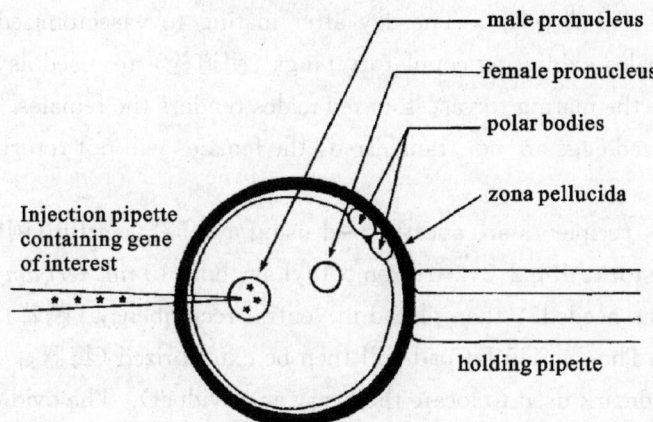

Fig. 36.6 **Diagram of DNA microinjection into a pronuclear murine zygote.** A pronuclear egg is held by a large-bore pipette using gentile suction. A small-bore injection pipette containing the DNA solution is inserted into the male pronucleus. The DNA solution is then slowly expelled into the pronucleus, which expands approximately twofold. The diameter of the mouse egg is about 75 μm. (Modified from H. G. Polites and C. A. Pinkert)

4. Treatment after Microinjection. Regardless of the skill and speed of the injectionist, the eggs should be left on the slide no longer than 30 min. Once the injection slide is completed, the pipettes are raised, the slide carefully returned to the warming stage on the dissecting microscope, and the injected eggs transferred through two "wash" dishes of culture medium (without cytochalasin B) and then placed in an appropriately labeled microdrop dish.

Routinely 10%-25% of eggs will lyse as a result of microinjection, but this figure is highly dependent on the skill of the injector, the egg background strain, the injection pipette shape, and the quality and concentration of the DNA preparation. If less than 10% of the microinjected eggs lyse, such results would suggest that the volume of DNA delivered into the egg pronucleus was insufficient to generate transgenic mice efficiently.

In Vitro Culture of Microinjected Eggs

Zygotes can be routinely cultured to the two-cell or blastocyst (胚泡) stage, then transferred to the oviduct of day 1 pseudopregnant mice (假孕鼠). Using outbred or hybrid mice, typically 40%-60% and 60%-80% of injected eggs (60%-80% and 75%-100% of noninjected eggs), respectively, develop overnight to the two-cell stage in our experience. A high proportion of eggs going to the two-cell stage will usually proceed to the blastocyst stage. The significant decrease in egg viability following microinjection is due to the trauma associated with the ionic and physical environment of the zygote. Egg culture provides a reasonable assessment of microinjection proficiency.

Egg Transfer and production of transgenic mice

Briefly, a pool of female mice are maintained and observed for visual evidence indicative of proestrus (发情前期). The day after mating to vasectomized (切除输精管) males, those females exhibiting copulatory plugs (阴道栓) are used as egg transfer recipients. Because the mating to vasectomized males renders the females "pseudopregnant (假孕)", if fertilized eggs are not transferred, the females will not return to estrus for about 11 days.

For transfer, recipients are anesthetized using a 2.5% avertin (阿佛丁，一种麻醉剂) i.p. (100% stock: 10g 2,2,2-tribromoethyl alcohol, 10 mL *tert-amyl* alcohol, then diluted in water as needed), then placed in ventral recumbency (躺着) under a dissecting microscope. The ovarian fat pad will then be exteriorized (将器官从腹中取出)(the fat pad is the landmark used to locate the ovary and oviduct). The oviduct is maintained in place using paper sponges around the uterus or by placing a 4-cm-long, 3-mm-wide sponge through the mesometrium(子宫系膜)or by grasping the fat pad with a small surgical clamp. The bursa(囊)surrounding the ovary is then either cut or torn to expose the ostium (口)of the oviduct.

Once these steps are accomplished, eggs are obtained using mouth-pipetting under a second dissecting microscope (Fig. 36.7). Between 20 and 30 injected ova are then expelled into the ostium of the oviduct of the recipient and are blown through to the ampulla(壶腹部)by gentle pressure through the pipette. The reproductive tract is then carefully replaced into the abdomen, and the body wall is closed with one or two sutures followed by skin closure with wound clips.

Production of Transgenic Golden Fish by DNA microinjection into Oocytes

As shown in the Fig. 36.8, it is much easier to produce transgenic fish than that done in mouse.

1. Prepare the microinjection needles with a diameter not more than 10 μm.

2. Prepare the DNA solution with a final concentration of 40 $\mu g/\mu L$ in Holtfreter's solution (3.5 g/L NaCl, 0.2 g/L $NaHCO_3$, 0.05 g/L KCl, 0.2 g/L $MgSO_4$, 0.3 g/L $CaCl_2$, pH 7-7.5)

PART Ⅱ CYTOTECHOLOGY IN ANIMALS（动物细胞工程）

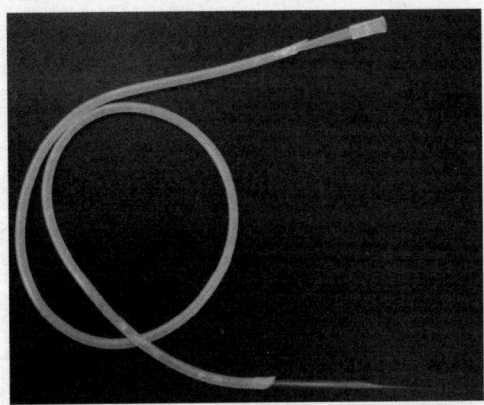

Fig. 36.7 Mouth-pipetting assembly with 200μL-pipette tip, plastic solft tube and glass needle.

3. Catch the sex mature female golden fishes, when they are chased by males, and dissect the abdomen to isolate the oocytes.

4. Incubate the oocytes in fertilized tap water containing 17α-20β-dihydroprogesterone (17α-20β-二氢孕酮) for more than 1 hour at room temperature.

5. When the germinal vesical (GV) moves right under the fertilization pore, inject the DNA solution (ranging from a few pL to 100 nL, $10^4 \sim 10^7$ copies) into the GV.

6. Three hours after microinjection, the GV will break down. The injected oocytes need a further development of 8-9 hours to be ready to fertilization.

7. Remove the follicle cells with a fine forceps and add the sperms.

8. Fertilized eggs then develop into larvae and adult fish.

9. Assay and characterization of transgenic fishes.

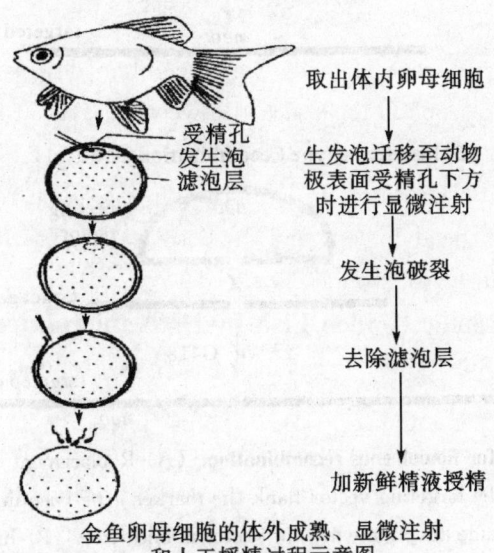

Fig 36.8 Workflow for production of transgenic fish by microinjection of exogenous DNA into the germinal vesical of oocytes.

GENE TARGETING IN EMBRYONIC STEM CELLS BY HOMOLOGOUS RECOMBINATION

Design Vectors for Homologous Recombination

The vast majority of work being done on ES cells involves preplanned genetic modifications by homologous recombination between exogenous targeting DNA and an endogenous target gene. Fig. 36.9 shows two basic schemes of gene targeting: replacement type and insertion type recombination. Because the targeting vector must contain sequences closely homologous to the target gene, these sequences should be cloned from DNA "isogenic" to that of the ES cell line. The use of isogenic DNA in the targeting vector increases targeting efficiency from 4 to 5 fold. The degree of homology in the targeting vector is another important consideration that has also been systematically tested for its effect on targeting efficiency. In general, it is advantageous to have 4-10 kb of homology between targeting sequences and target gene, with no homologous arm less than 1 kb.

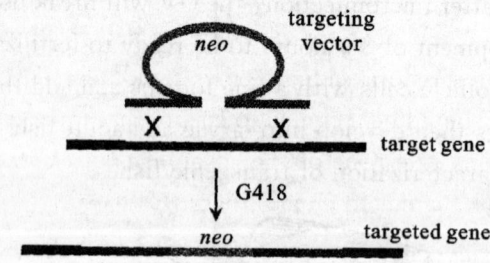

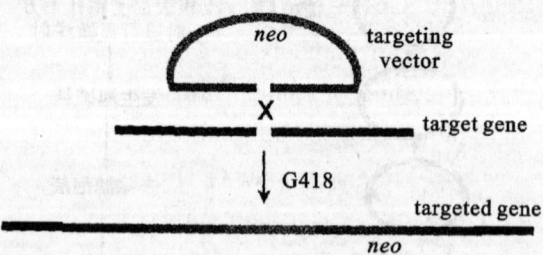

Fig. 36.9 **Basic schemes for homologous recombination.** (A) Replacement or "Ω"-type scheme in which the homologous arms of the targeting vector flank the marker gene (*neo* in this case). The homologous arms are oriented in the same direction when the vector is linearized. (B) Insertional or "O"-type scheme in which the targeting vector is linearized within one large homologous region or in which two homologous arms are oriented in the opposite direction with the marker gene between them. (By T. Doetschman)

A sequence in the targeting vector represented in multiple genomic loci such as repetitive sequence elements would be less likely to recombine homologously with a specific target locus. Therefore, choosing targeting fragments containing few mouse repetitive sequences may improve efficiencies. Because intronic sequences, which often contain repetitive elements, usually constitute much of the homologous sequence in a targeting vector, large regions of homology may occasionally decrease targeting efficiency. An early review of homology versus targeting efficiency suggested that greater than 10 kb of homology could be detrimental (有害).

It is unclear whether cotransfection of the targeting vector with enhancing sequences is of use. The orientation of the selectable marker gene usually does not affect the ability to target a gene. In one report, the neo^r gene only expressed when it was in the opposite orientation to the target gene.

While making the targeting vector it is important to think through all, of the possible transcriptional and translational products that the targeted cell might be able to generate from the planned targeted allele. For example, if the plan is to inactivate a functionally critical exon, either by excision or interruption, then one must consider whether the products of a splice over the alteration might have some other function or whether an alternative splice will render the transcript out of the reading frame. Potential fusion proteins between target and marker genes must also be considered, especially if the marker gene is promoterless or lacking a poly-A-addition signal.

A diagnostic test should be included to ensure that there is only one integration site-namely, the homologous site. One of the advantages of electroporation is that if the amount of targeting DNA is not too high, there will usually be only one integration site, but one should always make this determination just to be sure. It would be unfortunate to ablate (切除) both the target gene and a neighboring gene without knowing it and then report a phenotype, thinking it resulted from a single gene knockout.

Isolation of the Targeted Cells by Positive-Negative Selection (PNS)

Positive-negative selection is a commonly used approach to facilitate the isolation of targeted cells and avoid the blending of random integrants. In the PNS scheme (Fig. 36.10), it is assumed that nonhomologous integration events will retain the negative selector gene and confer sensitivity to gancyclovir (胸苷激酶对新底物), whereas homologous events will remove the negative selector gene and confer resistance. In most experiments, the neo^R gene is the positive selector and the HSV-tk (单纯疱疹病毒胸苷激酶) gene is the negative selector, but other positive selectors, such as the $hygromycin^R$ (潮霉素抗性) or $puromycin^R$ (嘌呤霉素抗性) genes have been used. Because of concerns about mutagenicity and toxicity of gancyclovir, other negative selectors, such as the *diphtheria toxin-A* (白喉毒素 A) gene or the HPRT (次黄嘌呤-鸟嘌呤磷酸核糖转移酶), GPT (谷丙转氨酶), or *ricin toxin* (蓖麻毒素) genes could be used. The enrichment for the *neo/tk* PNS system usually falls in a range of 4-to 10 fold. The *diphtheria toxin-A* gene yielded an enrichment of 25-fold, and the *HPRT* minigene yielded an enrichment of 6-to 7-fold. It should be noted that these enrichment factors are likely to vary from gene to gene. Nonetheless, it is clear that negative selection provides a significant enrichment for the number of targeted clones per positively selected clones.

Other enrichment schemes have used a promoterless positive selector gene. In this scheme, the selector gene will be expressed only if inserted in the correct orientation relative to the promoter of an endogenous gene, thereby reducing substantially the number of integrants capable of expressing the selector gene. The use of this targeting scheme has yielded enrichments estimated to range from 20-120. Similarly, another scheme relies on the integration site providing the poly-A-addition signal for the positive selector gene. Here, an integration in intergenic regions or in the wrong orientation within a gene would prevent its expression. Such experiments have yielded an enrichment of from 2-to 5-fold.

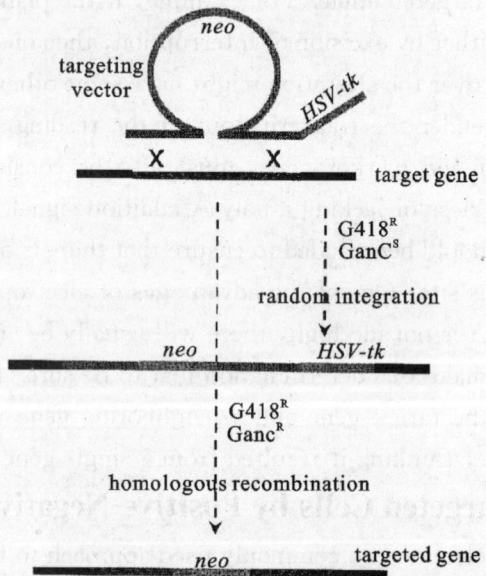

Fig. 36. 10 Positive-negative selection scheme. This scheme is constructed similarly to the replacement scheme shown in Fig. 36. 8, with the addition of a negative selector gene (*HSV-tk* in this case) ligated to the outside of one of the homologous arms. If the vector inserts nonhomologously in a random site, the *HSV-tk* gene may remain intact, in which case the cell will be killed by the antiviral drug gancyclovir (胸苷激酶对新底物). If the vector is inserted by homologous recombination, the *HSV-tk* will not become incorporated because it is outside of the region of homology. (By T. Doetschman)

Screening for Targeted Colonies

Pick G418^R colonies in their entirety using a Pipetman pipette and a blue tip. Transfer each picked colony to one well of a 24-well tray seeded earlier that day with G418^R

primary embryonic feeder cells. Keep G418 selection on the colonies at this point. Any colonies that were not actually G418 resistant will be eliminated by keeping the selective pressure on. Every 3-5 days, as required to prevent differentiation but maximize growth, retrypsinize each well and replate, as before, into the same well. There is no need to add new feeders. Repeat this procedure until the ES cell colonies cover about two/thirds of the entire surface of the well. At this time, the colonies are ready to harvest for DNA and for seeding new plates either for expansion and blastocyst injection or for freezing.

Blastocyst Injection

A typical blastocyst injection is shown in Fig. 36.11. As soon as the blastocyst injections start yielding chimeric animals, the animals will have to be mated with normal C57BL/6 females to test whether the targeted ES cells that were injected have colonized the germ line. The offspring of germline chimeras will then be used to establish a breeding colony and produce experimental animals from that strain.

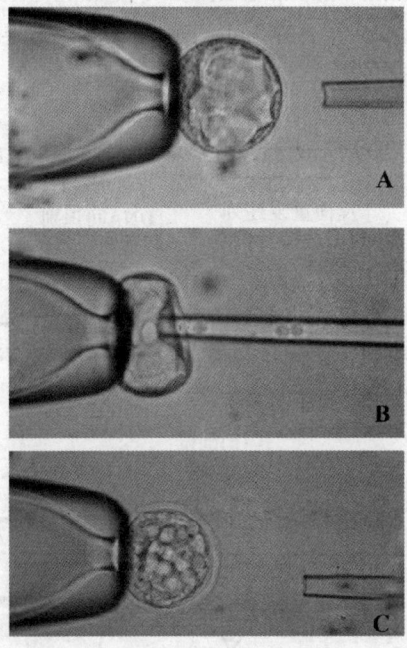

Fig. 36.11 ES cell transfer into mouse blastocysts. A. Blastocyst is mobilized by holding needle before transplantation. B. ES cells are transferred into the host blastocyst. C. After transfer, ES cells have been transferred into the blastocyst and reside by the inner cell mass.

GENE TARGETING IN EMBRYONIC STEM CELLS BY CRE-LOXP SYSTEM

Today, the P1 bacteriophage *Cre/Lox/P* recombination system is the method of choice for targeting subtle mutations into the genome. The Cre (causes recombination)-

*Lox*P (locus of crossing-over) recombination system requires only the cre enzyme and its *Lox*P recognition site on both partner molecules. Sequence analysis shows that the *Lox*P site consists of two symmetrical 13-bp protein-binding regions separated by an 8-bp spacer region, while the Cre recombinase is a 35-kDa integrase protein.

Briefly, Cre recombinase mediates recombination between two 34-bp *Lox*P sites. The *Lox*P sites consist of 13-bp inverted repeats flanking an 8-bp sequence that determines the orientation of the recombination. If the 8-bp sequences are similarly oriented in two *Lox*P sites, the sequences between those sites will be deleted; if the orientation is opposite, the sequences between the *Lox*P sites will be inverted Consequently, if *Lox*P sites flank the marker gene and a subtle mutation is incorporated into one of the homologous arms, then the marker gene can be removed from the targeted allele by transient transfection with a *Cre recombinase* expression vector or by mating the targeted mouse with a transgenic mouse expressing *Cre* in the oocyte or zygote. Many other types of modification such as inversions and replacements, translocations, large chromosomal deletions, and the construction of mouse balancer chromosomes have been made using the *Cre/LoxP* system. (Fig. 36.12).

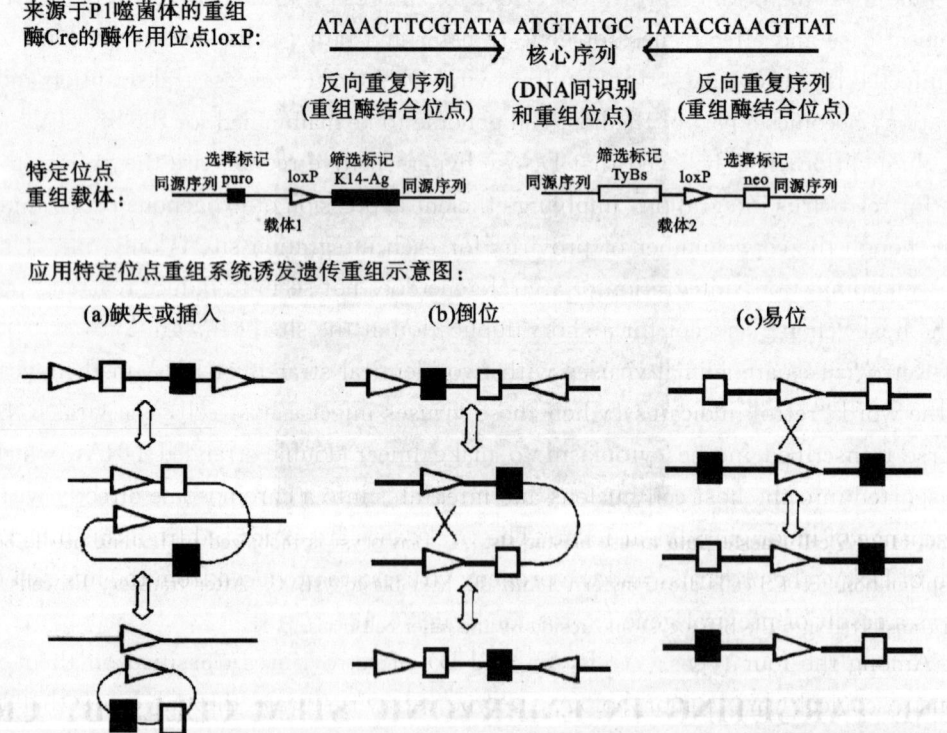

Fig. 36.12 Summary of Cre-mediated recombination events. An endogenous locus is targeted in embryonic stem cells by conventional gene targeting approaches using a positive-negative selection scheme. Cre recombinase can affect deletion or insertion (a), inversion (b) and translocation (c) events, depending on the placement and orientation of *Lox*P sites.

In summary, the power of site-specific recombination systems, with components harnessed from bacteria, phage, and yeast, has ushered(引领)in a new generation of molecular biology techniques. Chromosomal rearrangements for genetic engineering can be performed with precision at the nucleotide level. Second-generation knockout strategies using these systems allow for inducible-and developmental-specific DNA recombination at the single cell level.

Site-specific recombination systems have greatly expanded the repertoire of induced mutations that are possible and conceivable within the mouse genome. As the shared resources of transgenic Cre and floxed mice continue to expand, they will provide necessary research tools for investigators to explore crucial answers to biological questions in a whole-animal setting. These modifications continue to bridge our knowledge about genes, gene products, protein functionality, and phenotypes. With respect to long-term applications, information gleaned from these approaches will lead to significant advancements in biomedicine and agriculture.

RETROVIRUS-MEDIATED GENE TRANSFER

The most important features of retroviruses in regard to their use as vectors are the technical ease and effectiveness of gene transfer and target cells specificity. Once cells are infected by retroviruses, the resultant viral DNA, after reverse transcription and integration, becomes a part of the host cell genome to be maintained for the life of the host cell. In addition, it is believed that DNase hypersensitive regions are the preferred targets for retrovirus integration, implying efficient expression of exogenous proviral genes even though the copy number of provirus for each integration site is only one. Unlike DNA microinjection, integration of a viral gene does not seem to induce rearrangements of the host genome, except for a short duplication at the site of integration.

Retroviruses are animal viruses with two identical strands of RNA in their virion. As the word "retro" indicates, when these viruses infect a host cell, the viral RNA is reverse transcribed in the cytoplasm to make linear double-stranded DNA, which is transported into the host cell nucleus and integrates into a chromosome directly without any change of its original linear form. The viral genes are expressed from this integrated form of DNA, the provirus, and the progeny viruses are produced from the infected host cell as a result of proviral gene expression (Fig. 36.13)

Among the four types (A, B, C, and D) of retroviruses classified on their electromicroscopical(超微) morphology, C-type retroviruses are the only ones that have been used in vector construction as they are the most extensively studied. The most commonly used C-type retroviruses are murine leukemia virus (MLV,鼠白血病病毒), reticuloendotheliosis virus (REV, 网状内皮组织增殖病毒), and Rous sarcoma virus (RSV,劳氏肉瘤病毒); REV and RSV are avian retroviruses.

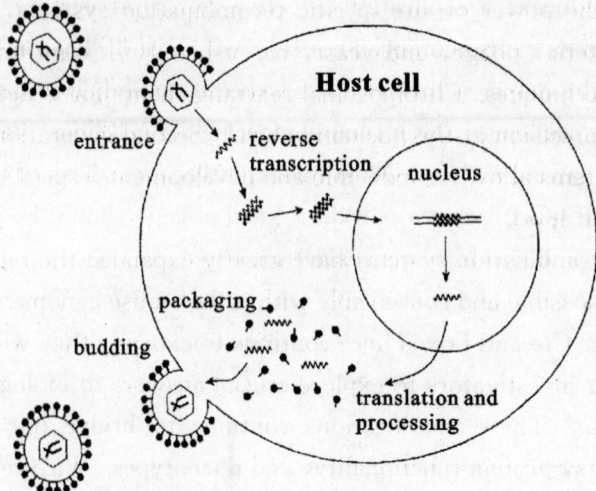

Fig. 36.13 The life cycle of a retrovirus. (By T. Kim)

Retroviral Gene Transfer Systems

Retroviruses have three trans-acting protein-coding genes: *gag*, *pol*, and *env*. The proteins encoded by the *gag* gene are responsible for virus internal structure. The *pol* gene encodes enzymes, including the carboxyl portion of protease, for posttranslational cleavage of viral proteins; reverse transcriptase, to transfer information from RNA into DNA; and integrase, for the integration of reversetranscribed viral DNA into the host cell chromosome. The *env* gene encodes viral envelope glycoproteins, which are part of the outside of the progeny virus and are the primary determinant of the retrovirus host range.

Cis-acting sequences of a retrovirus are clustered at the ends of the viral genome: (a) long terminal repeat (LTR) sequences for enhancer/promoter function, trans-acting factor response, integration, and polyadenylation of viral RNA; (b) primer binding site and polypurine sequence for reverse transcription; (c) posttranscriptional splicing sites, including splicing donor and acceptor sites along with two short fragments within the viral intron for *env* mRNA production; and (d) ψ or E signal in the case of MLV and REV, respectively, for encapsidation(包装). (see Fig. 36.14)

Packaging Cells

Helper cells (or packaging cells) are usually constructed by the transfection of viral transacting sequences (*gag*, *pol*, and *env*) to an appropriate eukaryote cell line. Because the viral RNAs produced from helper cells are devoid of retroviral cis-acting sequences, especially the encapsidation sites, the RNAs from helper cells cannot be encapsidated into retroviral virions. The purpose of a packaging cell is, therefore, to provide Gag, Pol, and Env proteins to the retroviral vector having no trans-acting sequences.

Unlike the construction of a replication-defective retroviral vector, the promoters for *gag*, *pol*, and *env* gene expression need not be a retroviral LTR promoter. Any promoter that is active in the helper cell can be used, because the helper cell does not require retrovirus cis-acting sequences, even though most helper cell lines available use the LTR promoter due to its strong activity. Most importantly, regardless of promoter, the (sequence (between the splicing donor and the 5' end of gag) should be removed to protect RNAs of *gag*, *pol*, and *env* from encapsidation.

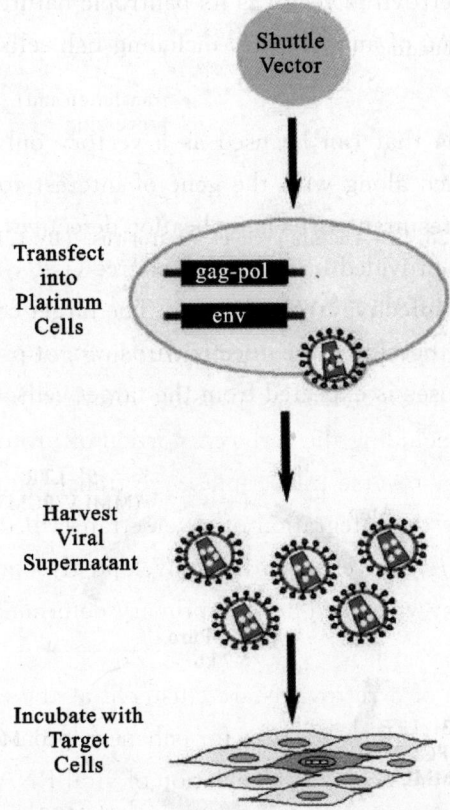

Fig. 36.14 Workflow for retrovirus production in packaging cells and infection of target cells.

Among retrovirus production systems for gene transfer in mammalian cells, NIH3T3 cells transformed with appropriate MLV genes are the most popular, due largely to the relatively simple structure and good characterization of MLV, as well as the permissiveness of NIH3T3 cells for MLV gene expression. The three kinds of MLV are ecotropic（单嗜性）, amphotropic（双嗜性）, and xenotropic（异嗜性）. Ecotropic MLV is able to infect only cells of mice and rats, amphotropic MLV can infect not only murine cells but also other cells, and xenotropic MLV can infect only non-murine cells. The host range difference is due primarily to differences in the envelope protein, which interacts with specific receptors on host cell membrane. For most currently available packaging cell lines for transgenic animal production, the *gag* and *pol* genes are from ecotropic MoMLV, while the *env* gene is from either ecotropic or amphotropic.

The significance of the 293GP/VSV G packaging cell line is that it overcomes the instability of retroviruses in centrifugal force. The stock of viral vector packaged with vesicular stomatitis virus (疱疹性口炎病毒) glycoprotein G (VSV-G) instead of MLV envelope protein can be centrifugally concentrated as much as 2000-fold, resulting in a titer of more than 10^9 transforming units per milliliter. Another characteristic of the MLV and VSV hybrid retrovirus vector is its pantropic nature, which enables the virus to infect almost every kind of animal cells, including fish cells.

Retroviral Vectors

To construct a virus that can be used as a vector, only the cis-acting DNA sequences should be retained along with the gene of interest to be transferred (Fig. 36. 15). Consequently, the resultant virus is replication defective, but with the coordination of trans-acting functions provided from another source (i. e., a helper cell or packaging cell), the virus becomes infective to target cells. The target cell has no trans-acting elements of the retrovirus; therefore, the infected virus cannot produce its progeny, and no spread of pathogenic viruses is expected from the target cells.

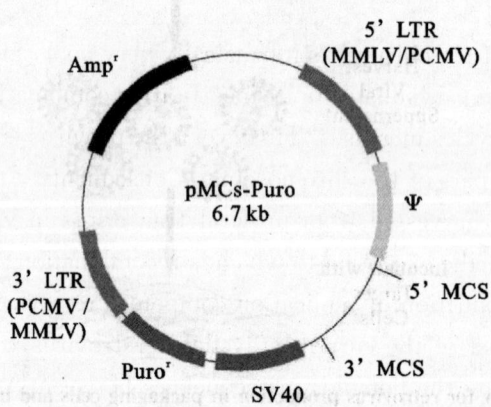

Fig. 36. 15 Schematic representation of pMCs-Puro retroviral vector. Multiple cloning site(MCS); Puromycin (Puro); Long terminal reapet (LTR); packaging signal (ψ); Moloney murine leukemia virus (MMLV); Porcine Cytomegalovirus (PCMV); Ampicilin (Amp).

Virus-producing cells

Virus-producing cells are constructed by either transfection of retroviral vector plasmid to helper cells or infection of the helper cells with the viruses produced from other virus-producing cells. The advantage of the latter approach over the former is higher virus productivity in general because of lower susceptibility to methylation of the provirus in the host celluar genome. One thing to be noted in the latter approach is that

Env protein produced from the helper cell should not be identical to that of the virus intended to infect the helper cell. Retroviruses cannot infect cells producing envelope proteins of the same class because the proteins block the interaction between cellular receptors and specific envelope proteins before infection. This phenomenon is called *superinfection interference*.

Problems in the Use of Retroviral Vectors in Gene Transfer

Although retroviral vectors have several advantages, some drawbacks include size limitation, a high rate of recombination, low titer, etc. Because the most significant limitation of the retroviral vector systems is low titer of the viruses, the following discussion focuses mainly on titer.

Size

The maximum permissible size for efficient encapsidation or reverse transcription of each retroviral vector is approximately 10 kb. Because introns of foreign genes carried by retrovirus vectors are spliced out during replication, the insert does not have to contain introns. Considering that an insert of permissible size (around 8 kb) can code for a protein as big as 300 kDa, a maximum size limitation does not seem to be a significant obstacle in most cases.

Recombination

Genetic stability of retroviruses is intrinsically very poor, and the genetic variations including deletion, base-pair substitution, insertion, recombination, etc. increase as the number of replication cycles increases. However, the provirus is as stable as the cellular genome because it is a part of the chromosome. Consequently, the genetic variations in retroviral vector stock harvested from proviral transcripts in virus-producing cells are very low and tolerable, the exception to which is recombination. The most serious effect of recombination is production of replication-competent retrovirus from virus-producing cells. Fortunately, most of the currently available retrovirus vector systems deal with this problem by reducing the homologous sequences between DNAs for packaging cells and vector and by using different plasmids to separate *gag*, *pol*, and *env* genes.

Titer

Because, in most cases, the titer of viruses is measured by the expression level of progeny virions on the target cells, factors affecting the titer are reviewed in the following four sections.

(1) Virus Productivity. When determining virus productivity from packaging cells, at least two factors are involved: (a) permissiveness of a specific cell line in regard to its use as a virus-producing cell line for the expression of *gag*, *pol*, and *env* genes as well as the genes of the retroviral vector, and (b) nonviral, negative cis-acting sequences in the retroviral vector.

(2) Infectivity of Virus. Because the susceptibility of host cells to retroviral infection is determined by specific interactions between the viral envelope glycoproteins and cell-surface receptors, it has been sometimes very difficult to find a suitable *env* gene whose product is recognized favorably by cellular receptors. Another critical problem in transgenic animal production is low infectivity of the amphotropic and xenotropic retrovirus vector viruses especially to the cells derived from ungulates（有蹄类）. One solution for this problem is conjugation of the virion with polybrene（聚凝胺）, which enhances gene transfer efficiency up to 10-fold. Another solution is the use of a pantropic （泛嗜性）VSV-G pseudotyped retrovirus vector system, from which the progeny viruses produced are encapsidated with VSV-G instead of MLV envelope protein. The biggest advantage of this hybrid vector system is that the virus titer can be increased more than 1000-fold by centrifugation of the virus stock.

One major drawback of this VSV-G pseudotyped packaging cell, however, is that it is impossible to make stable virus-producing cells because expression of the VSV-G gene in the mammalian cell is cytotoxic. Therefore, virus-producing cells can be made only by transient transfection of the VSV G gene to the 293 cells already expressing MoMLV *gag* and *pol* genes and the retrovirus vector sequence. Solutions for establishing permanent packaging or virus-producing cells are made by controlling VSV G expression employing a tetracycline（四环素）-regulatable system or an ecdysone（蜕皮激素）-inducible system.

(3) Integration into the Host Cell Genome. One of the significant advantages of the retroviral vector system is integration of reverse-transcribed retroviral vector DNA into the host cell genome. Without genomic integration, however, the transgene is destined to degrade in a short period of time. It has been reported that the intracellular half life of MLV-derived retroviral vector is in the range of 5.5-7.5 h and that the integration occurs only in mitotically active cells during the S phase. Therefore, gene transfer to nondividing cells has been a problem in a retrovirus vector system. Recently, it has been reported that using human immunodeficiency virus type 1 (HIV-1)-based or bovine immunodeficiency virus (BIV)-based retrovirus vector could overcome this nonintegration problem in the mitotically dormant cells.

(4) Expression of the Transgene in the Target Cells. It has been generally believed that the MoMLV LTR becomes inactive after introduction into preimplantation stage mouse embryo. This problem can be overcome either by modification of the LTR promoter or by introduction of an internal promoter or IRES (internal ribosome entry site) sequence into the retrovirus vector. The next problem, which is more critical in determining the success of transgenic livestock production, is how to optimally control the expression of the transgene *in vivo*. The best solution is the use of tissue-specific promoter or inducible promoter to achieve regulated expression. Some promoters used in

the retrovirus vector system for these purposes include CD4 minipromoter/enhancer for T-cell-specific expression.

APPLICATION AND SAFETY OF TRANSGENIC ANIMALS

Transgenic animals have given great benefit to the fields of agriculture, industry and medicine economically.

Disease resistance in humans and animals

From a basic research and ethical standpoint, it is imperative that we develop models for enhancing characteristic well-being.

Gene therapy

Models for growth, immunological, neurological, reproductive, and hematological （血液学）disorders have been developed. Circumvention and correction of genetic disorders are now possible to address using a variety of experimental methods.

Drug and product/testing screening

Toxicological screening protocols have already included the transgenic animal systems. For preclinical drug development, from a fundamental research perspective, a whole-animal model for screening is essential to understanding disease etiology, investigating drug pharmacokinetics, and evaluating therapeutic efficacy. A comparable and validated need is crucial to product safety testing as well.

Novel product development through "molecular farming"

In domestic animals, biomedical proteins have been targeted to specific organs and body fluids with reasonable production efficiencies. Tissue plasminogen（血纤维蛋白溶酶原）activator (TPA), human factor IX, and human α1 antitrypsin are a few products produced in transgenic animals that are currently in different stages of validation and commercialization.

Production agriculture

Long term, it may become possible to produce animals with enhanced characteristics that will have profound influences on the food we eat, influences ranging from production efficiency to the inherent safety of our food supply.

Although gene transfer technology continues to open new and unexplored biological frontiers, it also raises questions concerning regulatory and commercialization issues. A number of issues exist and will continue to plague the development of many of the systems described herein. Major aspects of the regulation of this technology will focus on the following issues as we begin the twenty-first century: (a) Environmental impact following "release" of transgenic animals; (b) Public perceptions; (c) Ethical considerations; (d) Legislation; (e) Safety of transgenic foodstuffs; (f) Patent aspects and product uniformity/economics.

Therefore, the central questions will revolve around the proper safeguards to employ and the development of a coherent and unified regulation of the technology. Can new animal reservoirs of fatal human diseases be created? Can more virulent pathogens be artificially created? What is the environmental impact of the "release" of genetically engineered animals? Do the advantages of a bioengineered product outweigh potential consequences of its use? These are but a few of the questions that researchers cannot ignore and must approach. They are not alone, however, as the many regulatory hurdles that exist today will challenge not only scientists and policymakers, but sociologists, ethicists, and legal scholars as well.

QUESTIONS FOR DISCUSSION

1. What is a transgenic animal? List some applications of transgenic animals.
2. How to produce transgenic mice by microinjection?
3. How to carry out gene targeting in ES cells by homologous recombination?
4. How to carry out gene targeting in ES cells using Cre-Lox P system?
5. How to produce a transgenic animal by retrovirus-mediated gene transfer?

知识要点

基因组中稳定地整合有外源基因的动物称为转基因动物。转生长激素基因的超级小鼠的成功,在学术界掀起了转基因动物研究的热潮。将外源DNA显微注射到动物的受精卵中是最早最常用的转基因动物技术,但基因整合效率低,而且需要熟练掌握显微操作技术。将体外培养ES细胞通过同源重组技术或Cre重组酶技术,进行基因打靶,并筛选出稳定转染ES细胞,然后通过细胞移植或细胞核移植来制备转基因动物,可大大提高转基因动物的成功率,缩短制备周期。反转录病毒介导的基因转移技术可实现基因的稳定整合,也大大提高了转基因动物的制备效率。

CHAPTER 37 CLONED ANIMAL(克隆动物)

A clone can be defined as a set of genetically identical animals. Small clones of two or occasionally up to four identical animals can be obtained by embryo splitting or blastomere separation(胚胎分割). Embryo cloning by nuclear transfer(核移植)involves the transfer of genetic material from a donor cell (karyoplast,核体) to the cytoplasm of an oocyte or zygote from which the genetic material has been removed (cytoplast,胞质体). In farm animals, metaphase II oocytes are most widely used as cytoplasts. There are now many factors known to influence the efficiency of embryo cloning by nuclear transfer. These include stage of development and cell cycle of donor cells, the choice of the recipient cell, the methods for activation of oocytes, the cell cycle coordination between donor cell and recipient cytoplast, and the method for fusion between nuclear donor and recipient cytoplast. Recent progress in cloning embryos and animals from cultured cells of embryonic, fetal, or adult origin offers a wide spectrum of potential applications of nuclear transfer, such as the unlimited multiplication of elite (优良) embryos or animals from selected matings and the potential for precise genetic modification of farm animals for gene farming or xenotransplantation (Fig. 37.1).

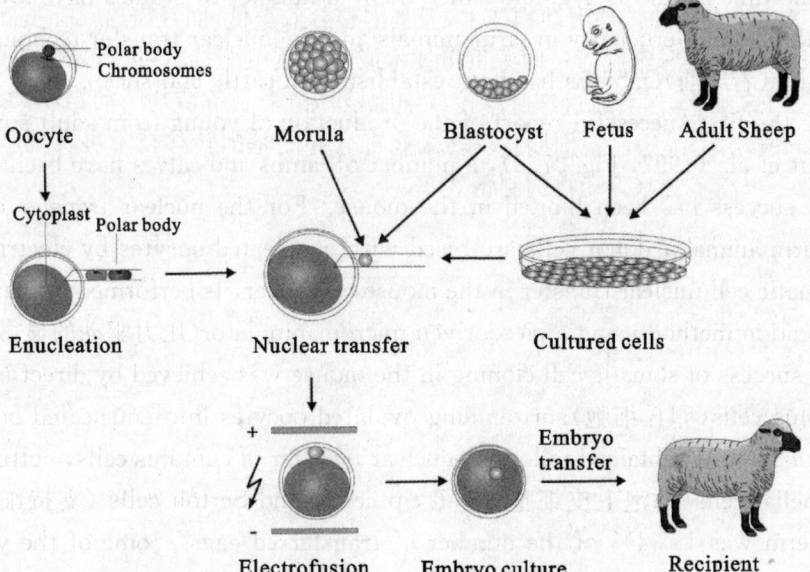

Fig. 37.1 Nuclear transfer using blastomeres or cultured cells of embryonic, fetal, and adult origin as nuclear donors. (By E. Wolf et al., 1998)

Since the first successful nuclear transfer in animals by Briggs and King (1952), who reported the development of enucleated eggs receiving blastula nuclei, nuclear transfer has been used extensively to examine the developmental potency of embryonic and somatic cell nuclei. A number of studies have demonstrated that at least some nuclei of embryos at different stages and from various somatic cells of tadpoles still have the potential to develop into fertile adults.

Another great advancement in nuclear transfer of somatic cells was reported by Shangqin Wu in 1982. She successfully obtained an adult karyblast-cytoplast hybrid fish by serial transplantation of the nuclear of erythrocytes of carp into the enucleated eggs of golden fish. Of interesting, the obtained fish was both good at jumping (like carp) and with beautiful fins (like golden fish). Erythrocytes are highly differentiated cells. This is the first and strongest evidence for the totipotential of somatic nuclear.

The development of nuclear transfer techniques in mammals has been delayed due to the lack of sufficient techniques and knowledge to manipulate mammalian oocytes, zygotes and embryos. Because the earlier studies on nuclear transfer in mammals were performed without enucleation(去核)of recipient eggs, nuclear-transferred eggs did not develop well. Later, enucleated zygotes receiving nuclei from early stage embryos can develop into blastocysts and even fertile young. In 1986, Willadsen reported that live lambs were produced after nuclear transfer of blastomeres from 8-to 16-cell-stage embryos to enucleated unfertilized oocytes but not to zygotes. The donor nuclei were considered to be reprogrammed because nuclear-transferred oocytes developed into blastocysts in the same time course as zygotes. Since then, a number of studies have attempted to increase the cloning efficiency in farm animals and the nuclear transfer technique of pre-implantation(着床前)embryos has been established in cattle and sheep.

Since the first successful report of the production of young from adult somatic cells by Wilmut et al. (1997, Fig. 37.2), a number of lambs and calves have been produced; however, success has been limited in the mouse. For the nuclear transfer of somatic cells in farm animals, donor cells are fused with enucleated oocytes by electric stimulation. Somatic cell nuclear transfer in the mouse, however, is performed primarily by the direct-injection method using a piezodriven micromanipulator(压力驱动的显微操作仪). The first success of somatic cell cloning in the mouse was achieved by direct injection of the cumulus cells(卵丘细胞)surrounding ovulated oocytes into enucleated oocytes. So far, mice have been obtained following nuclear transfer of cumulus cells, cultured follicular epithelial cells(滤泡上皮细胞), tail tip cells, and Sertoli cells (支持细胞). The yield to term was 1%-4% of the number of transferred eggs. Some of the young also died soon after birth due to respiration problems, similar to bovine somatic cell clones. Morphologic abnormalities of young were not observed in all pups (幼崽), except those obtained from Sertoli cells, but there is a significantly increased body weight in cloned mice. In most cases, placentas(胎盘)from somatic cell cloned mice were two to three times heavier than those of the controls, and there was also hypertrophy(肥大) of the placenta.

Fig. 37.2 Clonal lamb "Dolly" and his conceptus foster mother. Dolly is derived from nuclear transfer of adult somatic cells by Wilmut et al. (1997)

It was reported that, the enucleated oocytes receiving mouse embryonic stem (ES) cells, whose cell-cycle stage was not synchronized, developed into blastocysts but not into young. Recently, live mice were produced by direct injection of ES cells or gene-targeted ES cells, probably at the G_1 and G_2/M phase, into enucleated oocytes and also by fusing the ES cells at the M phase using Sendai virus.

NUCLEAR TRANSFER TECHNOLOGY

Preparation of Microtools

For nuclear transfer, three types of glass instruments are necessary: a holding pipette, an enucleation or injection pipette, and a fine glass needle. The holding pipette is used to hold the egg in place during microsurgery, the enucleation pipette is used to remove the chromosomes or nucleus from recipient oocytes or zygotes, the injection pipette is used for injection of the donor cell or nucleus into the perivitteline space(卵黄周隙)or ooplasm(卵质), and the fine glass needle is used to cut the zona pellucidae(透明带)of the eggs to facilitate nuclear transfer. All glass pipettes for micromanipulation can be prepared from fine glass capillaries using pipette pullers(拉针仪)of vertical design or horizontal design (Fig. 37.3) and further polished by needle grinder(磨针仪)and microforge(微煅仪)(Fig. 37.4). Practice is needed to create uniform tips. Once the tips are beveled, the microinjection pipettes may be bent to a 30° angle similar to the holding pipette, if desired, but the diameter of each is slightly different.

CYTOTECHNOLOGY (细胞工程技术)

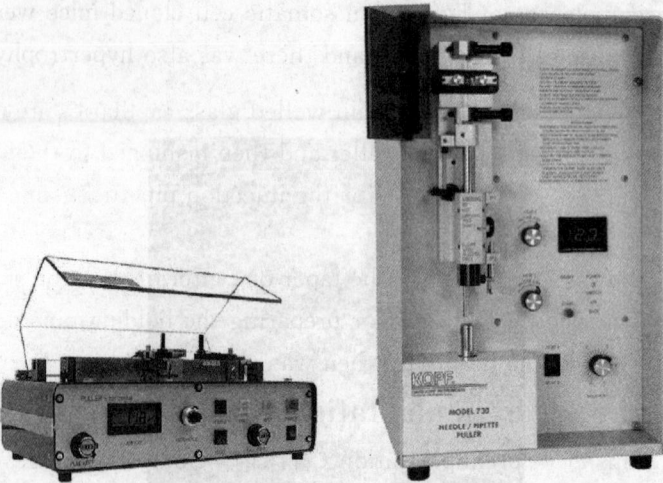

Fig. 37.3 Pipette puller. Uniform tapers and shapes of microneedles can be prepared using horizontal (up) or vertical pullers (down).

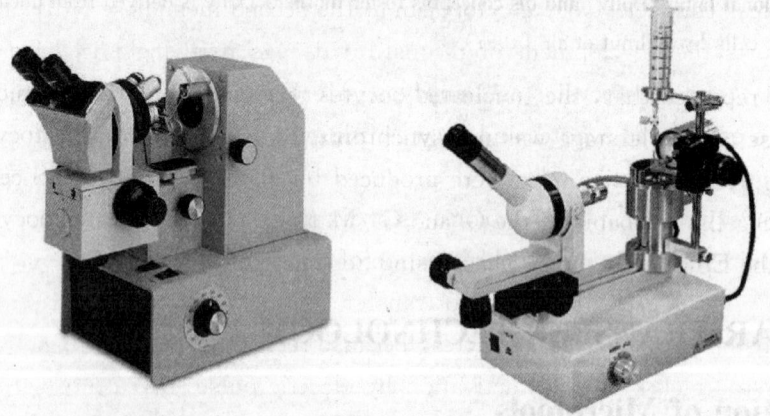

Fig. 37.4 Microforge (微煅仪) (left) and needle grinder(磨针仪)(right). Needles may be fire polished to produce a holding pipette, or bent to specification using the microforge mounted with microscope head. Needle tips can be beveled to desired shapes using the grinder incorporated with a microscope.

Holding Pipette

To make the holding pipettes, thick-walled glass tubes, are used. The most appropriate outer diameter for holding eggs is the same as or slightly greater than the diameter of the egg (80-100μm). The tip is examined under a microscope to determine if it is flat and whether the outer diameter is appropriate. It is useful to observe the tip under a microscope and fashion it on a microforge equipped with a platinum-iridium wire (铂铱丝) (Fig. 37.4). The glass capillary is mounted vertically against platinum-iridium filament wire on the microforge and is briefly heated. The tip of the glass capillary is reduced in size by melting the glass. The resulting tip is slightly bent to a 30° angle when working with dishes on a microscope.

Enucleation/Injection Pipette

For enucleation/injection pipettes, thin-walled glass capillaries are used. The glass capillary is first pulled using a pipette puller and then fashioned to the final shape using a microforge. A suitable outer diameter for the injection pipette is similar to or slightly smaller than that of the donor cell diameter.

Needle. For the fine glass needle, the taper or center of the capillary is first heated over a flame and then rapidly pulled as for preparing the holding pipette. The diameter is not critical (120 μm). The needle tip then was sharped finely.

Nuclear Transfer of Preimplantation Embryos

For nuclear transfer of preimplantation (着床前) embryos, enucleated zygotes and oocytes can be used for the recipient cytoplasm. Zygote cytoplasm can support development after nuclear transfer of karyoplasts from zygote or 2-cell-stage embryos used as donor nuclei. After the 4-cell stage, zygote cytoplasm does not support development following nuclear transfer. Oocyte cytoplasm at the second metaphase (MII phase), however, supports the development of reconstituted eggs receiving nuclei from 4-cell, 8-cell, morula, and blastocyst stage embryos. Here, we present the method of nuclear transfer into enucleated zygotes and oocytes using blastomeres from preimplantation embryos as donor nuclei.

In summary, nuclear transfer consists of six principal steps: enucleation of recipient eggs, isolation and cell-cycle synchronization of donor cells, fusion with donor cells, activation, *in vitro* culture, and embryo transfer. The activation process is necessary for nuclear transfer into enucleated oocytes, because recipient cytoplasm is not activated by sperm. Alternatively, artificial activation by electric pulses or culture with ethanol or strontium (锶) is effective for the further development of nuclear-transferred oocytes.

Nuclear Transfer into Enucleated Zygotes

1. Collection of Zygotes. Recipient zygotes for nuclear transfer are collected from hybrid mice (C57BL/6 CBA or C3H), because the viability of eggs from hybrid mice is higher and they are released from the developmental block (2-cell block) during *in vitro* culture. Briefly, mature female mice (6 weeks) are superovulated by injection of PMSG (pregnant mare serum gonadotropin, 5 IU) and HCG (human chorionic gonadotropin, 5 IU) 48 h apart, and oviducts are collected from females mated with same-strain males. Zygotes are collected from the dissected oviducts using a 27-gauge needle 20-22 h after HCG injection. Collected cumulus-oocyte complexes are treated with hyaluronidase for 1-3 min at 37℃ to dissociate the eggs from the cumulus cells. Denuded 1-cell eggs are picked up using a transfer pipette and washed three times in fresh M2 medium. One-cell eggs with two pronuclei and the second polar body are selected and incubated in M16

medium in a 5% CO_2 incubator.

Table 37.1 Strontium-Supplemented M16 Medium (By Y. Tsunoda and Y. Kato)

		M16 (10 mL)
Stock solution A (10×)	g/100 mL	1.0 mL
NaCl	5.534	
KCl	0.356	
KH_2PO_4	0.162	
$MgSO_4$ $7H_2O$	0.293	
Lactate	2.61	
Glucose	1.0	
Penicillin and streptomycin		
Stock solution B (10×)	(g/100 mL)	1.0 mL
$NaHCO_3$	2.101	
Phenol red	0.01	
Stock solution C (100×)	(g/100 mL)	0.1 mL
Sodium pyruvate	0.036	
Stock solution D (100×)	(g/100 mL)	0.1 mL
$SrCl_2 \cdot 6H_2O$	2.666	
H_2O (2×distilled)		7.8 mL
BSA		40mg

2. Enucleation of Zygotes. Collected zygotes are incubated in M2 medium supplemented with cytoskeleton inhibitors (cytochalasin B, 5 μg/mL; nocodazole(噻氨酯哒唑), 1-3 μg/mL) prior to microsurgery. For microsurgery, zygotes are manipulated in a flat droplet of M2 medium in a micromanipulation chamber under an inverted microscope using a micromanipulator. A cut (10-20% of the length of the zygote) is made in the zona pellucida with a fine glass needle. Zona-cut zygotes are held with the holding pipette placed opposite to the slit in the zona pellucida (Fig. 37.5A). Through the slit of the zona pellucida, the tip of the enucleation/injection pipette is inserted into the perivitelline space (Fig. 37.5B). The tip is then advanced to a point adjacent to one of the two pronuclei. One pronucleus with a small volume of cytoplasm is sucked into the injection pipette and this step is then repeated for the remaining pronucleus (Fig. 37.5C). For this step, it is necessary to avoid penetrating the ooplasm membrane by using cytoskeleton inhibitors. After zygotes are enucleated (Fig. 37.5D), they are rinsed with M2 medium and kept at room temperature or 37°C until use.

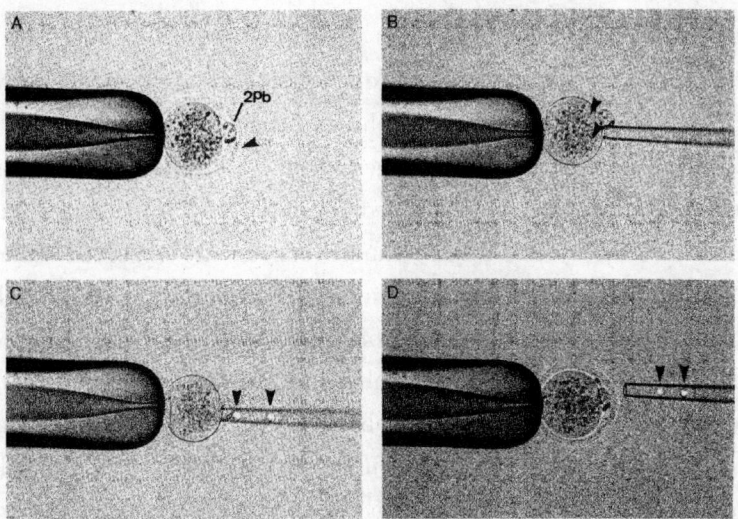

Fig. 37. 5 Enucleation of a zygote (see text). A. The arrowhead shows the slit of the zona. B, C and D. The arrowheads show two pronuclei. 2pb is the second polar body. (By Y. Tsunoda and Y. Kato)

3. Preparation of Donor Embryos. Zygotes are collected as described above. Two-cell-stage embryos are collected from oviducts of superovulated females by flushing the oviducts with M2 medium 42-44 h after HCG injection. Collected embryos are incubated in M2 medium until use.

4. Fusion Procedure. Fusion of donor nuclei with recipient cytoplasm is induced by electric pulses. Briefly, reconstituted zygotes are transferred to a chamber with two wire electrodes mounted 1 mm apart on a glass slide (Fig. 37. 8). Reconstituted zygotes are given two direct current (DC) pulses of 50 V/mm for 50 μsec three times at 20-min intervals. DC pulses are given after the donor cell is oriented parallel to the wires and recipient cytoplasm by hand. It is also useful to give alternating current (AC) pulses (100-500 kHz, 5 V/mm) before the DC pulses instead of orienting the cells by hand.

After electrofusion, reconstituted zygotes are cultured in M16 medium for 15-60 min and the success of fusion is verified. Fused eggs are further cultured for 4 days *in vitro* in a CO_2 incubator at 37℃, and embryos that develop into the blastocyst stage are transferred to oviducts on day 0. 5 or uteri on day 3. 0-3. 5 in recipient females previously mated with vasectomized (输精管结扎) males. Embryo transfer to recipient females is not recommended until the embryos have reached the compacted morula stage, because blastomeres of precompacted embryos often fall out of the slit in the zona, and such embryos do not develop further. Recipient females are allowed to go to term or are dissected on day 19.

Fusion of donor nuclei with recipient cytoplasm can also be induced by injecting the inactivated Sendai virus and donor cells into the recipient cytoplasm.

Nuclear Transfer into Enucleated Oocytes

1. Collection of Oocytes. Female mice (3 weeks of age) are superovulated using PMSG and HCG, and oocytes at the metaphase of the second meiosis (MII) are collected by dissecting the oviducts with 27-gauge needles. Morphologically normal MII oocytes are selected and incubated in M16 medium in a 5% CO_2 incubator until use.

2. Enucleation of Oocytes. When denuded oocytes are observed under an inverted microscope, a small area of swelling containing MII chromosomes can be observed. The zonae pellucidae around the swollen areas of collected MII oocytes are cut as shown above in M2 medium without cytoskeleton inhibitors at 37℃. After cutting the zonae pellucidae of the oocytes, the oocytes are incubated with cytochalasin B (5 μg/mL) in M2 medium. The swelling areas become difficult to distinguish after cytochalasin B treatment. Oocytes are held opposite the slit in the zona pellucida, and the enucleation pipette is inserted into the perivitteline space through the slit. An area of translucent cytoplasm containing MII chromosomes moves when the enucleation pipette touches the ooplasm, and then the area is aspirated into the enucleation pipette while visually confirming the presence of chromatin. After enucleation of all the oocytes, the eggs are rinsed with M2 medium and kept at room temperature or 37℃ until use.

If it is difficult to find the areas containing the MII chromosomes, they can first be stained using Hoechst 33342 to visualize them with fluorescence microscopy during the enucleation step. Briefly, collected oocytes are incubated with M2 medium containing Hoechst 33342 (0.5 μg/mL) for 3 min at 37℃. Stained oocytes are washed three times with M2 medium without Hoechst dye, and MII chromosomes are removed with an enucleation/injection pipette under a fluorescence microscope (blue-violet, UV excitation filter with 400-450 nm). UV irradiation exposure time is critical to the continued viability of the oocytes and should be less than 15 sec.

3. Preparation of Donor Cells.

Collection of Donor Embryos. Two-cell-stage embryos are collected from oviducts of superovulated females by flushing the oviducts with M2 medium 42-44 hours after HCG injection and mating. Four-cell and eight-cell embryos are collected from *in vitro* culture of two-cell embryos. Impacted morula-stage embryos are obtained after 24-hr culture of 8-cell-stage embryos.

Cell-Cycle Synchronization. When MII-phase oocytes are used for recipient cytoplasm, the cell cycle of the donor embryos must be synchronized at the G_1, G_2, or M phase because MII oocytes have high maturation promoting factor (MPF) activity in the cytoplasm. When donor cells are fused with MII oocytes, premature chromosome condensation (PCC) occurs due to MPF activity in the ooplasm. When the oocytes with donor nucleus in PCC are artificially activated, the nuclear envelope forms again and the pronuclear-like formation starts from the beginning of the G_1 phase of the cell cycle. If S-phase donor cells are fused with MII oocytes, the DNA will be abnormal due to PCC and subsequent activation. Thus, donor cells must be in G_1, G_2, or M phase. After synchronization at the G_1 phase, as described below, the medium must be supplemented with aphidicoline（阿非迪霉素）until fusion with the oocyte cytoplasm because the G_1 phase is very short in preimplantation embryos. If donor cells are at the G_2 phase, emission of the polar body after activation is necessary.

4. Fusion Procedure. Using the injection pipette, karyoplasts or disaggregated single blastomeres at the G_1 phase are injected into the perivitteline space of enucleated oocytes with inactivated Sendai virus. When donor cells are fused with recipient oocytes using electric pulses, calcium ions must be removed from the fusion medium to avoid oocyte activation.

5. Activation. Following fusion procedure, reconstituted oocytes are artificially activated by electric pulses or ethanol treatment. For electric pulses, reconstituted oocytes are given two DC pulses of 50V/mm for 50 sec three times at 20-min intervals. For ethanol treatment, reconstituted oocytes are cultured in M16 medium supplemented with 7% ethanol for 7 min to activate artificially.

6. *In Vitro* Culture and Embryo Transfer. After activation, reconstituted oocytes are cultured in M16 medium for 15-60 min and checked for fusion. Fused eggs are further cultured for 4 days *in vitro*, and embryos that develop into the blastocyst stage are transferred to day 0.5 oviducts or day 3.0-3.5 uterine horns of recipient females.

7. Serial Nuclear Transfer. After nuclear-transferred oocytes divide into the 2-cell stage during overnight culture *in vitro*, their nuclei are again transferred into enucleated blastomeres of 2-cell embryos that were derived from *in vivo* or *in vitro* fertilization. Serial nuclear transfer into fertilized embryo cytoplasm promotes the development of nuclear transplants.

NUCLEAR TRANSFER OF EMBRYONIC STEM CELLS AND SOMATIC CELLS

Due to the smaller size of embryonic stem (ES) cells and somatic cells, the nucleus can be directly injected into ooplasm. Here, we present the protocol for direct injection of donor nuclei into oocytes, in addition to the standard nuclear transfer method by cell fusion. Successful nuclear transfer of ES cells and somatic cells requires appropriate microtools, such as injection pipettes with the proper tip diameter. The cell size varies widely, depending on the cell cycle. Cells at the G_0 phase are smaller than those at the G_2 or M phase. If an injection pipette with a larger tip is used, inactivated Sendai virus is diluted in the pipette, and the fusion rate might decrease. When the nucleus is injected directly into the ooplasm, the injection pipette diameter must be much smaller (10-

20 μm); otherwise, the donor membrane may be destroyed during the pipetting, damaging the ooplasm membrane.

Collection of Oocytes

The materials and equipment are the same as described in the section on nuclear transfer of preimplantation embryos, except for the collection time and mouse strain of oocytes. When somatic cells are used as donor cells, oocytes are often collected earlier, such as 13-14 h after HCG injection.

Enucleation of Oocytes

The oocyte enucleation procedure is the same as described in the section on nuclear transfer of preimplantation embryos.

Preparation of Donor Cells

1. Embryonic Stem Cells. The basic protocol of maintenance of embryonic stem cells is described in chapter 31. Briefly, ES cells are cocultured with a primary culture of mouse embryonic fibroblast cells (MEFs) that were previously inactivated by mitomycin C (10 μg/mL for 2.5 h). At present, it is difficult to synchronize the cell cycle of ES cells at the G_o and G_l phases. In some reports, the cell cycles of ES cells are distinguished by the size of the ES cells. When ES cells in the M phase are used as donor cells, nuclear-transferred oocytes develop into the blastocyst stage at a high rate and some of them develop to term. For the synchronization of ES cells at the M phase, the cells are incubated in nocodazole (噻氨酯哒唑, 抗肿瘤药)-containing medium (1-3 μg/mL) for 3 h prior to use as nuclear transfer donors. The M-phase ES cells are round and easy to distinguish.

2. Somatic Cells. Somatic cells have successfully developed into full-term fetuses or adults by nuclear transfer into enucleated oocytes. But the developmental potential of cloned embryos into fetuses at full term or adults is extremely low. More basic research studies are necessary to characterize transgenic mouse development after nuclear transfer of somatic cells.

Procedure for Donor Nucleus Incorporation

Basically the same methods as for nuclear transfer of preimplantation embryos are used. Because the cell size is smaller than blastomeres or karyoplasts of preimplantation embryos, an injection pipette must be constructed with the same diameter as the donor cells (10-20μm).

Activation

1. Electric Pulses. Reconstituted oocytes are given two DC pulses of 50 V/mm for 50 μsec three times at 20-min intervals. After the electric pulses, reconstituted oocytes are cultured in M16 medium supplemented with CB (5 μg/mL) for 5 h and examined for

a pronuclear-like formation in the ooplasm. When M-phase cells are used as donor cells, fused oocytes are activated in M16 without CB to release the polar body.

2. Strontium(锶)treatment. Fused oocytes are cultured in M16 medium containing 10 mM $SrCl_2$ and cytochalasin B (5 $\mu g/mL$) for 5 h. After treatment with $SrCl_2$-containing medium, nuclear transfer oocytes are released from the drug by washing with M2 medium three times. They are then cultured further in standard M16 medium without $SrCl_2$. When M-phase cells are used as donor cells, fused oocytes are activated with $SrCl_2$ but not with cytochalasin B.

In Vitro Culture and Embryo Transfer

The protocol is basically the same as described previously.

QUESTIONS FOR DISCUSSION

1. What is a cloned animal? In what ways can we obtain cloned animals?
2. How to produce cloned animals by nuclear transfer of embryo cells, ES cells and somatic cells?

知识要点

遗传背景相同的动物称为克隆动物。早期胚胎分割可得到克隆动物,但可得到的数量有限。克隆动物还可以通过细胞核移植技术来获得,即将胚胎细胞、胚胎干细胞或体细胞的细胞核移植到去核的卵母细胞或受精卵,或者去核的 2-细胞期胚胎细胞中,核质杂交重构胚可发育成完整个体。

GLOSSARY

Part I Cytotechnology in plant

Acclimatization. Plants must gradually become accustomed to in vitro conditions when tissue cultures are transplanted in field.

Adenine. Aminopurine; exhibits cytokinin activity in bud initiation.

Adventitious. Initiation of a structure out of its usual place, i. e. , arising sporadically. Adventitious roots may originate from leaf or stem tissue.

Auxin. Plant growth regulator stimulating shoot cell elongation and resembling IAA in physiological activity.

Callus. Disorganized meristematic or tumor-like mass of plant cells formed in vitro; also refers to a meristematic growth arising at a wound site in vivo.

Chemically defined medium. A medium devoid of any natural plant or animal undefined organic supplement.

Clones. Genetically identical plants vegetatively propagated from a single individual.

Cryopreservation. Freeze preservation of plant material; typically maintained at the temperature of liquid nitrogen ($-196°C$).

Cybrid. Cytoplasmic hybrid; heteroplast.

Cytokinin. Plant growth regulator stimulating cell division and resembling kinetin in physiological activity; mainly N^6-substituted aminopurines.

Cytodifferentiation. Cell differentiation; morphological and biochemical specialization of a cell.

Determination. Resumption of reprogramming by which a cell becomes restricted to a new pathway of specialization.

Embryogenesis. Initiation of embryo formation.

Embryoid. Embryo-like structure formed in vitro. With proper care, an embryoid will develop into a normal plant.

Explant. Excised fragment of plant tissue or organ used to start a tissue culture; primary explant.

Habituation. Changes in exogenous nutritional requirements occurring during culture, particularly independent of external growth regulators.

Hairy root cultures. Genetic transformation by *Agrobacterium rhizogenes* that results in cultured roots showing a high degree of branching, a profusion of root hairs, and a high rate of biosynthesis of secondary metabolites.

Haploid. Having a single set of chromosomes; monoploid.

Heterokaryon. Fusion of unlike cells with dissimilar nuclei present; heterokaryocyte.

Homokaryon. Fusion of similar cells.

Hyperhydricity. Formerly called vitrification; A common cultural disorder that occurs in the multiplication phase. The leaves are brittle, malformed, and have a water-soaked or glassy appearance. This disorder is associated with too high a level of cytokinins, poor ventilation, choice of gelling agents and possibly ammonium in cultures.

Meristemoid. Cluster of meristematic cells within a callus with the potential to form a primordium.

Micropropagation. Clonal multiplication of plants originating from cultured shoot tips. Cytokinin-induced proliferation of shoots occurs during subcultures.

Organogenesis. The regeneration of whole plants adventitiously from callus or directly from plant tissue such as leaf explants.

Osmoticum. Isotonic plasmolyticum; external medium of low osmotic potential approximating the concentration of solutes within the cell, thereby preventing bursting of protoplasts due to excessive water uptake.

Passage. Subculture; transfer of inoculums from a culture to a fresh medium.

Passage number. Number of subcultures completed.

Passage time. Interval between successive subcultures.

Primordium (**pl.** *primordia*). Earliest detectable stage of development of an organ, usually pertaining to leaf and bud primordia.

Protoplast. Living isolated plant cell following removal of cell wall either by enzymatic or mechanical method.

Secondary metabolite. Metabolic compound unique to a limited number of species or cultivars.

Shoot-apex culture. Explant consisting of the apical dome plus a few subjacent leaf primordia.

Somatic embryogenesis. Formation of embryos in vitro from vegetative (asexual) cells.

Somatic hybrid. Cell or plant derived from the fusion of two genetically different vegetative (somatic) protoplasts.

Subculture. See "passage"; transfer of inoculum to a fresh medium.

Totipotency. Retention by a single cell of the genetic information for the recreation of the adult organism.

Transdifferentiation. Differentiation of a cell into a new cell type without undergoing cell division.

Transformation. Insertion of foreign DNA into a plant cell resulting in a phenotypic modification of the regenerated plant.

Transgenic plant. Plants possessing a single or multiple genes from a different species, and the forgein DNA is stably integrated into the genome.

Part Ⅱ Cytotechnology in animal

Adaptation. Induction or repression of synthesis of a macromolecule (usually a protein) in response to a stimulus; e. g. , enzyme adaptation-an alteration in enzyme activity brought about by an inducer or repressor and involving an altered rate of enzyme synthesis or degradation.

Anchorage dependent. Requiring attachment to a solid substrate for the survival or growth of cells.

Aneuploid. Not an exact multiple of the haploid chromosome number. (See *haploid.*)

Aseptic. Free of microbial infection.

Autocrine. Receptor-mediated response of a cell to a factor produced by the same cell.

Autograft. A graft from one individual transplanted back to the same individual.

Balanced salt solution. An isotonic solution of inorganic salts present in approximately the correct physiological concentrations; may also contain glucose, but is usually free of other organic nutrients.

Carcinoma. A tumor derived from epithelium, usually endodermally or ectodermally derived cells.

Cell culture. Growth of cells dissociated from the parent tissue by spontaneous migration or mechanical or enzymatic dispersal.

Cell fusion. Formation of a single cell body by the fusion of two other cells, either spontaneously or, more often, by induced fusion with inactivated Sendai virus or polyethylene glycol.

Cell hybridization. See *hybrid cell.*

Cell line. A propagated culture after the first subculture.

Cell strain. A characterized cell line derived by selection or cloning.

Clone. A population of cells derived from one cell.

Commitment. Irreversible progression from a stem cell to a particular defined lineage endowing the cell with the potential to express a limited repertoire of properties.

Conditioned medium. Used medium with small molecular metabolites and growth factors released by cultured cells.

Confluent. A monolayer of cells in which all cells are in contact with other cells all around their periphery, and no available substrate is left uncovered.

Constitutive. Expressed by a cell in the absence of external regulation.

Contact inhibition. Inhibition of plasma membrane ruffling and cell motility when cells are in complete contact with other adjacent cells, as in a confluent culture; often precedes, but is not necessarily causally related to, cessation of cell proliferation.

Continuous cell line or cell strain. Cell line or strain having the capacity for infinite survival. Previously known as "established" and often referred to as "immortal."

Cytokine. A factor, released by cells, that will induce a receptor-mediated effect on the proliferation, differentiation, or inflammation of other cells; usually a short range paracrine, rather than systemic, effect.

Deadaptation. Reversible loss of a specific property due to the absence of the appropriate inducer (not always defined).

Dedifferentiation. Irreversible loss of the specialized properties that a cell would have expressed *in vivo*.

Density limitation of growth. Mitotic inhibition correlated with an increase in cell density at confluence.

Dome. A hemicystic or blister-like structure in a confluent epithelial monolayer implying ion transport across the monolayer and resulting in the accumulation of water below the monolayer.

Ectoderm. The outer germ layer of the embryo, giving rise to the epithelium of the skin.

Endoderm. The innermost germ layer of the embryo, giving rise to the epithelial component of organs such as the gut, liver, and lungs.

Endothelium. An epithelial-like cell layer lining spaces within mesodermally derived tissues, such as blood vessels, and derived from the mesoderm of the embryo.

Epithelial. Cells derived from epithelium but often used more loosely to describe any cells of a polygonal shape with clear, sharp boundaries between them. More correctly, the latter should be referred to as epithelial-like or epithalioid.

Epithelium. A covering or lining of cells, as in the surface of the skin or lining of the gut, usually derived from the embryonic endoderm or ectoderm, but sometimes derived from mesoderm, as with kidney tubules and mesothelium lining body cavities.

Embryonic stem cells. Pluripotent cell lines that have been derived from the inner cell mass (ICM) of blastocyst stage embryos. They are characterized by their ability to be propagated indefinitely in culture as undifferentiated cells with a normal karyotype.

Euploid. Exact multiple of the haploid chromosome set. Otherwise we should say "euploid, with some chromosomal aberrations."

Explant. A fragment of tissue transplanted from its original site and maintained in an artificial medium.

Feeder layers. A layer of cells in arrested growth but can provide extracellular excretions to aid another cell in proliferation.

Fibroblast. A proliferating precursor cell of the mature differentiated fibrocyte.

Fibroblastic. Resembling fibroblasts (i. e. , spindle shaped (bipolar) or stellate (multipolar)); usually arranged in parallel arrays at confluence if contact is inhibited. More correctly, fibroblast-like or fibroblastoid.

Finite cell line. A culture that has been propagated by subculture, but is capable of only a limited number of cell generations *in vitro* before dying out.

Genotype. The total genetic characteristics of a cell.

Growth curve. A semi logarithmic plot of the cell number on a logarithmic scale against time on a linear scale, for a proliferating cell culture; usually divided into the lag phase (the phase before growth is initiated), the log phase (the period of exponential growth), and the plateau (a stable cell count achieved when the culture stops growing at a high cell density).

Growth factor. A factor, released by cells, that induces proliferation in other cells; mostly paracrine in effect, but may be released into the blood by platelets or endothelium.

Haploid. That chromosome number wherein each chromosome is represented once; in most higher animals, the number present in the gametes and half the number found in most somatic cells.

Heterokaryon. Cell containing two or more genetically different nuclei; usually derived by cell fusion.

Heteroploid. A culture in which the cells have chromosome numbers other than diploid and differing from each other.

Histotypic. A culture resembling a tissue-like morphology *in vivo*. Usually, a three-dimensional culture re-created from a dispersed cell culture that attempts to regain, by cell proliferation and multi-layering or by reaggregation, a tissue-like structure. Organ cultures cannot be propagated, whereas histotypic cultures can.

Homograft. (***Allograft.***) A graft derived from a genetically different donor of the same species as the recipient.

Homokaryon. Cell containing two or more genetically identical nuclei; usually a product of cell fusion.

Hybrid cell. Mononucleate cell that results from the fusion of two different cells, leading to the formation of a synkaryon. (See *synkaryon.*)

Immortalization. The acquisition of an infinite life span. May be induced in finite cell lines by transfection with telomerase, oncogenes, or the large T-region of the SV40 genome, or by infection with SV40 (whole virus) or Epstein-Barr virus (EBV). Immortalization is not necessarily a malignant transformation, although it may be a component of malignant transformation.

Induced pluripotent stem cells (***iPS***). A type of pluripotent artificially derived from a non-pluripotent cell, typically an adult somatic cell, by inducing a "forced" expression of specific genes such as four transcriptional factors: *Oct 3/4*, *Sox2*, *c-Myc* and *KLF4*. iPS cells are similar to natural pluripotent stem cells, such as embryonic stem (ES) cells, in many respects, such as the expression of certain stem cell genes and proteins, chromatin methylation patterns, doubling time, embryoid body formation, teratoma formation, viable chimera formation, and potency and differentiability, but the full extent of their relation to natural pluripotent stem cells is still being assessed.

In ovo. In the egg-usually, the hen's egg.

In vitro. Literally, "in glass," but used conventionally to mean cultured out with the host as cell cultures, organ cultures, or short-term organ bath preparations; also used to indicate biochemical and molecular reactions carried out in a test tube, but these reactions are better referred to as *cell free*.

In vivo. In the living plant or animal.

Karyotype. The distinctive chromosomal complement of a cell.

Laminar-flow hood or cabinet. A workstation with filtered air flowing in a laminar (nonturbulent) manner parallel to or perpendicular to the work surface, to maintain the sterility of the work; the parallel flow is called *horizontal* laminar flow, the perpendicular flow *vertical* laminar flow.

Lipofection. Transfection of DNA by fusion with lipid-encapsulated DNA.

Malignant. Invasive or metastatic; (said of a tumor). Usually progressive, leading to the destruction of host cells and, ultimately, death of the host.

Malignant transformation. The development of the ability to invade normal tissue without regulation in space or time; may also lead to metastatic growth.

Medium. A mixture of inorganic salts and other nutrients capable of sustaining cell survival *in vitro* for 24 hours. ***Growth medium***: That medium which is used in routine culture such that the cell number increases with time. ***Maintenance medium*** (***or Holding medium***): A medium that will retain cell survival without cell proliferation (e. g. , a low-serum or serum-free medium used with serum-dependent cells). The *plural* of medium is *media*.

Mesoderm. A germ layer in the embryo arising between the ectoderm and endoderm and giving rise to mesenchyme, which, in turn, gives rise to connective tissue, etc.

Monoclonal. Derived from a single clone of cells.

Monoclonal antibody. Antibody produced by a clone of lymphoid cells either *in vitro* or *in vivo*. *In vitro*, the clone is usually derived from a hybrid of a sensitized spleen cell and a continuously growing myeloma cell.

Morphogenesis. The development of form and structure of an organism.

Myeloma. A tumor derived from myeloid cells; used in monoclonal antibody production when the myeloma cell can produce immunoglobulin.

Neoplastic. A new, unnecessary proliferation of cells giving rise to a tumor.

Neoplastic transformation. The conversion of a non-tumorigenic cell into a tumorigenic cell.

Nuclear reprogramming. A term that describes a switch in nuclear gene expression of one kind of cell to that of an embryo or other cell type.

Oncogene. A gene that, when transfected or infected into normal cells, induces malignant transformation; usually positively-acting genes coding for growth factors, receptors, signal transducers, or nuclear regulators.

Organ culture. The maintenance or growth of organ primordia or the whole or parts of an organ *in vitro* in a way that may allow differentiation and preservation of the architecture or function of the organ.

Organogenesis. The development of organs.

Organotypic. Histotypic culture involving more than one cell type to create a model of the cellular interactions characteristic of an organ *in vivo*. A reconstruction from dissociated cells or fragments of tissue is implied, as distinct from organ culture, in which the structural integrity of the explanted tissue is retained.

Osmolality. The concentration of osmotically active particles in an aqueous solution, expressed in osmols/ kg.

Osmolarity. The concentration of osmotically active particles in an aqueous solution, expressed in osmoles/L.

Osmole. The amount of a substance containing 1 mole of osmotically active particles.

Passage. The transfer or subculture of cells from one culture vessel to another; usually, but not necessarily, involves the subdivision of a proliferating cell population, enabling the propagation of a cell line or cell strain.

Passage number. The number of times a culture has been subcultured.

Pavement-like. Cells in a regular monolayer or polygonal cells. More correctly, epithelioid or epitheliallike.

Phenotype. The aggregate of all the expressed properties of a cell; the product of the interaction of the genotype with the regulatory environment.

Plating efficiency. The percentage of cells seeded at subculture that gives rise to colonies. If each colony can be said to be derived from one cell, plating efficiency is identical to cloning efficiency. Sometimes the plating efficiency is used loosely to describe the number of cells surviving after subculture, but this is better termed the *seeding efficiency*.

Ploidy. Relationship of chromosome number of a given type of cell to that found in normal somatic cells *in vivo*. See also *haploid*, *diploid*, *euploid*, *aneuploid*, and *heteroploid*.

Population density. The number of monolayer cells per unit area of substrate; for cells growing in suspension, the population density is identical to the cell concentration.

Population doubling time. The interval required for a cell population to double at the middle of the logarithmic phase of growth.

Primary culture. A culture started from cells, tissues, or organs taken directly from an organism, and before the first subculture.

Saturation density. Maximum number of cells attainable per cm^2 (in a monolayer culture) or per ml (in a suspension culture) under specified conditions.

Seeding efficiency. The percentage of the inoculum that attaches to the substrate within a stated period of time (implying viability, or survival, but not necessarily proliferative capacity).

Senescence. Normal cells can divide only for a limited number of times and will die out. This genetically determined event is known as senescence.

Split ratio. The divisor of the dilution ratio of a cell culture at subculture (e. g., one flask divided into four, or 100 mL up to 400 mL, would be a split ratio of 4).

Stem cells. Cells that are capable of dividing and renewing themselves for long periods

Subconfluent. Less than confluent; not all of the available substrate is covered.

Substrate. The matrix or solid underlay upon which a monolayer culture grows.

Suspension culture. A culture in which cells will multiply when suspended in growth medium.

Teratocarcinomas. Malignant germ cell tumors that comprise an undifferentiated embryonal carcinoma (EC) component and a differentiated component that can include all three germ layers.

Tissue culture. Properly, the maintenance of fragments of tissue *in vitro*, but now commonly applied as a generic term denoting tissue explant culture, organ culture, and dispersed-cell culture, including the culture of propagated cell lines and cell strains.

Transdifferentiation. Cells from one lineage acquiring the ability to differentiate into cells of a different lineage.

Transgenic animals. Animals carrying new genes (integrating foreign DNA segments into their genome).

Transfection. The transfer, by artificial means, of genetic material from one cell to another, when less than the whole nucleus of the donor cell is transferred. Transfection is usually achieved by transferring isolated chromosomes, DNA, or cloned genes.

Transformation. A permanent alteration of the cell phenotype, presumed to occur via an irreversible genetic change. May be spontaneous, as in the development of rapidly growing continuous cell lines from slow-growing early passage rodent cell lines, or may be induced by chemical or viral action. Usually produces cell lines that have an increased growth rate, an infinite life span, a lower serum requirement, and a higher plating efficiency and that are often (but not necessarily) tumorigenic.

Tumorigenesis. The cells have developed the capacity to generate tumors if implanted *in vivo* into an isologous host or if transplanted as a xenograft into an immune-deprived animal.

Xenograft. Transplantation of tissue to a species different from that from which it was derived; often used to describe the implantation of human tumors in athymic (nude), immune-deprived, or immune-suppressed mice.

REFERENCES

[1] Alvarez M C, jar J B, Chen S, Hong Y. Fish ES Cells and Applications to Biotechnology. Marine Biotechnology, 2007,9:117-127.

[2] Ball E. Development in sterile culture of stem tips and subjacent regions of *Tropaeolum majus* L. and *Lupinus albus* L. Am. J. Bot, 1946,33:301-318.

[3] Ball E. Studies on the nutrition of the callus culture of *Sequoia senpervirens*. Ann. Biol, 1955,31:281-305.

[4] Bergmann L. Growth and division of single cells of higher plants in vitro. J. Gen. Physiol, 1960,43:841-851.

[5] Biondi S, Thorpe T A. Requirements for a tissue culture facility. In Plant tissue culture: Methods and applications in agriculture. New York: Academic Press, 1981.

[6] Bottcher U F, Aviv D, Galun E. Complementation between protoplasts treated with either of two metabolic inhibitors results in somatic-hybrid plants. Plant Science, 1989,63:67-77.

[7] Boyer L A, Lee T I, Cole M F, Johnstone S E, Levine S S, Zucker J P, Guenther M G, Kumar R M, Murray H L, Jenner R G, Gifford D K, Melton D A, Jaenisch R and Young R A. Core transcriptional regulatory circuitry in human embryonic stem cells. Cell, 2005,122:947-956.

[8] Briggs R, King T J. Transplantation of living nuclei from blastula cells into enucleated frogs' eggs. Proc. Natl. Acad. Sci. USA, 1952,38:455-463.

[9] Brinster R L, Chen H Y, Trumbauer M E, Avarbock M R. Translation of globin messenger RNA by the mouse ovum. Nature, 1980, 283:499-501.

[10] Butcher D N. Secondary products in tissue cultures. In Applied and fundamental aspects of plant cell, tissue and organ culture. Berlin: Springer-Verlag, 1977.

[11] Carrel A. On the permanent life of tissues outside the organism. J. Exp. Med, 1912,15:516-528.

[12] Carl A, Pinkert A. Transgenic animal technology-a laboratory handbook. 2nd edition. Cademic press, California, USA, 2002.

[13] Cutter E G. Recent experimental studies of the shoot apex and shoot morphogenesis. Bot. Rev, 1965, 31:7-113.

[14] Devreux M. New possibilities for the in vitro cultivation of plant cells. EuroSpec-

tra, 1970, 9:105-110.

[15] Dodds J H, Reynolds T L. A scanning electron microscope study of pollen embryogenesis in Hyoscyamus niger. Z. Pflanzenphysiol, 1980,97:271-276.

[16] De Ropp R S. The growth and behavior in vitro of isolated plant cells. Proc. Roy. Soc. B, 1955,144:86-93.

[17] Eagle H. Amino acid metabolism in mammalian cell cultures. Science, 1959,130: 432.

[18] Evans M J and Kaufman M. Establishment in culture of pluripotential cells from mouse embryos. Nature, 1981, 292:154-156.

[19] Folkman J and Haudenschild C. Angiogenesis in vitro. Nature, 1980, 288:551-556.

[20] Freshney R I. Culture of animal cells-a manual of basic technique. 4th edition. WILEY-Liss inc. , New York, USA, 2000.

[21] Falkiner F R. The criteria for choosing an antibiotic for control of bacteria in plant tissue culture. Newsl. IAPTC, 1990, 60:13-23.

[22] Gautheret R J. Sur la possibilite de realiser la culture indefinie destissu de tubercules de carotte. C. R. Acad. Sci. (Paris), 1939, 208:118-21.

[23] Gautheret R J. Comparison entre factions de l'acide indoleacetique et celle du Phytomonas tumefaciens sur la croissance des tissus vegetaux. C. R. Soc. Biol. (Paris), 1946,140:169-71.

[24] Green H and Thomas J. Pattern formation by cultured human epidermal cells: development of curved ridges resembling dermatoglyphs. Science, 1978, 200:1385-1388.

[25] Graham F L and Van der Eb A J. A new technique for the assay of infectivity of human adenovirus 5 DNA. Virology, 1973, 52:456-461.

[26] Gurdon J B and Melton D A. Nuclear Reprogramming in Cells. Science, 2008, 322:1811-1815.

[27] Gordon J W, Ruddle F H. Integration and stable germ line transmission of genes injected into mouse pronuclei. Science, 1981, 214:1244-1246.

[28] Guha S, Maheshwari S C. Cell division and differentiation of embryos in the pollen grains of Datura in vitro. Nature, 1966, 212:97-98.

[29] Haberlandt G. Kulturversuche mit isolierten Pflanzenzellen. Sitzungsber. Akad. Wiss. Wien, Math. Nat. Classe, 1902,111(1):69-92.

[30] Harrison R G. Observations on the living developing nerve fiber. Proc. Soc. Exp. Biol. Med. , 1907, 4:140-143.

[31] Harvey A E, Grasham J L. Procedure and media for obtaining tissue cultures of twelve conifer species. Can. J. Bot. , 1969, 47:547-549.

[32] Illmensee K and Hoppe P C. Nuclear transplantation in Mus musculus: develop-

REFERENCES

mental potential of nuclei from preimplantation embryos. Cell, 1981, 23:9-18.

[33] John H Dodds and Lorin W Roberts. Experiments in plant tissue culture. 3rd edition, Cambridge University press, New York, USA. 1995.

[34] King P J. Cell proliferation and growth in suspension cultures. Int. Rev. Cytol. Suppl. 11A, ed. I. K. Vasil, 1980:25-54.

[35] Kleinsmith L J and Pierce G B. Multipotentiality of single embryonal carcinoma cells. Cancer Res. , 1964, 24:1544-1552.

[36] Kotte W. Wurzelmeristem in Gewebekultur. Ber. Dtsch. Bot. Ges. , 1922, 40: 269-72.

[37] Leifert C, Waites W M. Contaminants in plant tissue cultures. Newsl. IAPTC, 1990, 60:2-13.

[38] Loo, S. W. Cultivation of excised stem-tips of asparagus in vitro. Am. J. Bot. , 1945, 32, 13-17.

[39] Marshak D R, Gardner R L and Gottlieb D. Stem cell biology. Cold spring harbor labotory press. USA, 2001.

[40] Martin G R. Isolation of a pluripotent cell line from early mouse embryos cultured in medium conditioned by teratocarcinoma stem cells. Proc. Natl. Acad. Sci. , 1981, 78:7634-7638.

[41] McGrath J and Solter D. Nuclear transplantation in the mouse embryo by microsurgery and cell fusion. Science, 1983, 220:1300-1302.

[42] Melchers G, Labib G. Somatic hybridisation of plants by fusion of protoplasts: I. Selection of light resistant hybrids of "haploid" light sensitive varieties of tobacco. Mol. Gen. Genet. , 1974, 135:277-294.

[43] Mertz J E, Gurdon J B. Purified DNAs are transcribed after microinjection into Xenopus oocytes. Proc. Natl. Acad. Sci. U S A, 1977, 74(4):1502-1506.

[44] Morel G. Sur la culture des tissus de deux Monocotyledons. C. R . Acad. Sci. (Paris), 1950, 230:1099-1101.

[45] Melchers G, Sacristan M D, Holder A A. Somatic hybrid plants of potato and tomato regenerated from fused protoplasts. Carlsberg Res. Commun, 1978, 43:203-218.

[46] Morel G. Tissue culture-a new means of clonal propagation in orchids. Am. Orchid Soc. Bull. , 1964, 33:473-478.

[47] Morel G. Producing virus-free Cymbidiums. Am. Orchid Soc. Bull. , 1960, 29: 4957.

[48] Morel G. Tissue culture-a new means of clonal propagation in orchids. Am. Orchid Soc. Bull. , 1964, 33:473-478.

[49] Muir W H. Culture conditions favoring the isolation and growth of single cells from higher plants in vitro. Ph. D. thesis, Dept. of Plant Pathology, University of

Wisconsin, Madison, 1953.

[50] Muir W H, Hildebrandt A C, Riker A J. Plant tissue cultures produced from single isolated plant cells. Science, 1954, 119:877-887.

[51] Murashige T. Plant propagation through tissue culture. Annu. Rev. Plant Physiol. , 1974, 25:135-166.

[52] Murashige T. Clonal crops through tissue culture. In Plant tissue culture and its bio-technological application, ed. W. Barz, E. Reinhard, & M. H. Zenk, 1977: 392-403. Berlin: Springer-Verlag.

[53] Murashige T, Skoog F. A revised medium for rapid growth and bioassays with tobacco tissue cultures. Physiol. Plant, 1962, 15:473-97.

[54] Nagy A, Gocza E, Merentes Diaz E, Prideaux V R, Ivanyi E, Markkula M and Rossant J. Embryonic stem cells alone are able to support fetal development in the mouse. Development, 1991, 110:815-821.

[55] Nagy A, Rossant J, Nagy R, Abramow-Newerly W and Roder J C. Derivation of completely cell culture-derived mice from early-passage embryonic stem cells. Proc. Natl. Acad. Sci. , 1993, 90:8424-8428.

[56] Nickell L G. Submerged growth of plant cells. Adv. Appl. Microbiol. , 1962, 4: 213-36.

[57] Nichols J, Zevnik B, Anastassiadis K, Niwa H, Klewe-Nebenius D, Chambers I, Scholer H, Smith A. Formation of pluripotent stem cells in the mammalian embryo depends on the POU transcription factor Oct4. Cell, 1998, 95:379-391.

[58] Odorico J S, Kaufman D S, Thomson J A. Multilineage differentiation from human embryonic stem cell lines. Stem Cells, 2001, 19:193-204.

[59] Owen H R, Miller A R. An examination and correction of plant tissue culture basal medium formulations. Plant Cell, Tissue & Organ Cult. , 1992, 28:147-50.

[60] Ojima K, Ohira K. Nutritional requirements of callus and cell suspension cultures. In Frontiers of plant tissue culture 1978, ed. T. A. Thorpe, pp. 1978: 265-75. Calgary: IAPTC.

[61] Palmiter R D, Norstedt G, Gelinas R E, Hammer R E, Brinster R L. Metallothionein-human GH fusion genes stimulate growth of mice. Science, 1983, 222 (4625):809-814.

[62] Power J B, Cummins S E, Cocking E C. Fusion of isolated plant protoplasts. Nature, 1970, 225:1016-18.

[63] Pei D. Regulation of Pluripotency and Reprogramming by Transcription Factors. The Journal of Biological Chemistry, 2009, 284(6):3365-3369.

[64] Raghavan V. Origin and development of pollen embryoids and pollen calluses in cultured anther segments of Hyoscyamus niger. Am. J. Bot. , 1978, 65: 984-1002.

[65] Rechinger C. Untersuchungen fiber die Grenzen der Teilbarkeit im Pflanzenreich. Abh. Zool. -Bot. Ges. (Vienna), 1893, 43:310-334.

[66] Reinert J. Uber die Kontrolle der Morphogenese and die Induktion von Adventiveembryonen an Gewebekulturen aus Karotten. Planta, 1959, 53:318-33.

[67] Reinert J. Yeoman M M. Plant cell and tissue culture-a laboratory manual. Springer-Verlag, Berlin Heidelberg, New York, 1982.

[68] Robbins W J. Cultivation of excised root tips and stem tips under sterile conditions. Bot. Gaz. , 1922, 73:376-390.

[69] Robbins W J, Maneval W E. Further experiments on growth of excised root tips under sterile conditions. Bot. Gaz. , 1923, 76:274-87.

[70] Scholer H R, Ruppert S, Suzuki N, Chowdhury K and Gruss P. New type of POU domain in germ line-specific protein Oct-4. Nature, 1990, 344:435-439.

[71] Smith R H, Murashige T. In vitro development of the isolated shoot apical meristem of angiosperms. Am. J. Bot. , 1970, 57:562-568.

[72] Staba E J. Tissue culture and pharmacy. In Applied and fundamental aspects of plant cell, tissue, and organ culture, ed. J. Reinert & Y. P. S. Bajaj, 1977:694-702. Berlin: Springer-Verlag.

[73] Sunderland N. Towards more effective anther culture. Newsl. IAPTC, 1979, 27: 10-12.

[74] Schwann Th. Mikroskepische Untersuchungen fiber die Ubereinstimmung in der struktur and dem Waschstume der Tiere and Pflanzen. Leipzig: W. Englemann, No. 1839:176, Oswalds Klassiker der exakten Wissenschaften, 1910.

[75] Skoog F, Miller C O. Chemical regulation of growth and organ formation in plant tissues cultured in vitro. Symp. Soc. Exp. Biol. , 1957, 11:118-130.

[76] Steward F C. Growth and development of cultivated cells. III. Interpretations of the growth from free cell to carrot plant. Am. J. Bot. , 1958, 45:709-713.

[77] Steward F C, Mapes M O, Mears K. Growth and organized development of cultured cells: II. Organization in cultures grown from freely suspended cells. Am. J. Bot. , 1958, 45:705-708.

[78] Street H E. Introduction. In Plant tissue and cell culture, ed. H. E. Street, pp. 1-10. Oxford: Blackwell Scientific Publications, 1977.

[79] Slater A, Scott N W, Fowler M R. Plant biotechnology-the genetic manipulation of plants. Oxford university press inc. , New York, USA, 2003.

[80] Sweet H C, Bolton W E. The surface decontamination of seeds to produce axenic seedlings. Am. J. Bot. , 1979, 66:692-698.

[81] Takahashi K, Yamanaka S. Induction of pluripotent stem cells from mouse embryonic and adult fibroblast cultures by defined factors. Cell, 2006, 126:663-676.

[82] Takahashi K, Tanabe K, Ohnuki M, Narita M, Ichisaka T, Tomoda K and Ya-

manakal S. Induction of Pluripotent Stem Cells from Adult Human Fibroblasts by Defined Factors. Cell, doi:10.1016/j.cell, 2007.11.019.

[83] Thomson J A, Itskovitz-Eldor J, Shapiro S S, Waknitz M A, Swiergiel J J, Marshall V S and Jones J M. Embryonic stem cell lines derived from human blastocysts. Science, 1998, 282:1145-1147.

[84] Tsunoda Y and Kato Y. Not only inner cell mass cell nuclei but also trophectoderm nuclei of mouse blastocysts have a developmental totipotency. J. Reprod. Fertil., 1998, 113:181-184.

[85] Tulecke W. A tissue derived from the pollen of Ginkgo biloba. Science, 1953, 117:599-600.

[86] Tulecke W. The pollen of Ginkgo biloba: In vitro culture and tissue formation. Am. J. Bot., 1957, 44:602-608.

[87] Tulecke W. The pollen cultures of C. D. La Rue: A tissue from pollen of Taxus. Bull. Torrey Bot. Club, 1959, 86:283-289.

[88] Van Overbeek J, Conklin M E and Blakeslee A F. Factors in coconut milk essential for growth and development of very young Datura embryos. Science, 1941, 94:350-351.

[89] Vasil V, Hildebrandt A C. Differentiation of tobacco plants from single, isolated cells in microculture. Science, 1965, 150:889-892.

[90] Wetmore R H, Wardlaw C W. Experimental morphogenesis in vascular plants. Annu. Rev. Plant Physiol., 1951, 2:269-292.

[91] Wakamatsu Y, Ozato K, Sasado T. Establishment of a pluripotent cell line derived from a medaka (Oryzias latipes) blastula embryo. Mol. Mar. Biol. Biotechnol., 1994, 3:185-191.

[92] White P R. Plant tissue culture: Results of preliminary experiments on the culturing of isolated stem-tips of Stellaria media. Protoplasma, 1933, 19:97-116.

[93] White P R. Potentially unlimited growth of excised tomato root tips in a liquid medium. Plant Physiol., 1934, 9:585-600.

[94] White P R. Nutritional requirements of isolated plant tissues and organs. Annu. Rev. Plant Physiol., 1951, 2:231-244.

[95] White P R. The cultivation of animal and plant cells. Ronald Press, New York, USA. 1954.

[96] White P R. The cultivation of animal and plant cells, 2d Ed. New York: Ronald Press. 1963.

[97] White P R, Braun A C. A cancerous neoplasm of plants. Autonomous bacteria-free crown-gall tissue. Cancer Res., 1942, 2:597-617.

[98] Willadsen S M. Nuclear transplantation in sheep. Nature, 1986, 320: 6365.

[99] Wilmut I, Schnieke A E, McWhir J, Kind A J and Campbell K H. Viable off-

spring derived from fetal and adult mammalian cells. Nature, 1997, 385:810-813.

[100] Wolf E, Zakhartchenko V and Brem G. Nuclear transfer in mammals: Recent developments and future perspectives. Journal of Biotechnology, 1998, 65(2-3):99-110.

[101] Yu J and Thomson J A. Pluripotent stem cell lines. Genes & Development, 2008, 22:1987-1997.

[102] Zaehres H and Sch lerl H R. Induction of Pluripotency: From Mouse to Human. Cell, 2007, 131:834-835.